Oxford Applied Mathematics and Computing Science Series

General Editors
R. F. Churchouse, W. F. McColl, and A. B. Tayler

Oxford Applied Mathematics and Computing Science Series

I. Anderson: *A First Course in Combinatorial Mathematics (Second Edition)*
D. W. Jordan and P. Smith: *Nonlinear Ordinary Differential Equations (Second Edition)*
D. S. Jones: *Elementary Information Theory*
B. Carré: *Graphs and Networks*
A. J. Davies: *The Finite Element Method*
W. E. Williams: *Partial Differential Equations*
R. G. Garside: *The Architecture of Digital Computers*
J. C. Newby: *Mathematics for the Biological Sciences*
G. D. Smith: *Numerical Solution of Partial Differential Equations (Third Edition)*
J. R. Ullmann: *A Pascal Database Book*
S. Barnett and R. G. Cameron: *Introduction to Mathematical Control Theory (Second Edition)*
A. B. Tayler: *Mathematical Models in Applied Mechanics*
R. Hill: *A First Course in Coding Theory*
P. Baxandall and H. Liebeck: *Vector Calculus*
D. C. Ince: *An Introduction to Discrete Mathematics and Formal System Specification*
P. Thomas, H. Robinson, and J. Emms: *Abstract Data Types: Their Specification, Representation, and Use*
D. R. Bland: *Wave Theory and Applications*
R. P. Whittington: *Database Systems Engineering*
J. J. Modi: *Parallel Algorithms and Matrix Computation*
P. Gibbins: *Logic with Prolog*
W. L. Wood: *Practical Time-stepping Schemes*
D. J. Acheson: *Elementary Fluid Dynamics*
L. M. Hocking: *Optimal Control: An Introduction to the Theory with Applications*
S. Barnett: *Matrices: Methods and Applications*
P. Grindrod: *Patterns and Waves: The Theory and Applications of Reaction-Diffusion Equations*
O. Pretzel: *Error-Correcting Codes and Finite Fields*

D. J. ACHESON

Jesus College, Oxford

Elementary Fluid Dynamics

CLARENDON PRESS · OXFORD

Oxford University Press, Walton Street, Oxford OX2 6DP
Oxford New York
Athens Auckland Bangkok Bombay
Calcutta Cape Town Dar es Salaam Delhi
Florence Hong Kong Istanbul Karachi
Kuala Lumpur Madras Madrid Melbourne
Mexico City Nairobi Paris Singapore
Taipei Tokyo Toronto
and associated companies in
Berlin Ibadan

Oxford is a trade mark of Oxford University Press

Published in the United States by
Oxford University Press Inc., New York

© D. J. Acheson 1990

First published 1990
Reprinted 1990, 1992, 1995

All rights reserved. No part of this publication may be
reproduced, stored in a retrieval system, or transmitted, in any
form or by any means, without the prior permission in writing of Oxford
University Press. Within the UK, exceptions are allowed in respect of any
fair dealing for the purpose of research or private study, or criticism or
review, as permitted under the Copyright, Designs and Patents Act, 1988, or
in the case of reprographic reproduction in accordance with the terms of
licences issued by the Copyright Licensing Agency. Enquiries concerning
reproduction outside those terms and in other countries should be sent to
the Rights Department, Oxford University Press, at the address above.

This book is sold subject to the condition that it shall not,
by way of trade or otherwise, be lent, re-sold, hired out, or otherwise
circulated without the publisher's prior consent in any form of binding
or cover other than that in which it is published and without a similar
condition including this condition being imposed
on the subsequent purchaser.

A catalogue record for this book is available from the British Library

Library of Congress Cataloging in Publication Data
Acheson, D. J.
Elementary fluid dynamics/D. J. Acheson.
p.   cm.—(Oxford applied mathematics and computing science series)
1. Fluid dynamics.   I. Title.   II. Series.
TA357.A276   1990
532—dc20   89–22947   CIP
ISBN 0 19 859679 0 (Pbk)

Printed and bound in Great Britain by Biddles Ltd, Guildford and King's Lynn

Preface

This book is an introduction to fluid dynamics for students of applied mathematics, physics, and engineering. The main mathematical requirements are the vector calculus and simple methods for solving differential equations. Exercises are provided at the end of each chapter, and extensive hints and answers are offered at the end of the book. In order to indicate how the text is organized it is first necessary to say a little about the subject itself.

It is a matter of common experience that some fluids are more *viscous* than others. No reader will be surprised to learn that the 'coefficient of viscosity' μ is much greater for syrup than it is for water. Many fluids, such as water and air, hardly seem to be viscous at all. It is natural, then, to construct a theory based on the concept of an *inviscid* fluid, i.e. one for which μ is precisely zero. This is how the subject first developed, and this is how we begin, in Chapter 1.

Yet inviscid theory has its dangers. Careful analysis of the equations of motion for a viscous fluid shows that strange things can happen in the limit $\mu \to 0$, so that *a fluid with very small viscosity may behave quite differently to a (hypothetical) fluid with no viscosity at all*. For this reason an elementary account of viscous flow appears very early in the book, in Chapter 2. The aim there, particularly in §§ 2.1 and 2.2, is to introduce some of the key ideas as simply as possible. In order to do this the viscous flow equations are merely stated; their derivation from first principles appears later.

While inviscid theory has to be used with caution there are major areas of fluid dynamics in which it is extremely successful, and one of these is wave motion (Chapter 3). Another is flow past a thin wing (Chapter 4), provided that the wing makes only a small angle of incidence with the oncoming stream. Inviscid theory has a further role in the study of vortex motion (Chapter 5), which turns out to be central to much of fluid dynamics, largely through the elegant theorems of Kelvin and Helmholtz.

In Chapter 6 we establish the equations of viscous flow from first principles, although some readers may wish to consult this chapter quite early. In Chapter 7 we explore very viscous flow, i.e. the case in which μ is large (in some appropriate sense). The flow problems here have some novel features and are the object of much current research. We return to fluids of low viscosity in Chapter 8, focusing on thin 'boundary layers', where viscous effects are of crucial importance, no matter how small μ happens to be. In the final chapter we examine the instability of fluid flow, which, together with boundary layer separation, gives rise to some of the deepest and most challenging problems in the subject.

I am extremely grateful to all the students who have tried out successive drafts of this book. I would also like to thank Brooke Benjamin, David Crighton, Raymond Hide, Tom Mullin, Hilary Ockendon, John Ockendon, Norman Riley, John Roe, Alan Tayler, and Robert Terrill for their comments on various chapters.

Finally, I take the opportunity to acknowledge all the help I received, when I was first learning the subject, from Raymond Hide at the Meteorological Office and from Norman Riley, Michael Glauert, and others at the University of East Anglia.

Jesus College, Oxford
April 1989   D. J. A.

Contents

1 **INTRODUCTION** 1
  1.1 An experiment 1
  1.2 Some preliminary ideas 2
  1.3 Equations of motion for an ideal fluid 6
  1.4 Vorticity: irrotational flow 10
  1.5 The vorticity equation 16
  1.6 Steady flow past a fixed wing 18
  1.7 Concluding remarks 22
  Exercises 23

2 **ELEMENTARY VISCOUS FLOW** 26
  2.1 Introduction 26
  2.2 The equations of viscous flow 30
  2.3 Some simple viscous flows: the diffusion of vorticity 33
  2.4 Flow with circular streamlines 42
  2.5 The convection and diffusion of vorticity 48
  Exercises 50

3 **WAVES** 56
  3.1 Introduction 56
  3.2 Surface waves on deep water 65
  3.3 Dispersion: group velocity 69
  3.4 Surface tension effects: capillary waves 74
  3.5 Effects of finite depth 78
  3.6 Sound waves 79
  3.7 Supersonic flow past a thin aerofoil 82
  3.8 Internal gravity waves 86
  3.9 Finite-amplitude waves in shallow water 89
  3.10 Hydraulic jumps and shock waves 100
  3.11 Viscous shocks and solitary waves 106
  Exercises 111

4 CLASSICAL AEROFOIL THEORY 120
4.1 Introduction 120
4.2 Velocity potential and stream function 122
4.3 The complex potential 124
4.4 The method of images 128
4.5 Irrotational flow past a circular cylinder 130
4.6 Conformal mapping 134
4.7 Irrotational flow past an elliptical cylinder 136
4.8 Irrotational flow past a finite flat plate 137
4.9 Flow past a symmetric aerofoil 138
4.10 The forces involved: Blasius's theorem 140
4.11 The Kutta–Joukowski Lift Theorem 143
4.12 Lift: the deflection of the airstream 145
4.13 D'Alembert's paradox 147
Exercises 151

5 VORTEX MOTION 157
5.1 Kelvin's circulation theorem 157
5.2 The persistence of irrotational flow 161
5.3 The Helmholtz vortex theorems 162
5.4 Vortex rings 168
5.5 Axisymmetric flow 172
5.6 Motion of a vortex pair 177
5.7 Vortices in flow past a circular cylinder 178
5.8 Instability of vortex patterns 184
5.9 A steady viscous vortex maintained by a secondary flow 187
5.10 Viscous vortices: the Prandtl–Batchelor theorem 189
Exercises 191

6 THE NAVIER–STOKES EQUATIONS 201
6.1 Introduction 201
6.2 The stress tensor 202
6.3 Cauchy's equation of motion 205
6.4 A Newtonian viscous fluid: the Navier–Stokes equations 207
6.5 Viscous dissipation of energy 216
Exercises 217

7 VERY VISCOUS FLOW 221

7.1 Introduction 221
7.2 Low Reynolds number flow past a sphere 223
7.3 Corner eddies 229
7.4 Uniqueness and reversibility of slow flows 233
7.5 Swimming at low Reynolds number 234
7.6 Flow in a thin film 238
7.7 Flow in a Hele-Shaw cell 241
7.8 An adhesive problem 243
7.9 Thin-film flow down a slope 245
7.10 Lubrication theory 248
Exercises 251

8 BOUNDARY LAYERS 260

8.1 Prandtl's paper 260
8.2 The steady 2-D boundary layer equations 266
8.3 The boundary layer on a flat plate 271
8.4 High Reynolds number flow in a converging channel 275
8.5 Rotating flows controlled by boundary layers 278
8.6 Boundary layer separation 287
Exercises 291

9 INSTABILITY 300

9.1 The Reynolds experiment 300
9.2 Kelvin–Helmholtz instability 303
9.3 Thermal convection 305
9.4 Centrifugal instability 313
9.5 Instability of parallel shear flow 320
9.6 A general theorem on the stability of viscous flow 325
9.7 Uniqueness and non-uniqueness of steady viscous flow 330
9.8 Instability, chaos, and turbulence 334
9.9 Instability at very low Reynolds number 341
Exercises 343

APPENDIX 348
HINTS AND ANSWERS FOR EXERCISES 356
BIBLIOGRAPHY 384
INDEX 390

1 Introduction

1.1. An experiment

Take a shallow dish and pour in salty water to a depth of 1 cm. Make a model wing with a length and span of 2 cm or so, ensuring that it has a sharp trailing edge. (One method is to cut the wing out of an india rubber with a knife.) Dip the wing vertically in the water and turn it to make a small angle of attack α with the direction in which it is to be moved. Put a blob of ink or food colouring around the trailing edge; a thin layer of this should then float on the salt water.

Now move the wing across the dish, giving it a clean, sudden start. If α is not too large there should be a strong anticlockwise vortex left behind at the point where the trailing edge started, as in Fig. 1.1.

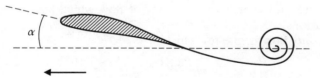

Fig. 1.1. The starting vortex.

A 'starting vortex' of this kind forms a crucial part of the mechanism by which an aircraft obtains lift, and we shall use aerodynamics in this chapter as a means of introducing some fundamental concepts of fluid flow.

Aerodynamics is, arguably, well suited to this purpose, but it goes without saying that the theory of fluid motion finds application in a wide variety of different fields. Within this book alone we may point to waves on a pond (§3.1), the instability of flow down a pipe (§9.1), the hydraulic jump in a kitchen sink

2 Introduction

(§3.10), the interaction of two smoke rings (§5.4), the jet stream in the atmosphere (§9.8), the motion of quantum vortices in liquid helium (§5.8), the flow of volcanic lava (§7.9), the swimming of biological micro-organisms (§7.5), and the spin-down of a stirred cup of tea (§8.5) as examples of the breadth and diversity of the subject.

1.2. Some preliminary ideas

The usual way of describing a fluid flow is by means of an expression

$$\boldsymbol{u} = \boldsymbol{u}(\boldsymbol{x}, t) \tag{1.1}$$

for the flow velocity $\boldsymbol{u}$ at any point $\boldsymbol{x}$ and at any time t. This tells us what all elements of the fluid are doing at any time; finding eqn (1.1) is usually the main task.

In general we must expect this task to be quite difficult. Let us take Cartesian coordinates, for example, and denote the three components of $\boldsymbol{u}$ by u, v, and w. Then eqn (1.1) is a convenient shorthand for

$$u = u(x, y, z, t), \qquad v = v(x, y, z, t), \qquad w = w(x, y, z, t).$$

There are, however, special classes of flow which have simplifying features.

A *steady* flow is one for which

$$\frac{\partial \boldsymbol{u}}{\partial t} = 0, \tag{1.2}$$

so that $\boldsymbol{u}$ depends on $\boldsymbol{x}$ alone. At any fixed point in space the speed and direction of flow are both constant.

A *two-dimensional (2-D) flow* is of the form

$$\boldsymbol{u} = [u(x, y, t), v(x, y, t), 0], \tag{1.3}$$

so that $\boldsymbol{u}$ is independent of one spatial coordinate (here selected to be z) and has no component in that direction.

A *two-dimensional steady flow* is thus of the form

$$\boldsymbol{u} = [u(x, y), v(x, y), 0]. \tag{1.4}$$

Introduction 3

These are idealizations. No real flow can be exactly two-dimensional, but in the case of flow past a fixed wing of long span and uniform cross-section we might reasonably expect a close approximation to 2-D flow, except near the wing-tips.

Before exploring such a flow more closely it is useful to introduce the concept of a *streamline*. This is, at any particular time t, a curve which has the same direction as $\boldsymbol{u}(\boldsymbol{x}, t)$ at each point. Mathematically, then, a streamline $x = x(s)$, $y = y(s)$, $z = z(s)$ is obtained by solving

$$\frac{\mathrm{d}x/\mathrm{d}s}{u} = \frac{\mathrm{d}y/\mathrm{d}s}{v} = \frac{\mathrm{d}z/\mathrm{d}s}{w} \qquad (1.5)$$

at a particular time t.

To imagine streamlines it can be convenient to consider a widely used experimental technique which involves putting tiny, neutrally buoyant polystyrene beads into the fluid. One particular plane of the fluid region is then illuminated by a collimated light beam, and the beads reflect this light to the camera, thus appearing as tiny pin-pricks of light if they are stationary. When the fluid is moving, however, the beads get carried around with it, so that a short-exposure-time photograph consists of short streaks, the length and direction of each one giving a measure of the fluid velocity at that particular point in space. As an example, we show in Fig. 1.2 a streak photograph for the flow (with uniform velocity at infinity) past a fixed wing. Because this is a steady flow the streamline pattern is the same at all times, and a fluid particle started on some streamline will travel along that

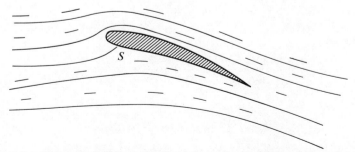

Fig. 1.2. Streamlines for steady flow past a fixed wing, as inferred from a streak photograph.

streamline as time proceeds. (In an unsteady flow, on the other hand, streamlines and particle paths are usually quite different; see Exercise 1.8.)

It is evident from Fig. 1.2 that even though the flow is steady, so that *u* is constant at a point fixed in space, *u* changes *as we follow any particular fluid element*. In particular—changes in direction of flow aside—an element riding over the top of the wing first speeds up and then slows down again.

Rate of change 'following the fluid'

This notion is of fundamental importance in fluid dynamics.

Let $f(x, y, z, t)$ denote some quantity of interest in the fluid motion. It could, for example, be one component of the velocity *u*, or it could be the density ρ. Note first that $\partial f/\partial t$ means the rate of change of f at fixed x, y, and z, i.e. at a fixed position in space.

In contrast, the rate of change of f 'following the fluid', which we denote by Df/Dt, is

$$\frac{Df}{Dt} = \frac{d}{dt} f[x(t), y(t), z(t), t],$$

where $x(t)$, $y(t)$, and $z(t)$ are understood to change with time at the local flow velocity *u*:

$$dx/dt = u, \qquad dy/dt = v, \qquad dz/dt = w,$$

so as to 'follow the fluid'. A simple application of the chain rule gives

$$\frac{Df}{Dt} = \frac{\partial f}{\partial x}\frac{dx}{dt} + \frac{\partial f}{\partial y}\frac{dy}{dt} + \frac{\partial f}{\partial z}\frac{dz}{dt} + \frac{\partial f}{\partial t},$$

whence

$$\frac{Df}{Dt} = \frac{\partial f}{\partial t} + u\frac{\partial f}{\partial x} + v\frac{\partial f}{\partial y} + w\frac{\partial f}{\partial z},$$

i.e.

$$\frac{Df}{Dt} = \frac{\partial f}{\partial t} + (\boldsymbol{u} \cdot \nabla)f. \tag{1.6}$$

By applying eqn (1.6) to the velocity components u, v, and w in turn it follows, in particular, that the *acceleration* of the fluid element at x is

$$\frac{D\boldsymbol{u}}{Dt} = \frac{\partial \boldsymbol{u}}{\partial t} + (\boldsymbol{u} \cdot \nabla)\boldsymbol{u}. \tag{1.7}$$

As an immediate check on eqn (1.7) consider fluid in uniform rotation with angular velocity Ω, so that

$$u = -\Omega y, \qquad v = \Omega x, \qquad w = 0.$$

Now $\partial \boldsymbol{u}/\partial t$ is zero, because the flow is steady, but

$$(\boldsymbol{u} \cdot \nabla)\boldsymbol{u} = \left(-\Omega y \frac{\partial}{\partial x} + \Omega x \frac{\partial}{\partial y}\right)(-\Omega y, \Omega x, 0)$$
$$= -\Omega^2(x, y, 0).$$

This is just as expected; it represents the familiar centrifugal acceleration $\Omega^2 r$ towards the rotation axis.

According to eqn (1.6) in any *steady* flow the rate of change of f following a fluid element is $(\boldsymbol{u} \cdot \nabla)f$, and it is quite easy to see why this should be so. Let $\boldsymbol{e}_s$ denote a unit vector which is always parallel to the streamlines and in the same sense as the flow. Then

$$\boldsymbol{u} \cdot \nabla f = |\boldsymbol{u}| \, \boldsymbol{e}_s \cdot \nabla f = |\boldsymbol{u}| \frac{\partial f}{\partial s},$$

where s denotes distance along a streamline. Now, $\partial f/\partial s$ is the rate of change of f with distance along a streamline, so multiplying it by the flow speed $|\boldsymbol{u}|$ evidently gives the rate of change with time as we follow a fluid element along that streamline.

The equation

$$(\boldsymbol{u} \cdot \nabla)f = 0, \tag{1.8}$$

which arises at some important stages in the following theory, thus implies that f is constant *along a streamline*. It should be emphasized that eqn (1.8) offers no information at all about whether f might be a different constant on different streamlines. Suppose, for instance, that the flow is everywhere in the x-direction, so that eqn (1.8) reduces to $\partial f/\partial x = 0$. This equation says that f is independent of x, but it contains no implication about how f might depend on y, z, or t.

6 Introduction

Likewise, the equation

$$\frac{Df}{Dt} = 0, \tag{1.9}$$

which also arises in the following theory, implies that f is a constant *for a particular fluid element*, and this follows directly from the definition of Df/Dt above. It does not preclude different elements having different values of f; it just implies that each such element will retain whatever value of f it started with.

Finally, it is worth remarking that there will be occasions on which we wish to follow not just an infinitesimal fluid element but a finite blob consisting always of the same fluid particles. Such a blob, which will of course deform as it moves about, is typically called a 'material' volume in the literature, but we shall freely describe it instead as 'dyed', with the understanding, of course, that no diffusion of this imaginary dye is envisaged. Such terminology can become rather colourful, but if it evokes a sharp mental picture of a moving and deforming blob of fluid, as opposed to some region fixed in space, it serves its purpose.

1.3. Equations of motion for an ideal fluid

In this text we define an *ideal fluid* as one with the following properties:

(i) It is *incompressible*, so that no 'dyed' blob of fluid can change in volume as it moves.

(ii) The density ρ (i.e. the mass per unit volume) is a constant, the same for all fluid elements and for all time t.

(iii) The force exerted across a geometrical surface element $\boldsymbol{n}\,\delta S$ within the fluid is

$$p\boldsymbol{n}\,\delta S, \tag{1.10}$$

where $p(x, y, z, t)$ is a scalar function, independent of the normal $\boldsymbol{n}$, called the *pressure*. (To be more precise, eqn (1.10) is the force exerted *on* the fluid into which $\boldsymbol{n}$ is pointing *by* the fluid on the other side of δS.)

There is, of course, no such thing in practice as an ideal fluid. All fluids are to some extent compressible, and all fluids are to some extent *viscous*, so that adjacent fluid elements exert both normal and tangential forces on one another across their common interface. For the time being, however, we explore some consequences of the assumptions (i)–(iii).

To examine the implications of (i), consider a fixed closed surface S drawn in the fluid, with unit outward normal $\boldsymbol{n}$. Fluid will be entering the enclosed region V at some places on S, and leaving it at others. The velocity component along the outward normal is $\boldsymbol{u} \cdot \boldsymbol{n}$, so the volume of fluid leaving through a small surface element δS in unit time is $\boldsymbol{u} \cdot \boldsymbol{n}\, \delta S$. The net volume rate at which fluid is leaving V is therefore

$$\int_S \boldsymbol{u} \cdot \boldsymbol{n}\, \mathrm{d}S.$$

But this must plainly be zero for an incompressible fluid, and on using the divergence theorem we find that

$$\int_V \nabla \cdot \boldsymbol{u}\, \mathrm{d}V = 0.$$

Now, this must be true for all regions V within the fluid. Suppose, then, that $\nabla \cdot \boldsymbol{u}$ is greater than zero at some point in the fluid. Assuming that it is continuous, $\nabla \cdot \boldsymbol{u}$ will be greater than zero in some small sphere around that point, and by taking V to be such a sphere we violate the above equation. The same applies if $\nabla \cdot \boldsymbol{u}$ is negative at some point. We thus conclude that

$$\nabla \cdot \boldsymbol{u} = 0 \qquad (1.11)$$

everywhere in the fluid.

This *incompressibility condition* is an important constraint on the velocity field $\boldsymbol{u}$ in virtually the whole of this book.†

To examine the implications of (iii) consider a surface S enclosing a 'dyed' blob of fluid. The force exerted by the surrounding fluid across any surface element δS is, by hypothesis, given by eqn (1.10), so that the net force exerted on the dyed blob is

$$-\int_S p\boldsymbol{n}\, \mathrm{d}S = -\int_V \nabla p\, \mathrm{d}V,$$

---

† Air is, of course, highly compressible, but it can behave like an incompressible fluid if the flow speed is much smaller than the speed of sound (see p. 58).

8 *Introduction*

where we have used the identity (A.14)—see the Appendix (the negative sign arises because n points out of S). Now, provided that ∇p is continuous it will be almost constant over a *small* blob of fluid of volume δV. The net force on such a small blob due to the pressure of the surrounding fluid will therefore be $-\nabla p \, \delta V$.

Euler's equations of motion

We are now in a position to apply the principle of linear momentum to a small 'dyed' blob of fluid of volume δV. Allowing for the presence of a gravitational body force per unit mass g, the total force on the blob is

$$(-\nabla p + \rho g) \, \delta V.$$

This force must be equal to the product of the blob's mass (which is conserved) and its acceleration, i.e. to

$$\rho \, \delta V \frac{\mathrm{D} u}{\mathrm{D} t}.$$

We thus obtain

$$\frac{\mathrm{D} u}{\mathrm{D} t} = -\frac{1}{\rho} \nabla p + g, \qquad (1.12)$$

$$\nabla \cdot u = 0,$$

as the basic equations of motion for an ideal fluid. They are known as *Euler's equations*, and written out in full they become

$$\frac{\partial u}{\partial t} + u \frac{\partial u}{\partial x} + v \frac{\partial u}{\partial y} + w \frac{\partial u}{\partial z} = -\frac{1}{\rho} \frac{\partial p}{\partial x},$$

$$\frac{\partial v}{\partial t} + u \frac{\partial v}{\partial x} + v \frac{\partial v}{\partial y} + w \frac{\partial v}{\partial z} = -\frac{1}{\rho} \frac{\partial p}{\partial y},$$

$$\frac{\partial w}{\partial t} + u \frac{\partial w}{\partial x} + v \frac{\partial w}{\partial y} + w \frac{\partial w}{\partial z} = -\frac{1}{\rho} \frac{\partial p}{\partial z} - g,$$

$$\frac{\partial u}{\partial x} + \frac{\partial v}{\partial y} + \frac{\partial w}{\partial z} = 0,$$

i.e. four scalar equations for four unknowns: u, v, w, and p. In dealing with the gravitational term we have momentarily taken the z-axis vertically upward, setting $\boldsymbol{g} = (0, 0, -g)$.

Now, the gravitational force, being conservative, can be written as the gradient of a potential:

$$\boldsymbol{g} = -\nabla \chi. \tag{1.13}$$

(In the above case, $\chi = gz$.) Using the expression (1.7) for the fluid acceleration we may rewrite eqn (1.12) in the form†

$$\frac{\partial \boldsymbol{u}}{\partial t} + (\boldsymbol{u} \cdot \nabla)\boldsymbol{u} = -\nabla\left(\frac{p}{\rho} + \chi\right),$$

where we have used the assumption that ρ is constant. Furthermore, it can be helpful to use the identity

$$(\boldsymbol{u} \cdot \nabla)\boldsymbol{u} = (\nabla \wedge \boldsymbol{u}) \wedge \boldsymbol{u} + \nabla(\tfrac{1}{2}u^2)$$

to cast the momentum equation into the form

$$\frac{\partial \boldsymbol{u}}{\partial t} + (\nabla \wedge \boldsymbol{u}) \wedge \boldsymbol{u} = -\nabla\left(\frac{p}{\rho} + \tfrac{1}{2}u^2 + \chi\right). \tag{1.14}$$

The Bernoulli streamline theorem

If the flow is steady, eqn (1.14) reduces to

$$(\nabla \wedge \boldsymbol{u}) \wedge \boldsymbol{u} = -\nabla H,$$

where

$$H = \frac{p}{\rho} + \tfrac{1}{2}u^2 + \chi. \tag{1.15}$$

On taking the dot product with $\boldsymbol{u}$ we obtain

$$(\boldsymbol{u} \cdot \nabla)H = 0, \tag{1.16}$$

---

† The way in which $p/\rho + \chi$ appears as a combination is significant; there will be many circumstances in this book in which gravity simply modifies the pressure distribution in the fluid and does nothing to change the velocity $\boldsymbol{u}$. Thus when we speak occasionally of 'ignoring' gravity, or of gravitational body forces being 'absent', what we often mean is that separate allowance may be made for gravity simply by subtracting $\rho\chi$ from the pressure field. This is emphatically not the case, however, if there is a free surface—as with water waves in Chapter 3—or if ρ is not constant—as in §3.8 and §9.3.

10 Introduction

so

> *If an ideal fluid is in steady flow,*
> *then H is constant along a streamline.*

In the absence of gravity it follows that $p + \frac{1}{2}\rho u^2$ is constant along a streamline in steady flow.

The above theorem says nothing about H being the same constant on different streamlines, only that it remains constant along each one. There is, however, one important circumstance in which H is constant throughout the whole flow field, and this now follows.

DEFINITION. An *irrotational* flow is one for which

$$\nabla \wedge u = 0. \qquad (1.17)$$

The Bernoulli theorem for irrotational flow

If the flow is steady and irrotational, then eqn (1.14) reduces to $\nabla H = 0$, so H is independent of x, y, and z, as well as t. Thus

> *If an ideal fluid is in steady irrotational flow,*
> *then H is constant throughout the whole flow field.*

Whether this result is of any value rests, evidently, on whether irrotational flows are of any real interest in practice. We address this matter in the next section.

1.4. Vorticity: irrotational flow

The *vorticity* ω is defined as

$$\omega = \nabla \wedge u, \qquad (1.18)$$

and it is a concept of central importance in fluid dynamics. The vorticity is, by definition, zero for an *irrotational* flow.

We consider vorticity first in the context of two-dimensional flow, for if

$$u = [u(x, y, t), v(x, y, t), 0]$$

then ω is $(0, 0, \omega)$, where

$$\omega = \frac{\partial v}{\partial x} - \frac{\partial u}{\partial y}. \qquad (1.19)$$

Interpretation of vorticity in 2-D flow

Consider two short fluid line elements AB and AC *which are perpendicular at a certain instant,* as in Fig. 1.3. Note that the *y*-component of velocity at B exceeds that at A by

$$v(x + \delta x, y, t) - v(x, y, t) \doteq \frac{\partial v}{\partial x} \delta x,$$

so that $\partial v/\partial x$ represents the instantaneous angular velocity of the fluid line element AB. Likewise, $\partial u/\partial y$ represents the instantaneous angular velocity (in the opposite sense) of the line element AC. Thus at any point of the flow field

$$\tfrac{1}{2}\omega = \tfrac{1}{2}\left(\frac{\partial v}{\partial x} - \frac{\partial u}{\partial y}\right)$$

represents the average angular velocity of two short fluid line elements that happen, at that instant, to be mutually perpendicular. In this precise sense the vorticity ω acts as a measure of the *local* rotation, or spin, of fluid elements.

We emphasize that vorticity has nothing directly to do with any global rotation of the fluid. Take, for example, the shear flow of

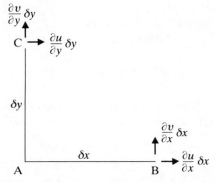

Fig. 1.3. Sketch for the interpretation of vorticity in 2-D flow. The velocity components shown are relative to the fluid particle at A.

12 Introduction

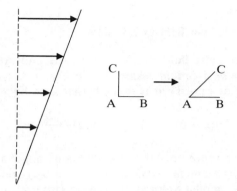

Fig. 1.4. Deformation of two short, momentarily perpendicular fluid line elements in a shear flow.

Fig. 1.4, in which

$$\boldsymbol{u} = (\beta y, 0, 0), \qquad (1.20)$$

where β is a constant. The fluid is certainly not rotating globally in any sense, but it has vorticity:

$$\omega = \frac{\partial v}{\partial x} - \frac{\partial u}{\partial y} = -\beta;$$

and two momentarily perpendicular line elements, AB and AC, orientated as shown plainly have an average angular velocity (in fact, of $-\tfrac{1}{2}\beta$), because while that of AB is zero that of AC is not.

A more colourful example of the distinction between vorticity and global rotation of the fluid is provided by the so-called *line vortex* flow given in cylindrical polar coordinates (r, θ, z) by

$$\boldsymbol{u} = \frac{k}{r} \boldsymbol{e}_\theta, \qquad (1.21)$$

where k is a constant. To find the vorticity of this flow we need the expression (A.32) for $\nabla \wedge \boldsymbol{u}$ in cylindrical polar coordinates:

$$\nabla \wedge \boldsymbol{u} = \frac{1}{r} \begin{vmatrix} \boldsymbol{e}_r & r\boldsymbol{e}_\theta & \boldsymbol{e}_z \\ \dfrac{\partial}{\partial r} & \dfrac{\partial}{\partial \theta} & \dfrac{\partial}{\partial z} \\ u_r & ru_\theta & u_z \end{vmatrix}.$$

Introduction   13

Plainly, then, the vorticity is zero except at $r = 0$, where neither u nor $\nabla \wedge u$ is defined. Thus although the fluid is clearly rotating in a global sense the flow is in fact *irrotational*, since $\nabla \wedge u = 0$, except on the axis. This is quite understandable if we consider two momentarily perpendicular fluid line elements, AB and AC, at $\theta = 0$ in Fig. 1.5. Clearly AC is rotating in an anticlockwise sense, because it will continue to lie along the circular streamline as time proceeds, but AB is rotating clockwise because of the decrease of u_θ with r in eqn (1.21). This particular fall-off of u_θ with r is, apparently, just the correct one—neither too slow nor too rapid—to ensure that AB has an equal and opposite angular velocity to AC at the instant they are perpendicular, so that their average angular velocity is zero.

We keep emphasizing the instantaneous nature of this conclusion about zero average angular velocity because two fluid line elements such as AB and AC in Fig. 1.5 will not remain perpendicular as they get carried about by the flow, and as soon as this happens we have no cause to conclude from the irrotationality of the flow that their average angular velocity should any longer be zero.

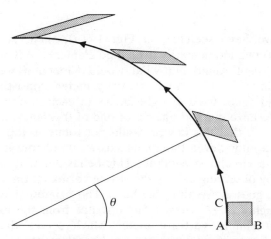

Fig. 1.5. The fate of a small square fluid element in a line vortex flow. The size of the element has been greatly exaggerated for the sake of clarity; an unfortunate consequence is that B does not look as if it is following a circular path.

14 *Introduction*

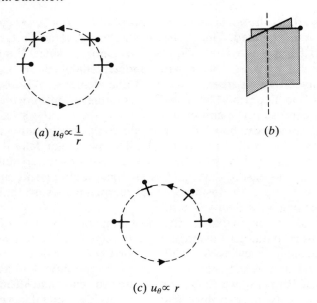

Fig. 1.6. A crude 'vorticity meter' (*b*), and its behaviour when immersed in a line vortex flow (*a*) and a uniformly rotating flow (*c*).

What we have sketched in Fig. 1.6(*a*), then, is not what happens to two momentarily dyed fluid elements, AB and AC, as they get swept round but what would happen if we were to immerse in the fluid a small 'vorticity meter' consisting of two short, rigid vanes fixed at right angles to each other, as in Fig. 1.6(*b*). We have marked one tip of one of the vanes, and in Fig. 1.6(*a*) we see that this device would not rotate in this particular (line vortex) flow, even though its axis would of course get swept round on a circular streamline. This behaviour may be seen in the bath by observing closely the strong vortex that may occur as the water goes down the plug-hole. The azimuthal velocity u_θ varies roughly as r^{-1} over a fair distance from the axis, and a crude but simple vorticity meter which serves the purpose consists of a pair of short wooden line elements shaved off a matchstick, sellotaped together at right angles and floated on the surface.

However, if such a vorticity meter were to be inserted in the flow

$$u = \Omega r e_\theta, \quad (1.22)$$

Ω being a constant, the result would of course be as in Fig. 1.6(c), because the device would get carried around just as if it were embedded in a rigid body. Its angular velocity would evidently be Ω, the same as the uniform angular velocity of the fluid as a whole, and the vorticity of the flow is therefore $(0, 0, 2\Omega)$, as may be confirmed by direct calculation of $\nabla \wedge \boldsymbol{u}$.

By putting the two flows in Fig. 1.6 together in the following way:

$$u_\theta = \begin{cases} \Omega r, & r < a, \\ \dfrac{\Omega a^2}{r}, & r > a, \end{cases}$$

$$u_r = u_z = 0, \quad (1.23)$$

we obtain a so-called 'Rankine vortex', which serves as a simple model for a real vortex such as that in Fig. 1.1. Real vortices are typically characterized by fairly small vortex 'cores' in which, by definition, the vorticity is concentrated, while outside the core the flow is essentially irrotational. The core is not usually exactly circular, of course; nor is the vorticity usually uniform within it. In these two respects the Rankine vortex of Fig. 1.7 is only an idealized model.

We have now said a fair amount about vorticity, albeit strictly

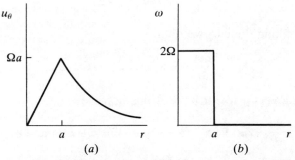

Fig. 1.7. Distribution of (a) azimuthal velocity u_θ and (b) vorticity ω in a Rankine vortex.

16 *Introduction*

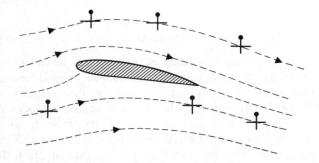

Fig. 1.8. The behaviour of a small 'vorticity meter' placed in the steady flow past a fixed wing at small angle of attack. The flow is clearly irrotational.

in the context of two-dimensional flow. We have discussed in particular detail the absence of vorticity, i.e. irrotational flow. At this stage, before the development seems to be getting rather a long way from our starting point (the experiment in §1.1), we should say that steady flow past a wing at small angles of incidence α *is* typically irrotational, as indicated in Fig. 1.8.

Why this should be so emerges from the Euler equations in a very elegant manner, as we now see.

1.5. The vorticity equation

In its form (1.14), Euler's equation may be written

$$\frac{\partial \boldsymbol{u}}{\partial t} + \boldsymbol{\omega} \wedge \boldsymbol{u} = -\nabla H,$$

and on taking the curl we obtain

$$\frac{\partial \boldsymbol{\omega}}{\partial t} + \nabla \wedge (\boldsymbol{\omega} \wedge \boldsymbol{u}) = 0. \qquad (1.24)$$

Using the vector identity (A.6) this becomes

$$\frac{\partial \boldsymbol{\omega}}{\partial t} + (\boldsymbol{u} \cdot \nabla)\boldsymbol{\omega} - (\boldsymbol{\omega} \cdot \nabla)\boldsymbol{u} + \boldsymbol{\omega}\nabla \cdot \boldsymbol{u} - \boldsymbol{u}\nabla \cdot \boldsymbol{\omega} = 0.$$

Now the fourth term vanishes because the fluid is incompressible, while the fifth term vanishes because div curl = 0. We therefore have

$$\frac{\partial \boldsymbol{\omega}}{\partial t} + (\boldsymbol{u} \cdot \nabla)\boldsymbol{\omega} = (\boldsymbol{\omega} \cdot \nabla)\boldsymbol{u},$$

or, alternatively,

$$\frac{D\boldsymbol{\omega}}{Dt} = (\boldsymbol{\omega} \cdot \nabla)\boldsymbol{u}. \tag{1.25}$$

This *vorticity equation* is extremely valuable. Note that the pressure has been eliminated; eqn (1.25) involves only $\boldsymbol{u}$ and $\boldsymbol{\omega}$, which are, of course, related by

$$\boldsymbol{\omega} = \nabla \wedge \boldsymbol{u}.$$

In particular, if the flow is *two-dimensional*, so that

$$\boldsymbol{u} = [u(x, y, t), v(x, y, t), 0] \tag{1.26}$$

and

$$\boldsymbol{\omega} = (0, 0, \omega),$$

then

$$(\boldsymbol{\omega} \cdot \nabla)\boldsymbol{u} = \omega \frac{\partial \boldsymbol{u}}{\partial z} = 0.$$

It then follows that

$$\frac{D\omega}{Dt} = 0, \tag{1.27}$$

and we thus conclude, referring back to eqn (1.9), that

> *In the two-dimensional flow of an ideal fluid subject to a conservative body force $\boldsymbol{g}$ the vorticity ω of each individual fluid element is conserved.* (1.28)

This result has important applications, which we discuss in Chapter 5. In the particular case of steady flow, eqn (1.27) reduces to

$$(\boldsymbol{u} \cdot \nabla)\omega = 0 \tag{1.29}$$

and consequently

> *In the steady, two-dimensional flow of an ideal fluid subject to a conservative body force $\boldsymbol{g}$ the vorticity ω is constant along a streamline.* (1.30)

18 Introduction

This, then, is the reason why the steady flow in Fig. 1.8 is irrotational. Note first that there are no regions of closed streamlines in the flow; all the streamlines can be traced back to $x = -\infty$. Now, the vorticity is constant along each one, and hence equal on each one to whatever it is on that particular streamline at $x = -\infty$. As the flow is uniform at $x = -\infty$ the vorticity is zero on all streamlines there. Hence it is zero throughout the flow field in Fig. 1.8.

1.6. Steady flow past a fixed wing

In Fig. 1.9 we show typical measured pressure distributions on the upper and lower surfaces of a fixed wing in steady flow. The pressures on the upper surface are substantially lower than the free-stream value p_∞, while those on the lower surface are a little higher than p_∞. In fact, then, the wing gets most of its lift from a suction effect on its upper surface.

But why is it that the pressures above the wing are less than those below? Well, because the flow is irrotational, the Bernoulli theorem tells us that $p + \tfrac{1}{2}\rho u^2$ is constant throughout the flow. Explaining the pressure differences, and hence the lift on the

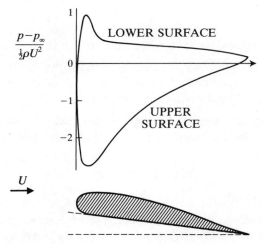

Fig. 1.9. Typical pressure distribution on a wing in steady flow.

wing, thus reduces to explaining why (as in Fig. 1.2) the flow speeds above the wing are greater than those below.

Let us first dispose of one bogus explanation that occasionally appears, namely that the air on the top of the wing flows faster 'because it has farther to go'. There are many woolly aspects to this argument, but it seems to turn principally on the notion that two neighbouring fluid elements, after parting to go their separate ways round the wing, meet up again at the trailing edge, and this is demonstrably false (see Fig. 2.4).

The right way forward to an explanation of the higher flow speeds above the wing is in terms of the concept of circulation.

Circulation

Let C be some closed curve lying in the fluid region. Then the circulation Γ round C is defined as

$$\Gamma = \int_C \boldsymbol{u} \cdot \mathrm{d}\boldsymbol{x}. \tag{1.31}$$

At first sight, perhaps, there cannot be any circulation in an irrotational flow, for Stokes's theorem gives

$$\int_C \boldsymbol{u} \cdot \mathrm{d}\boldsymbol{x} = \int_S (\nabla \wedge \boldsymbol{u}) \cdot \boldsymbol{n} \, \mathrm{d}S, \tag{1.32}$$

and an irrotational flow is, by definition, one for which $\nabla \wedge \boldsymbol{u}$ is zero. But such an argument holds only if the closed curve C in question can be spanned by a surface S which lies wholly in the region of irrotational flow. Thus in the two-dimensional context of Fig. 1.8, for example, for which eqn (1.32) reduces to

$$\Gamma = \int_C u \, \mathrm{d}x + v \, \mathrm{d}y = \int_S \left(\frac{\partial v}{\partial x} - \frac{\partial u}{\partial y} \right) \mathrm{d}x \, \mathrm{d}y, \tag{1.33}$$

it is true that Γ must be zero for any closed curve C not enclosing the wing, but the argument fails for any closed curve that does enclose the wing. The most that can be said about such circuits is that they all have the same value of Γ (Exercise 1.6).

Circulation round a wing is permissible, then, in a steady irrotational flow; but the question still arises as to why there should be any, and, in particular, why it should be negative, corresponding to larger flow speeds above the wing than below.

The Kutta–Joukowski hypothesis

In the case of a wing with a sharp trailing edge, one good reason for non-zero circulation Γ is that there would otherwise be a singularity in the velocity field. The irrotational flow past a wing with $\Gamma = 0$ is sketched in Fig. 1.10(*a*), but the velocity is infinite at the trailing edge where, loosely speaking, the fluid is having a hard time turning the corner. We show in Chapter 4 that only for one value of the circulation, Γ_K say, is the flow speed finite at the trailing edge, as in Fig. 1.10(*b*). It is natural to hope that this particular irrotational flow will correspond to the steady flow that is actually observed; this is the *Kutta–Joukowski hypothesis*.

This hypothesis is inevitably somewhat *ad hoc*, resting as it does on the unsatisfactory state of affairs that would otherwise arise because of the sharp trailing edge. (How are we to decide between all the different irrotational flows if the trailing edge is not sharp?) It is, nonetheless, one of the key steps in the development of aerodynamics, and gives results which are in excellent accord with experiment, as we shall shortly see.

The critical value Γ_K depends, naturally, on the flow speed at infinity U and on the size, shape, and orientation of the wing. In Chapter 4 we show that if the wing is thin and symmetrical, of length L, making an angle α with the oncoming stream, then

$$\Gamma_K \doteq -\pi UL \sin \alpha. \qquad (1.34)$$

Lift

According to ideal flow theory, the *drag* on the wing (the force parallel to the oncoming stream) is zero, but the *lift* (the force

Fig. 1.10. Irrotational flow past a fixed wing with (*a*) $\Gamma = 0$ and (*b*) $\Gamma = \Gamma_K < 0$.

perpendicular to the stream) is

$$\mathcal{L} = -\rho U \Gamma. \tag{1.35}$$

This *Kutta–Joukowski Lift Theorem* is proved in §4.11.

That negative Γ should give positive lift is entirely natural; we have argued as much in the preceding sections. As a precise theorem, however, eqn (1.35) is rather extraordinary, as it holds for irrotational flow (uniform at infinity) past a two-dimensional body of any size or shape; $\mathcal{L}$ depends on the size and shape of the body only inasmuch as Γ does. For the thin symmetrical wing of Fig. 1.10(*b*), for example, with Γ as in eqn (1.34) by the Kutta–Joukowski condition, the lift is

$$\mathcal{L} \doteqdot \pi \rho U^2 L \sin \alpha. \tag{1.36}$$

Agreement with experiment is good provided that α is only a few degrees (Fig. 1.11). Thereafter the measured lift falls dramatically and diverges substantially from the predictions of inviscid theory, for reasons to be discussed later. The angle α at which this divergence begins may be anywhere between about 6° and 12°, depending on the shape of the wing (see, e.g., Nakayama 1988, pp. 76–80).

Accounting for the flow past a wing at small angles of attack α is nevertheless one of the great, and practically important, successes of ideal-flow theory.

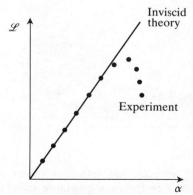

Fig. 1.11. Lift on a symmetric aerofoil.

1.7. Concluding remarks

In this chapter we have introduced some of the basic concepts of fluid dynamics and, at the same time, given some indication of how they figure in one particular branch of the subject, namely aerodynamics. Our treatment of this branch has inevitably been sketchy.

We have, for instance, focused wholly on 2-D aerodynamics, yet any real wing, no matter how long, has ends, and important new phenomena then arise. The circulation round a circuit such as C in Fig. 1.12(a) is essentially that predicted by the 2-D theory (i.e. eqn (1.34)), but plainly the flow cannot be everywhere irrotational, because C can now be spanned by a surface S which lies wholly in the fluid. Indeed, from Stokes's theorem (1.32) we deduce that there must be a positive flux of vorticity out of S, and this is in practice observed as a concentrated *trailing vortex* emanating from the wing-tip as shown. The higher the lift (and

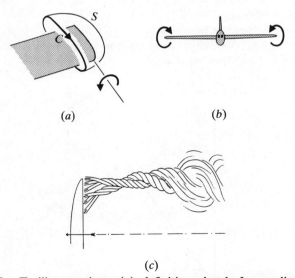

Fig. 1.12. Trailing vortices: (a) definition sketch for application of Stokes's theorem; (b) view from some distance ahead of the aircraft; (c) the original drawing from Lanchester's *Aerodynamics* (1907).

therefore the circulation), the stronger the trailing vortices. Furthermore, the presence of these trailing vortices results in a drag on the wing, even on ideal flow theory, for as they lengthen they contain more and more kinetic energy, and creating all this kinetic energy takes work.

But even within a purely two-dimensional framework we have left some key questions unanswered. We indicated how the Kutta–Joukowski hypothesis provides a rational, although *ad hoc,* basis for deciding the circulation round an aerofoil in steady flight, and we have noted that this gives good agreement with experiment. Yet we have given no account of the dynamical processes by which that circulation is *generated* when the aerofoil starts from a state of rest. It arises, in fact, in response to the 'starting vortex' in §1.1, but why this should be so is far from obvious, and rests on one of the deepest theorems in the subject (§5.1).

Again, the sceptical reader may even be asking: 'But what is all this business about a starting vortex? If the aerofoil and fluid are initially at rest, the vorticity ω is initially zero for each fluid element. By eqn (1.27) it remains zero for each fluid element, even when the aerofoil has been started into motion. Therefore there should not be a starting vortex.'

This is a legitimate conclusion—on the basis of ideal flow theory. While that theory accounts well for the steady flow past an aerofoil, the explanation of how that flow became established involves *viscous* effects in a crucial way.

If this provokes the response: 'But air isn't very viscous, is it?', the answer is, 'No, in some sense air is hardly viscous at all'. Yet, as we shall see, viscous effects are sufficiently subtle that the shedding of the vortex in §1.1, while being an essentially viscous process, would occur no matter how small the viscosity of the fluid happened to be.

Exercises

1.1. Whether a fluid is incompressible or not, each element must conserve its mass as it moves. Consider the rate of mass flow through a fixed closed surface S drawn in the fluid, and use an argument similar to that on p. 7 to show that this *conservation of mass* implies

$$\frac{\partial \rho}{\partial t} + \nabla \cdot (\rho \boldsymbol{u}) = 0, \qquad (1.37)$$

where $\rho(\mathbf{x}, t)$ denotes the (variable) density of the fluid. Show too that this equation may alternatively be written

$$\frac{D\rho}{Dt} + \rho \nabla \cdot \mathbf{u} = 0. \tag{1.38}$$

It follows that if $\nabla \cdot \mathbf{u} = 0$, then $D\rho/Dt = 0$. What does this mean, exactly, and does it make sense?

1.2. An ideal fluid is rotating under gravity g with constant angular velocity Ω, so that relative to fixed Cartesian axes $\mathbf{u} = (-\Omega y, \Omega x, 0)$. We wish to find the surfaces of constant pressure, and hence the surface of a uniformly rotating bucket of water (which will be at atmospheric pressure).

'By Bernoulli,' $p/\rho + \frac{1}{2}\mathbf{u}^2 + gz$ is constant, so the constant pressure surfaces are

$$z = \text{constant} - \frac{\Omega^2}{2g}(x^2 + y^2).$$

But this means that the surface of a rotating bucket of water is at its highest in the middle. What is wrong?

Write down the Euler equations in component form, integrate them directly to find the pressure p, and hence obtain the correct shape for the free surface.

1.3. Find the pressure p both inside and outside the core of the Rankine vortex (1.23). Show that the pressure at $r = 0$ is lower than that at $r = \infty$ by an amount $\rho\Omega^2 a^2$ (hence the very low pressure in the centre of a tornado). Deduce that if there is a free surface to the fluid and gravity is acting, then the surface at $r = 0$ is a depth $\Omega^2 a^2/g$ below the surface at $r = \infty$ (hence the dimples in a cup of tea accompanying the vortices that are shed by the edges of the spoon).

1.4. Take the Euler equation for an incompressible fluid of constant density, cast it into an appropriate form, and perform suitable operations on it to obtain the *energy equation*:

$$\frac{d}{dt}\int_V \tfrac{1}{2}\rho \mathbf{u}^2 \, dV = -\int_S (p' + \tfrac{1}{2}\rho \mathbf{u}^2)\mathbf{u} \cdot \mathbf{n} \, dS,$$

where V is the region enclosed by a fixed closed surface S drawn in the fluid, and p' denotes $p + \rho\chi$, the non-hydrostatic part of the pressure field.

1.5. For an inviscid fluid we have Euler's equation

$$\frac{\partial \mathbf{u}}{\partial t} + \boldsymbol{\omega} \wedge \mathbf{u} + \nabla(\tfrac{1}{2}\mathbf{u}^2) = -\frac{1}{\rho}\nabla p - \nabla \chi,$$

and, whether or not the fluid is incompressible, we also have conservation of mass (Exercise 1.1):

$$\frac{D\rho}{Dt} + \rho \nabla \cdot \boldsymbol{u} = 0.$$

Show that

$$\frac{D}{Dt}\left(\frac{\boldsymbol{\omega}}{\rho}\right) = \left(\frac{\boldsymbol{\omega}}{\rho} \cdot \nabla\right)\boldsymbol{u} - \frac{1}{\rho}\nabla\left(\frac{1}{\rho}\right) \wedge \nabla p. \qquad (1.39)$$

Deduce that, if p is a function of ρ alone, the vorticity equation is exactly as in the incompressible, constant density case, except that $\boldsymbol{\omega}$ is replaced by $\boldsymbol{\omega}/\rho$.

1.6. Show that the circulation is the same round all simple closed circuits enclosing the wing in Fig. 1.8. (Hint: sketch two such circuits, and then make a construction so as to create a single closed circuit that does not enclose the wing.)

1.7. Sketch the streamlines for the flow

$$u = \alpha x, \qquad v = -\alpha y, \qquad w = 0,$$

where α is a positive constant.

Let the concentration of some pollutant in the fluid be

$$c(x, y, t) = \beta x^2 y e^{-\alpha t},$$

for $y > 0$, where β is a constant. Does the pollutant concentration for any particular fluid element change with time?

An alternative way of describing any flow is to specify the position $\boldsymbol{x}$ of each fluid element at time t in terms of the position $\boldsymbol{X}$ of that element at time $t = 0$. For the above flow this 'Lagrangian' description is

$$x = X e^{\alpha t}, \qquad y = Y e^{-\alpha t}, \qquad z = Z.$$

Verify by direct calculation that

$$\left(\frac{\partial \boldsymbol{x}}{\partial t}\right)_X = \boldsymbol{u}, \qquad \left(\frac{\partial \boldsymbol{u}}{\partial t}\right)_X = \frac{D\boldsymbol{u}}{Dt}$$

in this particular case. Why are these results true in general?

Write c as a function of X, Y, and t.

1.8. Consider the unsteady flow

$$u = u_0, \qquad v = kt, \qquad w = 0,$$

where u_0 and k are positive constants. Show that the streamlines are straight lines, and sketch them at two different times. Also show that any fluid particle follows a parabolic path as time proceeds.

2 Elementary viscous flow

2.1. Introduction

Steady flow past a fixed aerofoil may seem at first to be wholly accounted for by inviscid flow theory. The streamline pattern seems right, and so does the velocity field. In particular, the fluid in contact with the aerofoil appears to slip along the boundary in just the manner predicted by inviscid theory. Yet close inspection reveals that there is in fact no such slip. Instead there is a very thin *boundary layer,* across which the flow velocity undergoes a smooth but rapid adjustment to precisely zero—corresponding to *no slip*—on the aerofoil itself (Fig. 2.1). In this boundary layer inviscid theory fails, and viscous effects are important, even though they are negligible in the main part of the flow.

To see why this should be so we must first make precise what we mean by the term 'viscous'. To this end, consider the case of simple shear flow, so that $\boldsymbol{u} = [u(y), 0, 0]$. The fluid immediately above some level y = constant exerts a stress, i.e. a force per unit area of contact, on the fluid immediately below, and vice versa. For an inviscid fluid this stress has no tangential component τ, but for a viscous fluid τ is typically non-zero. In this book we shall be concerned with *Newtonian* viscous fluids, and in this case the shear stress τ is proportional to the velocity gradient du/dy, i.e.

$$\tau = \mu \frac{du}{dy}, \qquad (2.1)$$

where μ is a property of the fluid, called the *coefficient of viscosity*. Many real fluids, such as water or air, behave according to eqn (2.1) over a wide range of conditions (although there are many others, including paints and polymers, which are non-Newtonian, and do not; see Tanner (1988)).

From a fluid dynamical point of view the so-called *kinematic viscosity*

$$\nu = \mu/\rho \qquad (2.2)$$

Elementary viscous flow 27

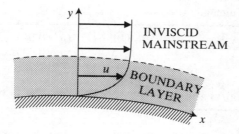

Fig. 2.1. A boundary layer.

is often more significant than μ itself, and some typical values of ν are given in Table 2.1. These values can vary quite substantially with temperature, but throughout much of this book we shall concentrate on a simple model of fluid flow in which μ, ρ, and ν are all constant.

We can see now, in general terms, why viscous effects become important in a boundary layer. The reason is that the velocity gradients in a boundary layer are much larger than they are in the main part of the flow, because a substantial change in velocity is taking place across a very thin layer. In this way the viscous stress (2.1) becomes significant in a boundary layer, even though μ is small enough for viscous effects to be negligible elsewhere in the flow.

But why are boundary layers so important that we begin this chapter with them? The answer is that in certain circumstances

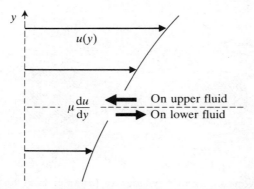

Fig. 2.2. Viscous stresses in a simple shear flow.

28   Elementary viscous flow

Table 2.1.   Kinematic viscosity v (cm^2 s^{-1}) at 15°C.

Water	0.01	($\mu = 0.01$ c.g.s. units)
Air	0.15	($\mu = 0.0002$ c.g.s. units)
Olive oil	1.0	
Glycerine	18	
Golden syrup/treacle	~1200	($v \sim 200$ at 27°C)

they may *separate* from the boundary, *thus causing the whole flow of a low-viscosity fluid to be quite different to that predicted by inviscid theory.*

Consider, for example, the flow of a low-viscosity fluid past a circular cylinder. In the first instance it is natural to assume that viscous effects will be negligible in the main part of the flow, which will therefore be irrotational, by the argument of §1.5. If we solve the problem of irrotational flow past a circular cylinder (§4.5) we obtain the streamline pattern of Fig. 2.3(a). This 'solution' is not wholly satisfactory, for it predicts slip on the surface of the cylinder. We might then suppose that a thin viscous boundary layer intervenes to adjust the velocity smoothly to zero on the cylinder itself. But this turns out to be wishful thinking; the observed flow of a low-viscosity fluid past a circular cylinder is, instead, of an altogether different kind, with massive separation of the boundary layer giving rise to a large vorticity-filled *wake* (Fig. 2.3(b)).

Why does separation occur? The answer lies in the variation of pressure p along the boundary, as predicted by inviscid theory.

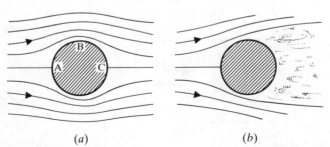

(a)   (b)

Fig. 2.3.   Flow past a circular cylinder for (a) an inviscid fluid and (b) a fluid of small viscosity.

Elementary viscous flow 29

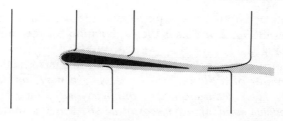

Fig. 2.4. Flow past an aerofoil: the fate of successive lines of fluid particles.

In Fig. 2.3(a), inviscid theory predicts that p has a local maximum at the forward stagnation point A, falls to a minimum at B, then increases to a local maximum at C, with $p_A = p_C$. This implies that between B and C there is a substantial increase in pressure along the boundary in the direction of flow. It is this *severe adverse pressure gradient along the boundary* which causes the boundary layer to separate, for reasons which are outlined in §§8.1 and 8.6 (see especially Fig. 8.2.)

An aerofoil, on the other hand, is deliberately designed to avoid such large-scale separation, the key feature being its slowly tapering rear. In Fig. 1.9, for example, the substantial fall in pressure over the first 10% or so of the upper surface is followed by a very *gradual* pressure rise over the remainder. For this reason the boundary layer does not separate until close to the trailing edge, and there is only a very narrow wake (Fig. 2.4). This state of affairs persists as long as the angle of attack α is not too large; if α is greater than a few degrees, the pressure rise over the remainder of the upper surface is no longer gradual,

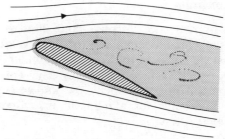

Fig. 2.5. Separated flow past an aerofoil.

30 Elementary viscous flow

large-scale separation takes place, and the aerofoil is said to be *stalled*, as in Fig. 2.5. This is the explanation for the sudden drop in lift in Fig. 1.11.

The most important overall message of this introduction is that *the behaviour of a fluid of small viscosity μ may, on account of boundary layer separation, be completely different to that of a (hypothetical) fluid of no viscosity at all.* From a mathematical point of view, what happens in the limit $\mu \to 0$ may be quite different to what happens when $\mu = 0$.

2.2. The equations of viscous flow

So far we have considered the motion of fluids of small viscosity. Yet there is more to the subject than this, including the opposite extreme of very viscous flow (Chapter 7). It is time, then, to take a more balanced—if brief—look at viscous flow as a whole.

The Navier–Stokes equations

Suppose that we have an incompressible Newtonian fluid of constant density ρ and constant viscosity μ. Its motion is governed by the *Navier–Stokes equations*†

$$\frac{\partial \boldsymbol{u}}{\partial t} + (\boldsymbol{u} \cdot \nabla)\boldsymbol{u} = -\frac{1}{\rho}\nabla p + \nu \nabla^2 \boldsymbol{u} + \boldsymbol{g}, \qquad (2.3)$$

$$\nabla \cdot \boldsymbol{u} = 0.$$

These differ from the Euler equations (1.12) by virtue of the viscous term $\nu \nabla^2 \boldsymbol{u}$, where ∇^2 denotes the Laplace operator $\partial^2/\partial x^2 + \partial^2/\partial y^2 + \partial^2/\partial z^2$.

The no-slip condition

Observations of real (i.e. viscous) fluid flow reveal that both normal and tangential components of fluid velocity at a rigid boundary must be equal to those of the boundary itself. Thus if the boundary is at rest, $\boldsymbol{u} = 0$ there. The condition on the tangential component of velocity is known as the *no-slip condition*, and it holds for a fluid of any viscosity $\nu \neq 0$, no matter how small ν may be.

---

† The Navier–Stokes equations are derived from first principles in Chapter 6.

The Reynolds number

Consider a viscous fluid in motion, and let U denote a typical flow speed. Furthermore, let L denote a characteristic length scale of the flow. This is all somewhat subjective, but in dealing with the spin-down of a stirred cup of tea, for instance, 4 cm and 5 cm s^{-1} would be reasonable choices for L and U, while 10 m and 100 m s^{-1} would not. Having thus chosen a value for L and for U we may form the quantity

$$R = \frac{UL}{\nu}, \qquad (2.4)$$

which is a pure number known as a *Reynolds number*.

To see why R should be important, note that derivatives of the velocity components, such as $\partial u/\partial x$, will typically be of order U/L—assuming, that is, that the components of $\boldsymbol{u}$ change by amounts of order U over distances of order L. Typically, these derivatives will themselves change by amounts of order U/L over distances of order L, so second derivatives such as $\partial^2 u/\partial x^2$ will be of order U/L^2. In this way we obtain the following order of magnitude estimates for two of the terms in eqn (2.3):

inertia term: $\quad |(\boldsymbol{u} \cdot \nabla)\boldsymbol{u}| = O(U^2/L),$
viscous term: $\quad |\nu\nabla^2\boldsymbol{u}| = O(\nu U/L^2).$ $\qquad (2.5)$

Provided that these are correct we deduce that

$$\frac{|\text{inertia term}|}{|\text{viscous term}|} = O\left(\frac{U^2/L}{\nu U/L^2}\right) = O(R). \qquad (2.6)$$

The Reynolds number is important, then, because it can give a rough indication of the relative magnitudes of two key terms in the equations of motion (2.3). It is not surprising, therefore, that high Reynolds number flows and low Reynolds number flows have quite different general characteristics.

High Reynolds number flow

The case $R \gg 1$ corresponds to what we have hitherto called the motion of a fluid of small viscosity. Equation (2.6) suggests that viscous effects should on the whole be negligible, and flow past a

32 Elementary viscous flow

thin aerofoil at small angle of attack provides just one example where this is indeed the case. Even then, however, viscous effects become important in thin boundary layers, where the unusually large velocity gradients make the viscous term much larger than the estimate in eqn (2.5). We show in §§8.1 and 8.2 that the typical thickness δ of such a boundary layer is given by

$$\delta/L = O(R^{-\frac{1}{2}}). \tag{2.7}$$

The larger the Reynolds number, then, the thinner the boundary layer.

A large Reynolds number is *necessary* for inviscid theory to apply over most of the flow field, but it is not sufficient. As we have seen, boundary layer separation can lead to a quite different state of affairs. A further complication at high Reynolds number is that steady flows are often *unstable* to small disturbances, and may, as a result, become *turbulent*. It was in fact in this context that Reynolds first employed the dimensionless parameter that now bears his name (see §9.1).

Low Reynolds number flow

Consider a laboratory experiment in which golden syrup occupies the gap between two circular cylinders, the inner one rotating and the outer one at rest. For reasonable rotation rates of the inner cylinder the Reynolds number might be in the region of 10^{-2} or so; it will certainly be much less than 1. At such Reynolds numbers there is no sign of turbulence, and the flow is extremely well ordered.

The flow is so well ordered, in fact, that if the rotation of the

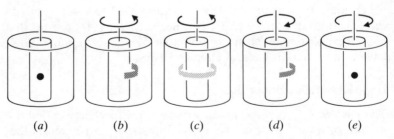

Fig. 2.6. The reversibility of a very viscous flow.

inner cylinder is stopped after a few revolutions, and the inner cylinder is then rotated back through the correct number of turns to its original position, a dyed blob of syrup, which has been greatly sheared in the meantime, will return almost exactly to its original configuration as a concentrated blob (Fig. 2.6).

This near *reversibility* is characteristic of low Reynolds number flows, and helps account, in fact, for the unusual swimming techniques that are adopted by certain biological microorganisms such as the Spermatozoa (§7.5).

2.3. Some simple viscous flows: the diffusion of vorticity

We now turn to some elementary exact solutions of the Navier–Stokes equations. There is, in addition, a major theme running through §§2.3 and 2.4, and that theme is the *viscous diffusion of vorticity*, an important mechanism which was wholly absent in Chapter 1, where v was zero.

Plane parallel shear flow

Suppose that a viscous fluid is moving so that relative to some set of rectangular Cartesian coordinates

$$\boldsymbol{u} = [u(y, t), 0, 0]. \tag{2.8}$$

Such a flow is termed a plane parallel shear flow. It automatically satisfies $\nabla \cdot \boldsymbol{u} = 0$, as u is independent of x, and in the absence of gravity† the Navier–Stokes equations (2.3) become, in component form:

$$\frac{\partial u}{\partial t} = -\frac{1}{\rho}\frac{\partial p}{\partial x} + v\frac{\partial^2 u}{\partial y^2},$$
$$\frac{\partial p}{\partial y} = \frac{\partial p}{\partial z} = 0. \tag{2.9}$$

The pressure p is thus a function of x and t only. But from eqn (2.9) $\partial p/\partial x$ is equal to the difference between two terms which are independent of x. Thus $\partial p/\partial x$ must be a function of t alone. As we shall see shortly, there are important circumstances in which this fact enables us to deduce that $\partial p/\partial x$ must be zero.

---

† See footnote on p. 9.

34 Elementary viscous flow

First, however, it is instructive to see how eqn (2.9) may be obtained by a simple and direct application of the expression (2.1).

An ad hoc derivation of the equations of motion for a viscous fluid in plane parallel shear flow

First note that in the absence of viscous forces the corresponding Euler equation

$$\rho \frac{\partial u}{\partial t} = -\frac{\partial p}{\partial x} \qquad (2.10)$$

may be deduced by considering an element of fluid of unit length in the z-direction and of small, rectangular cross-section in the x–y plane, with sides of length δx and δy (see Fig. 2.7). The net pressure force on the element in the x-direction is

$$p(x)\,\delta y - p(x + \delta x)\,\delta y \doteq -\frac{\partial p}{\partial x}\,\delta x\,\delta y,$$

and this is equal to the product of the element's mass $\rho\,\delta x\,\delta y$ and its acceleration

$$\frac{Du}{Dt} = \frac{\partial u}{\partial t} + u \frac{\partial u}{\partial x},$$

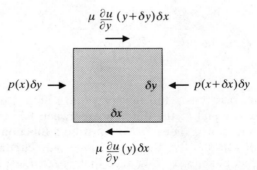

Fig. 2.7. The forces in the x-direction on a small rectangular blob in a plane parallel shear flow.

which reduces simply to $\partial u/\partial t$ because u is independent of x.

In a similar manner we may use eqn (2.1) to deduce that viscous forces on the top and bottom of the element give rise to a net contribution in the x-direction of

$$\mu \frac{\partial u}{\partial y}\bigg|_{y+\delta y} \delta x - \mu \frac{\partial u}{\partial y}\bigg|_{y} \delta x \doteq \mu \frac{\partial^2 u}{\partial y^2} \delta x\, \delta y, \qquad (2.11)$$

whence eqn (2.10) becomes modified to

$$\rho \frac{\partial u}{\partial t} = -\frac{\partial p}{\partial x} + \mu \frac{\partial^2 u}{\partial y^2},$$

i.e. to eqn (2.9).

This equation is, of course, valid only for a very restricted class of flows, but the brevity of the above derivation does have its merits. In particular, it brings out rather clearly, via eqns (2.1) and (2.11), why the viscous term in the equation of motion (2.3) involves the *second* derivatives of the velocity field.

The flow due to an impulsively moved plane boundary

Suppose that viscous fluid lies at rest in the region $0 < y < \infty$ and suppose that at $t = 0$ the rigid boundary $y = 0$ is suddenly jerked into motion in the x-direction with constant speed U. By virtue of the no-slip condition the fluid elements in contact with the boundary will immediately move with velocity U. We wish to find how the rest of the fluid responds.

It is natural to look for a flow of the form (2.8), and eqn (2.9) then applies. We assume that the flow is being driven only by the motion of the boundary, i.e. not by any externally applied pressure gradient. This experimental consideration corresponds to asserting that the pressures at $x = \pm\infty$ are equal, and as $\partial p/\partial x$ is independent of x (so that p is a linear function of x) it follows that $\partial p/\partial x$ is zero.

The velocity $u(y, t)$ thus satisfies the classical one-dimensional *diffusion equation*

$$\frac{\partial u}{\partial t} = \nu \frac{\partial^2 u}{\partial y^2}, \qquad (2.12)$$

together with the initial condition

$$u(y, 0) = 0, \qquad y > 0,$$

36 Elementary viscous flow

and the boundary conditions

$$u(0, t) = U, \quad t > 0, \qquad u(\infty, t) = 0, \quad t > 0.$$

This whole problem is in fact identical with the problem of the spreading of heat through a thermally conducting solid when its boundary temperature is suddenly raised from zero to some constant.

We may proceed most easily, on this occasion, by seeking a *similarity solution*. We postpone a more rational discussion of this method until §8.3; for the time being we simply observe that the equation is unchanged by the transformation of variables $y \Rightarrow \alpha y$, $t \Rightarrow \alpha^2 t$, α being a constant. This suggests the possibility that there are solutions to eqn (2.12) which are functions of y and t simply through the single combination $y/t^{\frac{1}{2}}$, for this 'similarity' variable would itself be unchanged by such a transformation. Inspection of eqn (2.12) suggests that it may be more convenient still to take $y/(vt)^{\frac{1}{2}}$ as the similarity variable. Thus if we try

$$u = f(\eta), \qquad \text{where } \eta = y/(vt)^{\frac{1}{2}}, \qquad (2.13)$$

so that

$$\frac{\partial u}{\partial t} = f'(\eta)\frac{\partial \eta}{\partial t} = -f'(\eta)\frac{y}{2v^{\frac{1}{2}}t^{\frac{3}{2}}},$$

$$\frac{\partial u}{\partial y} = f'(\eta)\frac{\partial \eta}{\partial y} = f'(\eta)\frac{1}{v^{\frac{1}{2}}t^{\frac{1}{2}}}, \qquad \text{etc.},$$

we obtain, from eqn (2.12),

$$f'' + \tfrac{1}{2}\eta f' = 0.$$

Integrating,

$$f' = B e^{-\eta^2/4},$$

whence

$$f = A + B \int_0^{\eta} e^{-s^2/4}\, ds,$$

where A and B are constants of integration, to be determined from the initial and boundary conditions. By virtue of eqn (2.13) these reduce to

$$f(\infty) = 0, \qquad f(0) = U,$$

Elementary viscous flow 37

so that
$$u = U\left[1 - \frac{1}{\pi^{\frac{1}{2}}}\int_0^\eta e^{-s^2/4}\,ds\right] \quad (2.14)$$
is the solution of the problem, where $\eta = y/(vt)^{\frac{1}{2}}$.

The simple form of the initial and boundary conditions was essential to the success of the method. The underlying reason lies in the nature of the similarity solution (2.14) itself. As its name implies, the velocity profiles $u(y)$ are, at different times, all geometrically similar. At time t_1 the velocity u is a function of $y/(vt_1)^{\frac{1}{2}}$; at a later time t_2 the velocity u is the *same* function of $y/(vt_2)^{\frac{1}{2}}$. All that happens as time goes on is that the velocity profile becomes stretched out, as indicated in Fig. 2.8. We would not expect this to be the case if, for instance, an upper boundary were present, and the solution is, indeed, not then of similarity form (see eqn (2.21)).

At time t the effects of the motion of the plane boundary are largely confined to a distance of order $(vt)^{\frac{1}{2}}$ from the boundary; u is less than 1% of U at $y = 4(vt)^{\frac{1}{2}}$. In this way viscous effects gradually communicate the motion of the boundary to the whole fluid.

A more fundamental way of viewing this process, open to considerable generalization, is in terms of the diffusion of vorticity. The vorticity is
$$\omega = -\frac{\partial u}{\partial y} = \frac{U}{(\pi vt)^{\frac{1}{2}}} e^{-y^2/4vt}, \quad (2.15)$$

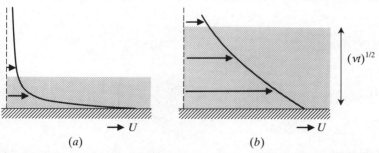

Fig. 2.8. The diffusion of vorticity from a plane boundary suddenly moved with velocity U. The solid line indicates the velocity profile at some early time (*a*) and some later time (*b*); the shading indicates the region of significant vorticity.

38 Elementary viscous flow

and this is exponentially small beyond a distance of order $(vt)^{\frac{1}{2}}$ from the boundary. The spreading of vorticity by viscous action thus smooths out what was, initially, a *vortex sheet*, i.e. an infinite concentration of vorticity at the boundary ($y = 0$, $t \to 0$) with none elsewhere ($y > 0$, $t \to 0$).

Finally we may state these broad conclusions in a slightly different way. Vorticity diffuses a distance of order $(vt)^{\frac{1}{2}}$ in time t. Equivalently, *the time taken for vorticity to diffuse a distance of order L is of the order*

$$\text{viscous diffusion time} = O(L^2/v). \qquad (2.16)$$

Steady flow under gravity down an inclined plane

This next solution of the Navier–Stokes equations serves to make one or two elementary points about technique.

It may be argued that the key step in solving any flow problem, having decided on a sensible coordinate system, is to decide the number of independent variables (e.g. x, y, z, t) on which $\boldsymbol{u}$ depends, and the rule is 'the fewer, the better'.

In the present problem $\boldsymbol{u}$ is zero on $y = 0$ (see Fig. 2.9), by virtue of the no-slip condition, so $\boldsymbol{u}$ must depend on y. In the absence of any a priori reason why $\boldsymbol{u}$ needs to depend on anything else we examine the possibility that there is a two-dimensional steady flow solution in which $\boldsymbol{u} = [u(y), v(y), 0]$.

Now, it is only common sense in any problem to *turn to the incompressibility condition at an early stage*, for of the two

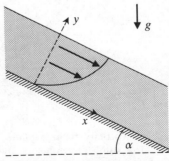

Fig. 2.9. Steady flow of a viscous fluid down an inclined plane.

equations (2.3) it is by far the simpler. In the present instance it tells us immediately that

$$dv/dy = 0,$$

i.e. that v is a constant. But $v = 0$ on $y = 0$, so v is zero everywhere.

Substituting $\boldsymbol{u} = [u(y), 0, 0]$ into the momentum equation (2.3), with the gravitational body force included, we obtain

$$0 = -\frac{1}{\rho}\frac{\partial p}{\partial x} + \nu\frac{d^2 u}{dy^2} + g \sin \alpha,$$
$$0 = -\frac{1}{\rho}\frac{\partial p}{\partial y} - g \cos \alpha. \qquad (2.17)$$

Integrating the second of these we find

$$p = -\rho g y \cos \alpha + f(x),$$

where $f(x)$ is an arbitrary function of x.

Now, the free surface must be $y = h$, where h is a constant, because all the streamlines are parallel to the plane. At this free surface the tangential stress must be zero and the pressure p must be equal to atmospheric pressure p_0 (see Exercise 6.3), so

$$p = p_0 \quad \text{and} \quad \mu\frac{du}{dy} = 0 \quad \text{at } y = h, \qquad (2.18)$$

by virtue of eqn (2.1). Consequently,

$$p - p_0 = \rho g (h - y) \cos \alpha,$$

whence $\partial p/\partial x$ is zero. Equation (2.17) then reduces to

$$\nu\frac{d^2 u}{dy^2} = -g \sin \alpha,$$

and we may easily integrate this twice, applying the boundary conditions

$$u = 0 \quad \text{at } y = 0, \quad \mu\frac{du}{dy} = 0 \quad \text{at } y = h,$$

to obtain

$$u = \frac{g}{2\nu} y (2h - y) \sin \alpha. \qquad (2.19)$$

40 Elementary viscous flow

The velocity profile is therefore parabolic, as shown in Fig. 2.9. The volume flux down the plane, per unit length in the z-direction, is

$$Q = \int_0^h u \, dy = \frac{gh^3}{3\nu} \sin \alpha.$$

Another example of vorticity diffusion

Consider the problem in Fig. 2.10, in which a lower rigid boundary $y = 0$ is suddenly moved with speed U, while an upper rigid boundary to the fluid, $y = h$, is held at rest. As in an earlier subsection, we argue that $\boldsymbol{u} = [u(y, t), 0, 0]$ will satisfy eqn (2.12):

$$\frac{\partial u}{\partial t} = \nu \frac{\partial^2 u}{\partial y^2}, \qquad (2.20)$$

subject to the initial condition

$$u(y, 0) = 0, \qquad 0 < y < h;$$

but this time the boundary conditions will be

$$u(0, t) = U, \quad t > 0, \qquad u(h, t) = 0, \quad t > 0.$$

The equation is homogeneous, but the boundary conditions are not. Before using the method of separation of variables and

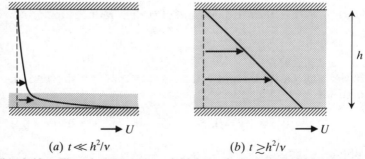

Fig. 2.10. Flow between two rigid boundaries, one suddenly moved with speed U and one held fixed. Shading indicates regions of significant vorticity.

Fourier series we therefore reformulate the problem by first seeking a *steady* solution that satisfies the boundary conditions; this is clearly $U(1 - y/h)$. We therefore write

$$u = U(1 - y/h) + u_1,$$

where

$$\frac{\partial u_1}{\partial t} = v \frac{\partial^2 u_1}{\partial y^2},$$

$$u_1(y, 0) = -U(1 - y/h), \qquad 0 < y < h,$$

$$u_1(0, t) = 0, \qquad t > 0, \qquad u_1(h, t) = 0, \qquad t > 0.$$

The boundary conditions are now homogeneous. By the method of separation of variables we find that the functions

$$\exp(-n^2\pi^2 vt/h^2)\sin(n\pi y/h), \qquad n = 1, 2, \ldots.$$

all satisfy the equation for u_1 and the boundary conditions for u_1 at $y = 0, h$. None of these individually satisfies the initial condition for u_1, but by writing

$$u_1 = \sum_{n=1}^{\infty} A_n \exp(-n^2\pi^2 vt/h^2)\sin(n\pi y/h),$$

we may use Fourier theory to determine the A_n such that

$$\sum_{n=1}^{\infty} A_n \sin(n\pi y/h) = -U(1 - y/h) \qquad \text{in } 0 < y < h,$$

thus satisfying the initial condition. In this way we find

$$A_n = -\frac{2}{h} \int_0^h U(1 - y/h)\sin(n\pi y/h)\, dy = -2U/n\pi,$$

and the solution is therefore

$$u(y, t) = U(1 - y/h) - \frac{2U}{\pi} \sum_{n=1}^{\infty} \frac{1}{n} \exp(-n^2\pi^2 vt/h^2)\sin(n\pi y/h).$$

(2.21)

The main feature of this solution is that for times $t \gtrsim h^2/v$ (cf. eqn (2.16)) the flow has almost reached its steady state, as in Fig. 2.10(b), and the vorticity is almost distributed uniformly throughout the fluid.

2.4. Flow with circular streamlines

The Navier–Stokes equations are

$$\frac{\partial \boldsymbol{u}}{\partial t} + (\boldsymbol{u} \cdot \nabla)\boldsymbol{u} = -\frac{1}{\rho}\nabla p + \nu\nabla^2\boldsymbol{u},$$

$$\nabla \cdot \boldsymbol{u} = 0,$$

and when written out in cylindrical polar coordinates they become

$$\frac{\partial u_r}{\partial t} + (\boldsymbol{u} \cdot \nabla)u_r - \frac{u_\theta^2}{r} = -\frac{1}{\rho}\frac{\partial p}{\partial r} + \nu\left(\nabla^2 u_r - \frac{u_r}{r^2} - \frac{2}{r^2}\frac{\partial u_\theta}{\partial \theta}\right)$$

$$\frac{\partial u_\theta}{\partial t} + (\boldsymbol{u} \cdot \nabla)u_\theta + \frac{u_r u_\theta}{r} = -\frac{1}{\rho r}\frac{\partial p}{\partial \theta} + \nu\left(\nabla^2 u_\theta + \frac{2}{r^2}\frac{\partial u_r}{\partial \theta} - \frac{u_\theta}{r^2}\right)$$

$$\frac{\partial u_z}{\partial t} + (\boldsymbol{u} \cdot \nabla)u_z = -\frac{1}{\rho}\frac{\partial p}{\partial z} + \nu\nabla^2 u_z,$$

(2.22)

$$\frac{1}{r}\frac{\partial}{\partial r}(ru_r) + \frac{1}{r}\frac{\partial u_\theta}{\partial \theta} + \frac{\partial u_z}{\partial z} = 0,$$

where

$$(\boldsymbol{u} \cdot \nabla) = u_r\frac{\partial}{\partial r} + \frac{u_\theta}{r}\frac{\partial}{\partial \theta} + u_z\frac{\partial}{\partial z},$$

$$\nabla^2 = \frac{1}{r}\frac{\partial}{\partial r}\left(r\frac{\partial}{\partial r}\right) + \frac{1}{r^2}\frac{\partial^2}{\partial \theta^2} + \frac{\partial^2}{\partial z^2},$$

(see eqn (A.35)).

Note the 'extra' terms that arise; the r-component of $(\boldsymbol{u} \cdot \nabla)\boldsymbol{u}$ is not $(\boldsymbol{u} \cdot \nabla)u_r$, for instance, but $(\boldsymbol{u} \cdot \nabla)u_r - u_\theta^2/r$ instead. This kind of thing occurs because $\boldsymbol{u} = u_r\boldsymbol{e}_r + u_\theta\boldsymbol{e}_\theta + u_z\boldsymbol{e}_z$, and *some of the unit vectors involved change with* θ:

$$\frac{\partial \boldsymbol{e}_r}{\partial \theta} = \boldsymbol{e}_\theta, \qquad \frac{\partial \boldsymbol{e}_\theta}{\partial \theta} = -\boldsymbol{e}_r, \qquad \frac{\partial \boldsymbol{e}_z}{\partial \theta} = 0, \qquad (2.23)$$

(see eqn (A.29)). When $(\boldsymbol{u} \cdot \nabla)\boldsymbol{u}$ and $\nu\nabla^2\boldsymbol{u}$ are expanded carefully using these expressions they may be seen to yield eqn (2.22).

Elementary viscous flow 43

Taking explicit account of the change in direction of unit vectors may alternatively be avoided by use of the identities

$$(\boldsymbol{u} \cdot \nabla)\boldsymbol{u} = (\nabla \wedge \boldsymbol{u}) \wedge \boldsymbol{u} + \nabla(\tfrac{1}{2}u^2), \qquad (2.24)$$

$$\nabla^2 \boldsymbol{u} = \nabla(\nabla \cdot \boldsymbol{u}) - \nabla \wedge (\nabla \wedge \boldsymbol{u}). \qquad (2.25)$$

For this purpose we recall

$$\nabla \wedge \boldsymbol{u} = \frac{1}{r} \begin{vmatrix} \boldsymbol{e}_r & r\boldsymbol{e}_\theta & \boldsymbol{e}_z \\ \dfrac{\partial}{\partial r} & \dfrac{\partial}{\partial \theta} & \dfrac{\partial}{\partial z} \\ u_r & ru_\theta & u_z \end{vmatrix} \qquad (2.26)$$

(see Exercise 2.13).

The differential equation for circular flow

Consider solutions to the Navier–Stokes equations of the form

$$\boldsymbol{u} = u_\theta(r, t)\boldsymbol{e}_\theta, \qquad (2.27)$$

so that the streamlines are circular. The incompressibility condition $\nabla \cdot \boldsymbol{u} = 0$ is automatically satisfied for any flow of the form (2.27).

Rather than use the remaining equations in the ready-made form (2.22) it is instructive to derive them, for the flow (2.27), using the expressions (2.23). Thus

$$(\boldsymbol{u} \cdot \nabla)\boldsymbol{u} = \frac{u_\theta}{r} \frac{\partial}{\partial \theta}[u_\theta(r, t)\boldsymbol{e}_\theta] = \frac{u_\theta^2}{r} \frac{\partial \boldsymbol{e}_\theta}{\partial \theta} = -\frac{u_\theta^2}{r}\boldsymbol{e}_r, \qquad (2.28)$$

while

$$\nu \nabla^2 \boldsymbol{u} = \nu\left(\frac{\partial^2}{\partial r^2} + \frac{1}{r}\frac{\partial}{\partial r} + \frac{1}{r^2}\frac{\partial^2}{\partial \theta^2} + \frac{\partial^2}{\partial z^2}\right)[u_\theta(r, t)\boldsymbol{e}_\theta],$$

and

$$\frac{1}{r^2}\frac{\partial^2}{\partial \theta^2}[u_\theta \boldsymbol{e}_\theta] = \frac{1}{r^2}\frac{\partial}{\partial \theta}\left(u_\theta \frac{\partial \boldsymbol{e}_\theta}{\partial \theta}\right) = \frac{-1}{r^2}\frac{\partial}{\partial \theta}(u_\theta \boldsymbol{e}_r) = -\frac{u_\theta}{r^2}\boldsymbol{e}_\theta,$$

so

$$\nu \nabla^2 \boldsymbol{u} = \nu\left(\frac{\partial^2 u_\theta}{\partial r^2} + \frac{1}{r}\frac{\partial u_\theta}{\partial r} - \frac{u_\theta}{r^2}\right)\boldsymbol{e}_\theta. \qquad (2.29)$$

When $\boldsymbol{u} = u_\theta(r, t)\boldsymbol{e}_\theta$ the Navier–Stokes equations therefore reduce to

$$-\frac{u_\theta^2}{r} = -\frac{1}{\rho}\frac{\partial p}{\partial r},$$

$$\frac{\partial u_\theta}{\partial t} = -\frac{1}{\rho r}\frac{\partial p}{\partial \theta} + \nu\left(\frac{\partial^2 u_\theta}{\partial r^2} + \frac{1}{r}\frac{\partial u_\theta}{\partial r} - \frac{u_\theta}{r^2}\right),$$

$$0 = -\frac{1}{\rho}\frac{\partial p}{\partial z},$$

as we might have deduced more quickly from eqn (2.22).

Now, u_θ is a function of r and t only, so from the second equation the same must be true of $\partial p/\partial \theta$, so $\partial p/\partial \theta = P(r, t)$, say. Integrating:

$$p = P(r, t)\theta + f(r, t),$$

as $\partial p/\partial z = 0$. We conclude that $P(r, t) = 0$, for otherwise p would be a multivalued function of position (different at $\theta = 0$ and at $\theta = 2\pi$, say). Thus

$$\frac{\partial u_\theta}{\partial t} = \nu\left(\frac{\partial^2 u_\theta}{\partial r^2} + \frac{1}{r}\frac{\partial u_\theta}{\partial r} - \frac{u_\theta}{r^2}\right) \tag{2.30}$$

is the evolution equation for a viscous flow with $\boldsymbol{u} = u_\theta(r, t)\boldsymbol{e}_\theta$.

Steady flow between rotating cylinders

For steady flow we have

$$r^2 \frac{d^2 u_\theta}{dr^2} + r\frac{du_\theta}{dr} - u_\theta = 0,$$

with general solution

$$u_\theta = Ar + \frac{B}{r}. \tag{2.31}$$

If the fluid occupies the gap $r_1 \leq r \leq r_2$ between two circular cylinders which rotate with angular velocities Ω_1 and Ω_2, then we may apply the no-slip condition at each cylinder to obtain

$$A = \frac{\Omega_2 r_2^2 - \Omega_1 r_1^2}{r_2^2 - r_1^2}, \qquad B = \frac{(\Omega_1 - \Omega_2)r_1^2 r_2^2}{r_2^2 - r_1^2}. \tag{2.32}$$

Elementary viscous flow 45

The most interesting thing about this flow is the manner in which it becomes unstable if Ω_1 is too large, so that superbly regular and axisymmetric *Taylor vortices* appear (see §9.4, especially Fig. 9.8).

Spin-down in an infinitely long circular cylinder

Suppose viscous fluid occupies the region $r \leq a$ within a circular cylinder of radius a, and suppose that both cylinder and fluid are initially rotating with uniform angular velocity Ω, so that

$$u_\theta = \Omega r, \qquad r \leq a, \qquad t = 0.$$

Suppose that the cylinder is then suddenly brought to rest. We need to solve

$$\frac{\partial u_\theta}{\partial t} = \nu \left(\frac{\partial^2 u_\theta}{\partial r^2} + \frac{1}{r} \frac{\partial u_\theta}{\partial r} - \frac{u_\theta}{r^2} \right)$$

with the above initial condition and the boundary condition

$$u_\theta = 0 \qquad \text{at } r = a, \quad t > 0.$$

The problem may be tackled in a Fourier-series type manner, as for eqn (2.21), but the separable solutions now involve Bessel functions, and

$$u_\theta(r, t) = -2\Omega a \sum_{n=1}^{\infty} \frac{J_1(\lambda_n r/a)}{\lambda_n J_0(\lambda_n)} \exp\left(-\lambda_n^2 \frac{\nu t}{a^2}\right). \qquad (2.33)$$

Here λ_n denote the positive values of λ at which $J_1(\lambda) = 0$, and J_k denotes the Bessel function of order k. All the terms of the series decay rapidly with t; the one that survives longest is the first one, and $\lambda_1 \doteq 3.83$. The 'spin-down' process is therefore well under way in a time of order $a^2/\nu\lambda_1^2$, i.e. in the classic viscous diffusion time (2.16).

If we apply this to a stirred cup of tea, with $a = 4$ cm and $\nu = 10^{-2}$ cm^2 s^{-1} for water, we obtain a 'spin-down' time of about 2 minutes. This is much too long; casual observation suggests that u_θ drops to about $1/e$ of its original value in about 15 s. The discrepancy arises because straightforward diffusion of (negative) vorticity from the side walls is not the key process by which a stirred cup of tea comes to rest; the bottom of the cup—wholly absent in the present model—plays a crucial role (see Fig. 5.6.)

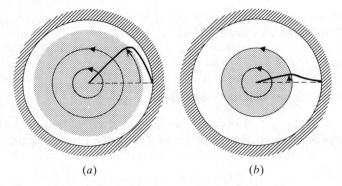

Fig. 2.11. 'Spin-down' in an infinitely long circular cylinder. Initially there is vorticity 2Ω everywhere, but negative vorticity diffuses inward from the stationary boundary $r = a$, so that the (shaded) region of significant vorticity shrinks with time.

Viscous decay of a line vortex

The line vortex

$$u = \frac{\Gamma_0}{2\pi r} e_\theta, \qquad (2.34)$$

where Γ_0 is a constant, has zero vorticity in $r > 0$ but infinite vorticity at $r = 0$. In a viscous fluid, then, this flow does not persist; the vorticity diffuses outward as time goes on.

To examine this process it is convenient to take the circulation

$$\Gamma(r, t) = 2\pi r u_\theta(r, t) \qquad (2.35)$$

as the dependent variable of the problem. In place of eqn (2.30) we then obtain

$$\frac{\partial \Gamma}{\partial t} = \nu \left(\frac{\partial^2 \Gamma}{\partial r^2} - \frac{1}{r} \frac{\partial \Gamma}{\partial r} \right). \qquad (2.36)$$

The initial condition is

$$\Gamma(r, 0) = \Gamma_0.$$

We require u_θ finite at $r = 0$ at any later time, so

$$\Gamma(0, t) = 0, \qquad t > 0.$$

Elementary viscous flow

This problem is very similar to that in which a plane rigid boundary is jerked into motion (see eqn (2.14)); we leave it as an exercise to seek, as in that case, a similarity solution in which

$$\Gamma = f(\eta), \quad \text{where } \eta = r/(vt)^{\frac{1}{2}}.$$

In this way we may discover that

$$\Gamma = \Gamma_0(1 - e^{-r^2/4vt}),$$

so

$$u_\theta = \frac{\Gamma_0}{2\pi r}(1 - e^{-r^2/4vt}). \tag{2.37}$$

At distances greater than about $(4vt)^{\frac{1}{2}}$ from the axis the circulation is almost unaltered, because very little vorticity has yet diffused that far out. At small distances from the axis, however, where $r \ll (4vt)^{\frac{1}{2}}$, the flow is no longer remotely irrotational; indeed

$$u_\theta \doteqdot \frac{\Gamma_0}{8\pi vt} r \quad \text{for} \quad r \ll (4vt)^{\frac{1}{2}}, \tag{2.38}$$

which corresponds to almost uniform rotation with angular velocity $\Gamma_0/8\pi vt$. The intensity of the vortex thus decreases with time as the 'core' spreads radially outward (Fig. 2.12).

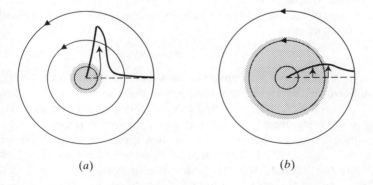

(a) (b)

Fig. 2.12. The viscous diffusion of a vortex.

2.5. The convection and diffusion of vorticity

If we take the curl of the momentum equation (2.3) we obtain

$$\frac{\partial \boldsymbol{\omega}}{\partial t} + (\boldsymbol{u} \cdot \nabla)\boldsymbol{\omega} = (\boldsymbol{\omega} \cdot \nabla)\boldsymbol{u} + \nu \nabla^2 \boldsymbol{\omega}, \qquad (2.39)$$

(cf. eqn (1.25)), and in the case of a 2-D flow this reduces to

$$\frac{\partial \omega}{\partial t} + (\boldsymbol{u} \cdot \nabla)\omega = \nu \left(\frac{\partial^2 \omega}{\partial x^2} + \frac{\partial^2 \omega}{\partial y^2} \right). \qquad (2.40)$$

In Chapter 1 we set the viscosity ν to zero from the outset; ω was then conserved by individual fluid elements in 2-D flow. Changes in ω at a particular point in space took place only by the *convection* of vorticity from elsewhere in the fluid, and this process is represented by the second term in eqn (2.40). In §§2.3 and 2.4, on the other hand, we looked at some simple viscous flow problems in which the term $(\boldsymbol{u} \cdot \nabla)\omega$ happened to be identically zero; in other words, we isolated *diffusion* of vorticity as a mechanism, this being represented by the third term in eqn (2.40).

In general, there is both diffusion and convection of vorticity in a viscous fluid flow, and we end this chapter with two examples.

2-D flow near a stagnation point

The main features of this exact solution of the Navier–Stokes equations (Exercise 2.14) are as follows. First, there is an inviscid 'mainstream' flow

$$u = \alpha x, \qquad v = -\alpha y, \qquad (2.41)$$

where α is a positive constant. This fails to satisfy the no-slip condition at the rigid boundary $y = 0$, but the mainstream flow speed $\alpha |x|$ increases with distance $|x|$ along the boundary. By Bernoulli's theorem, the mainstream pressure p *decreases* with distance along the boundary in the flow direction (Fig. 2.13), so we may hope for a thin, unseparated boundary layer which adjusts the velocity to satisfy the no-slip condition (see §2.1). This is indeed the case, as Exercise 2.14 shows, and the boundary layer, in which all the vorticity is concentrated, has thickness

Elementary viscous flow 49

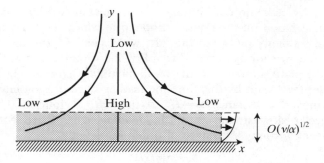

Fig. 2.13. Flow towards a 2-D stagnation point.

$\delta = O(\nu/\alpha)^{\frac{1}{2}}$. In this boundary layer there is a steady state balance between the viscous diffusion of vorticity from the wall and the convection of vorticity towards the wall by the flow. Thus if ν decreases the diffusive effect is weakened, while if α increases the convective effect is enhanced; in either case the boundary layer becomes thinner.

High Reynolds number flow past a flat plate

In uniform flow past a flat plate with a leading edge, as in Fig. 2.14, there is no flow component convecting vorticity towards the plate to counter the diffusion of vorticity from it, so the boundary layer becomes progressively thicker with downstream distance x. (In less formal terms, the layers of fluid closest to the centreline are the first to be slowed down as they pass the leading edge, and they in turn gradually slow down the layers of fluid which are further away.)

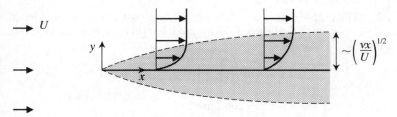

Fig. 2.14. The boundary layer on a flat plate.

Elementary viscous flow

We may estimate the boundary layer thickness δ by a simple argument based on the related problem in which the flat plate is instead suddenly pulled to the left, with speed U, through fluid which was previously at rest. From Fig. 2.8 we infer that at time t after the plate is moved vorticity will have diffused out a distance of order $(\nu t)^{\frac{1}{2}}$. But by this time the leading edge of the plate will have moved a distance $x = Ut$ to the left. It follows that at distance x downstream from the leading edge there will be significant vorticity a distance of order

$$\delta \sim (\nu x/U)^{\frac{1}{2}} \tag{2.42}$$

from the plate, but not beyond.

This crude estimate for the growth of the boundary layer with downstream distance x in Fig. 2.14 is indeed confirmed by the appropriate solution of the boundary layer equations (see §8.3). For a plate of finite length L the thickness (2.42) is in keeping with the claim (2.7) and is small compared with L at all points of the plate if $R = UL/\nu \gg 1$.

Exercises

2.1. Give an order of magnitude estimate of the Reynolds number for:

(i) flow past the wing of a jumbo jet at 150 m s^{-1} (roughly half the speed of sound);

(ii) the experiment in §1.1 with, say, $L = 2$ cm and $U = 5 \text{ cm s}^{-1}$;

(iii) a thick layer of golden syrup draining off a spoon;

(iv) a spermatozoan with tail length of 10^{-3} cm swimming at $10^{-2} \text{ cm s}^{-1}$ in water.

Give an order of magnitude estimate of the thickness of the boundary layer in case (i).

2.2. The problem of 2-D steady viscous flow past a circular cylinder of radius a involves finding a velocity field $\boldsymbol{u} = [u(x, y), v(x, y), 0]$ which satisfies

$$(\boldsymbol{u} \cdot \nabla)\boldsymbol{u} = -\frac{1}{\rho}\nabla p + \nu \nabla^2 \boldsymbol{u}, \qquad \nabla \cdot \boldsymbol{u} = 0,$$

together with the boundary conditions

$$\boldsymbol{u} = 0 \quad \text{on } x^2 + y^2 = a^2; \qquad \boldsymbol{u} \to (U, 0, 0) \quad \text{as } x^2 + y^2 \to \infty.$$

Rewrite this problem in *dimensionless form* by using the dimensionless variables

$$x' = x/a, \quad u' = u/U, \quad p' = p/\rho U^2$$

in places of x, u, and p. Without attempting to solve the problem, show that the streamline pattern can depend on ν, a, and U only in the combination $R = Ua/\nu$, so that flows at equal Reynolds numbers are geometrically similar.

2.3. (i) Viscous fluid flows between two stationary rigid boundaries $y = \pm h$ under a constant pressure gradient $P = -\mathrm{d}p/\mathrm{d}x$. Show that

$$u = \frac{P}{2\mu}(h^2 - y^2), \quad v = w = 0.$$

(ii) Viscous fluid flows down a pipe of circular cross-section $r = a$ under a constant pressure gradient $P = -\mathrm{d}p/\mathrm{d}z$. Show that

$$u_z = \frac{P}{4\mu}(a^2 - r^2), \quad u_r = u_\theta = 0.$$

[These are called *Poiseuille flows* (Fig. 2.15), after the physician who first studied (ii) in connection with blood flow. Their instability at high Reynolds number constitutes one of the most important problems of fluid dynamics (see §9.1).]

2.4. Two incompressible viscous fluids of the same density ρ flow, one on top of the other, down an inclined plane making an angle α with the horizontal. Their viscosities are μ_1 and μ_2, the lower fluid is of depth h_1 and the upper fluid is of depth h_2. Show that

$$u_1(y) = [(h_1 + h_2)y - \tfrac{1}{2}y^2]\frac{g \sin \alpha}{\nu_1},$$

so that the velocity of the lower fluid $u_1(y)$ is dependent on the depth h_2, but not the viscosity, of the upper fluid. Why is this?

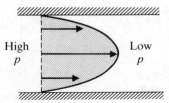

Fig. 2.15. Poiseuille flow.

52 Elementary viscous flow

2.5. Viscous fluid is at rest in a two-dimensional channel between two stationary rigid walls $y = \pm h$. For $t \geq 0$ a constant pressure gradient $P = -\mathrm{d}p/\mathrm{d}x$ is imposed. Show that $u(y, t)$ satisfies

$$\frac{\partial u}{\partial t} = \nu \frac{\partial^2 u}{\partial y^2} + \frac{P}{\rho},$$

and give suitable initial and boundary conditions. Find $u(y, t)$ in the form of a Fourier series, and show that the flow approximates to steady channel flow when $t \gg h^2/\nu$.

2.6. Viscous fluid flows between two rigid boundaries $y = 0$, $y = h$, the lower boundary moving in the x-direction with constant speed U, the upper boundary being at rest. The boundaries are porous, and the vertical velocity v is $-v_0$ at each one, v_0 being a given constant (so that there is an imposed flow across the system). Show that the resulting flow is

$$u = U\left(\frac{\mathrm{e}^{-v_0 y/\nu} - \mathrm{e}^{-v_0 h/\nu}}{1 - \mathrm{e}^{-v_0 h/\nu}}\right), \qquad v = -v_0.$$

Show that the horizontal velocity profile $u(y)$ is as in Fig. 2.16, so that when $v_0 h/\nu$ is large the downflow v_0 confines the vorticity to a very thin layer adjacent to $y = 0$.

[This is probably the mathematically simplest example of a steady boundary layer, but it is untypical in that the boundary layer thickness is proportional to ν, rather than to $\nu^{\frac{1}{2}}$ (see eqn (2.7)).]

2.7. Incompressible fluid occupies the space $0 < y < \infty$ above a plane rigid boundary $y = 0$ which oscillates to and fro in the x-direction with velocity $U \cos \omega t$. Show that the velocity field $\mathbf{u} = [u(y, t), 0, 0]$ satisfies

$$\frac{\partial u}{\partial t} = \nu \frac{\partial^2 u}{\partial y^2}$$

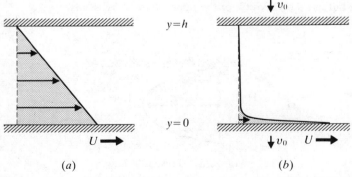

Fig. 2.16. Wall-driven channel flow with (a) $v_0 = 0$ and (b) $v_0 h/\nu \gg 1$.

(there being no applied pressure gradient), and by seeking a solution of the form
$$u = \mathcal{R}[f(y)e^{i\omega t}],$$
where $\mathcal{R}$ denotes 'real part of', show that
$$u(y, t) = Ue^{-ky}\cos(ky - \omega t),$$
where $k = (\omega/2\nu)^{\frac{1}{2}}$.

Sketch the velocity profile at some time t, and note that there is hardly any motion beyond a distance of order $(\nu/\omega)^{\frac{1}{2}}$ from the boundary.

2.8. A circular cylinder of radius a rotates with constant angular velocity Ω in a viscous fluid. Show that the line vortex flow
$$\mathbf{u} = \frac{\Omega a^2}{r}\mathbf{e}_\theta, \qquad r \geq a,$$
is an exact solution of the equations and boundary conditions. Describe roughly how the vorticity changes with time when the cylinder is suddenly started into rotation with angular velocity Ω from a state of rest. Likewise, discuss the case in which an outer cylinder $r = b$ is simultaneously given an angular velocity $\Omega a^2/b^2$.

2.9. A viscous flow is generated in $r \geq a$ by a circular cylinder $r = a$ which rotates with constant angular velocity Ω. There is also a radial inflow which results from a uniform suction on the (porous) cylinder, so that $u_r = -U$ on $r = a$. Show that
$$u_r = -Ua/r \qquad \text{for } r \geq a,$$
and that
$$r^2\frac{d^2 u_\theta}{dr^2} + (R+1)r\frac{du_\theta}{dr} + (R-1)u_\theta = 0,$$
where $R = Ua/\nu$.

Show that if $R < 2$ there is just one solution of this equation which satisfies the no-slip condition on $r = a$ and has finite circulation $\Gamma = 2\pi r u_\theta$ at infinity, but that if $R > 2$ there are many such solutions.

2.10. Show that, as claimed in eqn (2.37), a line vortex of strength Γ_0 decays by viscous diffusion in the following manner:
$$u_\theta = \frac{\Gamma_0}{2\pi r}(1 - e^{-r^2/4\nu t}).$$

Calculate and sketch the vorticity as a function of r at two different times.

54 Elementary viscous flow

2.11. Viscous fluid occupies the region $0 < z < h$ between two rigid boundaries $z = 0$ and $z = h$. The lower boundary is at rest, the upper boundary rotates with constant angular velocity Ω about the z-axis. Show that a steady solution of the full Navier–Stokes equations of the form

$$\boldsymbol{u} = u_\theta(r, z)\boldsymbol{e}_\theta$$

is not possible, so that any rotary motion $u_\theta(r, z)$ in this system must be accompanied by a *secondary flow* $(u_r, u_z \neq 0)$.

2.12. Viscous fluid is inside an infinitely long circular cylinder $r = a$ which is rotating with angular velocity Ω, so that $u_\theta = \Omega r$ for $r \leq a$. The cylinder is suddenly brought to rest at $t = 0$. Rewrite the evolution equation (2.30) in the form

$$\frac{\partial u_\theta}{\partial t} = \frac{\nu}{r}\frac{\partial}{\partial r}\left(r\frac{\partial u_\theta}{\partial r}\right) - \frac{\nu u_\theta}{r^2},$$

and thereby show that

$$\frac{dE}{dt} + \frac{2\nu}{a^2}E \leq 0,$$

where

$$E = \tfrac{1}{2}\int_0^a r u_\theta^2 \, dr,$$

which is proportional to the kinetic energy of the flow. Hence show that $E \to 0$ as $t \to \infty$.

[This may seem a little pointless, given that the exact solution (2.33) is available, but the above approach is in fact of very general value, and provides the basis for the proof, in §9.7, of an important uniqueness theorem.]

2.13. Re-derive the results (2.28) and (2.29) by the alternative route involving eqns (2.24), (2.25), and (2.26).

2.14. Consider in $y \geq 0$ the 2-D flow

$$u = \alpha x f'(\eta), \qquad v = -(\nu\alpha)^{\frac{1}{2}}f(\eta),$$

where

$$\eta = (\alpha/\nu)^{\frac{1}{2}}y.$$

Show that it is an exact solution of the Navier–Stokes equations which (i) satisfies the boundary conditions at the stationary rigid boundary $y = 0$ and (ii) takes the asymptotic form $u \sim \alpha x$, $v \sim -\alpha y$ far from the

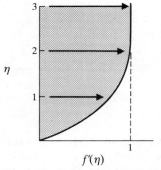

Fig. 2.17. The velocity profile in the boundary layer near a 2-D stagnation point.

boundary (see Fig. 2.13) if

$$f''' + ff'' + 1 - f'^2 = 0,$$

with

$$f(0) = f'(0) = 0, \qquad f'(\infty) = 1.$$

[The differential equation for $f(\eta)$ is solved numerically, and $f'(\eta)$ is shown in Fig. 2.17. Notably, $f'(3) = 0.998$, so beyond a distance of $3(\nu/\alpha)^{\frac{1}{2}}$ from the boundary the flow is effectively inviscid and irrotational, with $u \doteqdot \alpha x$ and $v \doteqdot -\alpha y$.]

2.15. If a flat plate is fixed between $(0, 0)$ and $(0, L)$ in Fig. 2.13, with $L \gg (\nu/\alpha)^{\frac{1}{2}}$, one might at first think that the flow would not be much affected, for the plate lies along one of the streamlines of the original flow. Why is it, then, that the observed flow is quite different, as in Fig. 2.18?

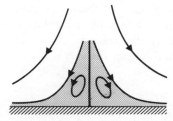

Fig. 2.18. High Reynolds number stagnation-point flow with a protruding flat plate.

3 Waves

3.1. Introduction

When a group of waves moves across the surface of a pond, each particular wavecrest travels faster than the group as a whole, and eventually passes through it. Thus new crests are continually being created at the back of the group while old crests are continually disappearing at the front. Suppose, for instance, that a snapshot of the wavetrain reveals ten crests, as in Fig. 3.1. A stationary observer at some fixed x will count substantially more than ten crests as the whole wavetrain passes by.

The reason for this curious behaviour is that water waves are *dispersive,* i.e. the different Fourier components that make up a general disturbance propagate at different speeds, depending on their wavelength. We show in fact, in §3.2, that a simple harmonic surface wave described by

$$\eta = A \cos(kx - \omega t), \tag{3.1}$$

where k and ω are both positive, has a wave speed

$$c = \omega/k = (g/k)^{\frac{1}{2}} \tag{3.2}$$

so that *waves of longer wavelength* $\lambda = 2\pi/k$ *travel faster.* (Contrast this with small amplitude waves on a stretched string, where all disturbances travel at a speed determined wholly by the tension in the string and its mass per unit length.) We show further, in §3.3, that while each individual wavecrest in Fig. 3.1 travels with speed c, the velocity of travel of the group as a whole is the so-called *'group velocity'*

$$c_g = \frac{d\omega}{dk}. \tag{3.3}$$

According to eqn (3.2) the frequency is given by

$$\omega = (gk)^{\frac{1}{2}} \tag{3.4}$$

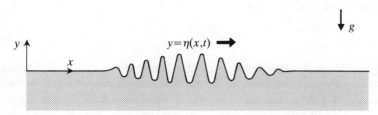

Fig. 3.1. A group of surface waves on deep water.

so the group velocity is

$$c_g = \tfrac{1}{2}(g/k)^{\tfrac{1}{2}} = \tfrac{1}{2}c. \tag{3.5}$$

Thus in the case of surface waves on (deep) water, individual wavecrests travel twice as fast as the group as a whole, and this is why they may be seen to be continually appearing at the back of the group and disappearing at the front.

The dispersive property of water waves is fundamental. It is responsible, for example, for the complicated wave pattern behind a moving ship (Fig. 3.2(a)). For very short waves, surface tension effects make the dispersion properties even more complicated (§3.4). If, for example, water flows past a fishing line, the stationary wave pattern contains both upstream and downstream disturbances (Fig. 3.2(b)).

We obtain a still broader class of wave if we consider disturbances to the interface between two fluids. If the lower fluid has density ρ_1 and the upper fluid has density $\rho_2 < \rho_1$, then

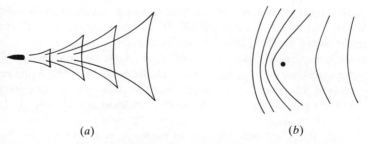

Fig. 3.2. Stationary wave patterns in (a) flow past a ship and (b) flow past a fishing line.

in place of eqn (3.2) we find

$$c^2 = \frac{g}{k}\left(\frac{\rho_1 - \rho_2}{\rho_1 + \rho_2}\right) \quad (3.6)$$

(Exercise 3.2). Buoyancy, i.e. the action of gravity on density differences, is here seen to be the mechanism responsible for the waves. We may view (3.2) as an extreme case, with $\rho_2 = 0$. If, on the other hand, the densities ρ_1 and ρ_2 are almost equal—as would be the case with a layer of fresh water overlying a layer of salty water—then the wave speed c is much reduced.

A rather different kind of wave can arise if the density of a fluid decreases continuously with height. Buoyancy forces are again responsible, but now *internal gravity waves* travel through the main body of the fluid, and possess some surprising properties, as we see in §3.8.

But buoyancy is not, of course, the only mechanism that enables a fluid to support wave motion. Another is compressibility, which permits the propagation of *sound waves*. We show in §3.6 that small-amplitude sound waves propagate through a gas at the *speed of sound*:

$$a_0 = (\gamma p_0/\rho_0)^{\frac{1}{2}}, \quad (3.7)$$

where p_0 and ρ_0 denote the background pressure and density respectively, while γ is a constant for the gas in question, approximately 1.4 for air. The most notable feature of eqn (3.7), given all the remarks above, is that it is independent of the wavelength of the waves. Sound waves are therefore non-dispersive.

In fact, eqn (3.7) also reveals why, elsewhere in the text, we so readily treat aerodynamic problems using incompressible flow theory. Isn't air compressible? It is, of course, but it may nevertheless be that, in a particular motion, the air is not being much compressed. By virtue of eqn (3.7) we may write the background pressure p_0 as $\rho_0 a_0^2/\gamma$. By inspection of the equations of motion, however, it is evident that the pressure fluctuations p_1 within the gas associated with the fluid motion are of order $\rho_0 U^2$, where U is a typical flow speed. Provided, therefore, that $U^2 \ll a_0^2$, the fractional change in pressure wrought by the fluid motion will be small, and will result in little expansion or compression of fluid elements. For this reason incompressible flow theory is aerodynamically useful, provided that the flow speed is much smaller than the speed of sound.

The above argument is borne out by the analysis in §3.7 of 2-D compressible flow past a thin aerofoil. Such a flow may be written in the form

$$u = U + \frac{\partial \phi}{\partial x}, \qquad v = \frac{\partial \phi}{\partial y}, \qquad (3.8)$$

where the *velocity potential* ϕ for the small disturbance to the uniform flow U satisfies

$$(1 - M^2) \frac{\partial^2 \phi}{\partial x^2} + \frac{\partial^2 \phi}{\partial y^2} = 0, \qquad (3.9)$$

the *Mach number M* being defined as the ratio of the speed of the free stream to the speed of sound:

$$M = U/a_0. \qquad (3.10)$$

Now, if M^2 is small compared with 1, then eqn (3.9) is approximately Laplace's equation, which is exactly the equation that arises if we treat the fluid as incompressible, as may be seen by substituting eqn (3.8) directly into $\nabla \cdot \boldsymbol{u} = 0$. In fact, provided that $M^2 < 1$, it is possible to infer certain properties of the flow from those of the corresponding incompressible flow by exploiting the change of variable $X = (1 - M^2)^{-\frac{1}{2}} x$. In this way it may be shown, for example, that for a steady, unseparated and *subsonic* flow past a thin aerofoil

$$\text{lift} = \text{lift}_{\text{incompressible}}/(1 - M^2)^{\frac{1}{2}}, \qquad \text{drag} = 0. \qquad (3.11)$$

It follows, too, that at subsonic speeds there is some disturbance to the oncoming flow at all distances from the aerofoil, even though that disturbance will be very small when the distance is large.

At supersonic speeds, however, the situation is quite different. The flow past a thin aerofoil when $M^2 > 1$ is as indicated in Fig. 3.3(*b*); there is no disturbance whatever to the oncoming stream except between the *Mach lines* extending from the ends of the aerofoil. These make an angle

$$\alpha = \sin^{-1}(1/M) \qquad (3.12)$$

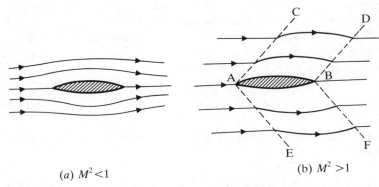

(a) $M^2 < 1$ (b) $M^2 > 1$

Fig. 3.3. Subsonic and supersonic flow past a thin symmetrical aerofoil at zero incidence.

with the uniform stream, and until an approaching parcel of air reaches the leading Mach line it is totally unaffected by the aerofoil's presence.

The reason for this may be appreciated in elementary terms by reference to Fig. 3.4. Imagine the fluid to be at rest at infinity, and a body moving through it to the left, from A to A'. In moving a small distance $U \, \delta t$ to the left of A it generates a sound wave which, after time t, will have radiated to the position shown in Fig. 3.4(a). This generation of sound waves continues all along the path from A to A'. Now, provided that $U < a_0$ the motion of the body clearly makes itself felt ahead of A', and, most

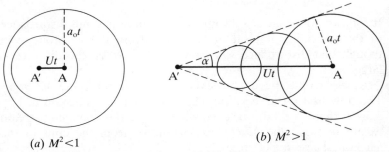

(a) $M^2 < 1$ (b) $M^2 > 1$

Fig. 3.4. Acoustic radiation by a body moving (a) subsonically and (b) supersonically.

significantly, throughout the entire fluid as $t \to \infty$. But contrast this case with the supersonic case in Fig. 3.4(b), where A' plainly lies outside the region affected by the initial sound wave; indeed, the disturbances resulting from each infinitesimal part of the body's motion are all confined to a wedge of angle α defined by eqn (3.12).

The final major distinction between subsonic and supersonic flow concerns the drag on the body. For $M^2 < 1$ the drag is precisely zero according to inviscid theory (see eqn (3.11)), but for $M^2 > 1$ it is not. In the case of a thin symmetric aerofoil, length L, at zero incidence, the drag is

$$D = \frac{2\rho_0 U^2}{(M^2 - 1)^{\frac{1}{2}}} \int_0^L [f'(x)]^2 \, dx, \qquad \text{for } M^2 > 1, \qquad (3.13)$$

where $y = f(x)$ denotes the shape of the upper surface (see §3.7). It arises because of the sound wave energy which the aerofoil radiates to infinity between the Mach lines in Fig. 3.3(b). (The ship and the fishing line in Fig. 3.2 experience a similar *wave drag*, but the dispersive property of water waves makes the corresponding theory more difficult.)

The above theory breaks down when M is very close to 1, but there is nonetheless a clear indication from both (3.11) and (3.13) that the aerofoil will then be subject to exceptionally large aerodynamic forces. This is indeed the case, and it was the destructive effect of these forces that led at one time to the notion of a 'sound barrier'.

Waves of finite amplitude

The formula (3.7) for the speed of small-amplitude sound waves was first given by Laplace, in a paper published in 1816. Even at that time there had already been one theoretical study of sound waves of arbitrary amplitude, at least in the 1-D case, with $\boldsymbol{u} = [u(x, t), 0, 0]$. This analysis, by Poisson, was based on an inappropriate thermodynamic assumption (constant temperature), but it suggested correctly that, for a disturbance travelling in the positive x-direction, larger values of u travel faster than smaller values, the difference being negligible only in the small-amplitude limit ($u \to 0$).

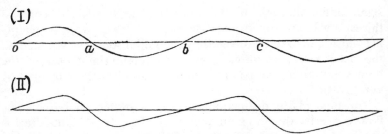

Fig. 3.5. Stokes's original sketches illustrating sound waves of finite amplitude. In each case the velocity $u(x, t)$ is plotted against x. (I) indicates the waveform at $t = 0$ and (II) indicates the waveform at a later time t. In order for the change in *shape* to be most evident (a finite-amplitude effect), the uniform displacement of the original curve by an amount $a_0 t$ in the x-direction has been omitted.

By 1848 Stokes had noticed an important consequence. In a paper entitled 'On a difficulty in the theory of sound' he remarks, with reference to Fig. 3.5:

... it is evident that in the neighbourhood of the points a, c the curve becomes more and more steep as t increases, while in the neighbourhood of the points $o, b, \ldots$ its inclination becomes more and more gentle.

He observes, too, that the steepening will lead, at some finite time, to an infinite value of $\partial u / \partial x$ at some point x, which is both mathematically and physically unacceptable.

What happens, then, beyond the time at which $\partial u / \partial x$ is predicted to be infinite? Stokes suggests an answer:

Perhaps the most natural supposition to make for trial is, that a surface of discontinuity is formed, in passing across which there is an abrupt change of density and velocity.

While this idea of a *shock wave* is now so familiar, if only because of sonic bangs from supersonic aircraft, it was not readily accepted at the time, and was advanced only hesitantly by Stokes himself. Yet in 1858 Riemann arrived independently at the notion that a compression wave would steepen into a shock (see §3.10), and during the latter half of the 19th century the conditions relating physical quantities on either side of a shock wave were clarified by Rankine and Hugoniot. By 1887 shock waves were to be seen in some of Mach's ingenious experiments (see in particular the photograph that serves as frontispiece to van Dyke (1982) and which appears as Fig. 22 of Reichenbach (1983)).

Stokes recognized that when $\partial u/\partial x$ becomes very large the neglect of viscosity may cease to be justified:

> Suppose now that a surface of discontinuity is very nearly formed, that is to say, that in the neighbourhood of a certain surface there is a very rapid change of density and velocity. It may be easily shown, that in such a case the rapid condensation or rarefaction implies a rapid sliding motion of the fluid; and this rapid sliding motion would call into play a considerable tangential force, the effect of which would be to check the relative motion of the parts of the fluid. It appears, then, almost certain that the internal friction would effectually prevent the formation of a surface of discontinuity, and even render the motion continuous again if it were for an instant discontinuous.

The limiting effect of viscosity does indeed give rise to a region of very rapid change, as opposed to a discontinuity, as Rayleigh and Taylor eventually showed in 1910. There is no telling, however, whether Stokes had any idea how thin such a region can be; a strong shock is in fact typically no thicker than the mean free path of the constituent gas molecules, and provides one well known instance where our whole treatment of the fluid as a continuous medium breaks down.

Early in the 19th century it was recognized that surface waves of finite amplitude in shallow water have a similar capacity to steepen. The consequence can in fact be something remarkably analogous to a shock wave, namely a *hydraulic jump*. This is a sudden change in water level—the bore on the River Severn being a well known example—with turbulent dissipation of energy in the jump itself being typically the limiting mechanism that prevents the occurrence of an actual discontinuity.

But in the case of surface waves on water there is another mechanism that can counteract finite-amplitude wave steepening, namely dispersion. The outstanding example of this came to light as the result of a chance observation by Russell on the Edinburgh–Glasgow canal in 1834:

> I happened to be engaged in observing the motion of a vessel at a high velocity, when it was suddenly stopped, and a violent and tumultuous

agitation among the little undulations which the vessel had formed around it attracted my notice. The water in various masses was observed gathering in a heap of a well-defined form around the centre of the length of the vessel. This accumulated mass, raising at last to a pointed crest, began to rush forward with considerable velocity towards the prow of the boat, and then passed away before it altogether, and, retaining its form, appeared to roll forward alone along the surface of the quiescent fluid, a large, solitary, progressive wave. I immediately left the vessel, and attempted to follow this wave on foot, but finding its motion too rapid, I got instantly on horseback and overtook it in a few minutes, when I found it pursuing its solitary path with a uniform velocity along the surface of the fluid. After having followed it for more than a mile, I found it subside gradually, until at length it was lost among the windings of the channel.

By 1844, when Russell reported this phenomenon to a meeting of the British Association for the Advancement of Science, he had produced such 'solitary waves' in the laboratory by dropping weights at one end of a water channel (Fig. 3.6), and he had also constructed an empirical formula for their wave speed. The matter was taken up by others, notably by Korteweg and de Vries (1895), and a solitary wave is now understood to arise from a precise balance between the steepening effects of finite amplitude and the smoothing effects of dispersion.

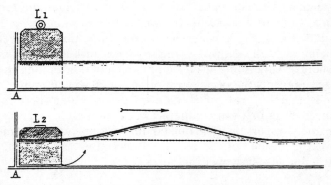

Fig. 3.6. Russell's original sketches of the generation of a solitary wave.

In the second part of this chapter (§§3.9–3.11) we give a brief account of all the finite-amplitude phenomena that have just been mentioned, and in discussing the interaction of two solitary waves in §3.11 we touch on one of the most unexpected developments of modern wave theory.

3.2. Surface waves on deep water

Let us investigate two-dimensional water waves, with

$$\boldsymbol{u} = [u(x, y, t), v(x, y, t), 0],$$

and suppose that the fluid motion is *irrotational*, so that

$$\frac{\partial v}{\partial x} - \frac{\partial u}{\partial y} = 0.$$

This would certainly be the case if, for example, the fluid were initially at rest. The vorticity of each fluid element would then be zero at $t = 0$, and would remain zero by virtue of the 2-D vorticity equation (1.27), provided viscous effects were negligible. The above condition implies the existence of a *velocity potential* $\phi(x, y, t)$ such that

$$u = \partial \phi / \partial x, \qquad v = \partial \phi / \partial y \qquad (3.14)$$

(see §4.2). By virtue of the incompressibility condition $\nabla \cdot \boldsymbol{u} = 0$ this velocity potential ϕ will satisfy Laplace's equation

$$\frac{\partial^2 \phi}{\partial x^2} + \frac{\partial^2 \phi}{\partial y^2} = 0. \qquad (3.15)$$

The fluid motion will arise from a deformation of the water surface, which is of major interest. We denote the equation of this free surface, as in Fig. 3.1, by

$$y = \eta(x, t). \qquad (3.16)$$

Kinematic condition at the free surface

Fluid particles on the surface must remain on the surface, as may be seen by imagining all the surface particles being marked with dye at some instant. If we define the quantity $F(x, y, t) = y - \eta(x, t)$ we may then claim that $F(x, y, t)$ remains constant (in fact, zero) for any particular particle on the free surface. It follows that $DF/Dt = 0$ on the free surface, i.e.

$$\frac{\partial F}{\partial t} + (\boldsymbol{u} \cdot \nabla)F = 0 \qquad \text{on } y = \eta(x, t). \tag{3.17}$$

Now

$$\frac{\partial F}{\partial t} = -\frac{\partial \eta}{\partial t}, \qquad u\frac{\partial F}{\partial x} = -u\frac{\partial \eta}{\partial x}, \qquad v\frac{\partial F}{\partial y} = v,$$

so eqn (3.17) is equivalent to

$$\frac{\partial \eta}{\partial t} + u\frac{\partial \eta}{\partial x} = v \qquad \text{on } y = \eta(x, t). \tag{3.18}$$

There are two special cases in which we can check this. If the free surface stays horizontal, then $\partial \eta/\partial x = 0$, so $v = d\eta/dt$, which is obviously correct. If, on the other hand, the free surface is stationary, so that $\eta = \eta(x)$ and $\partial \eta/\partial t = 0$, then eqn (3.18) reduces to $v/u = d\eta/dx$. This implies that the slope of the streamlines at $y = \eta(x)$ is equal to the slope of the free surface, as it should be in this particular case.

Bernoulli's equation for unsteady irrotational flow

Let us return for a moment to Euler's equation (1.14). If the flow is irrotational, so that $\boldsymbol{u} = \nabla \phi$ (see eqn (4.3)), the second term vanishes, and we are left with

$$\frac{\partial}{\partial t}\nabla \phi = -\nabla\left(\frac{p}{\rho} + \tfrac{1}{2}\boldsymbol{u}^2 + \chi\right),$$

where $\chi = gy$ in the present context. Integration gives

$$\frac{\partial \phi}{\partial t} + \frac{p}{\rho} + \tfrac{1}{2}\boldsymbol{u}^2 + \chi = G(t), \tag{3.19}$$

where $G(t)$ is an arbitrary function of time alone, which may in fact be chosen at our convenience, as it corresponds to adding a function of time alone to ϕ, and this makes no difference whatever to the flow velocity $\boldsymbol{u} = \nabla \phi$.

Equation (3.19) is a direct extension to *unsteady* irrotational flow of the second Bernoulli result in §1.3.

The pressure condition at the free surface

We are assuming that the fluid is inviscid, so the condition at the free surface is simply that the pressure p is equal to atmospheric pressure p_0 at $y = \eta(x, t)$. For practical purposes p_0 may be taken to be constant. By choosing $G(t)$ in a convenient manner, then, the pressure condition may be written

$$\frac{\partial \phi}{\partial t} + \tfrac{1}{2}(u^2 + v^2) + g\eta = 0 \qquad \text{on } y = \eta(x, t). \tag{3.20}$$

Small-amplitude waves: 'linearization' of the surface conditions

We shall now suppose that both the free surface displacement $\eta(x, t)$ and the associated fluid velocities u, v are small, in a sense to be made precise later. On this basis we will 'linearize' the problem by neglecting quadratic (and higher) terms in small quantities. In this way eqn (3.18) simplifies at once to

$$v(x, \eta, t) = \partial \eta / \partial t.$$

Furthermore, on expanding the left-hand side in a Taylor series,

$$v(x, 0, t) + \eta \frac{\partial v}{\partial y}(x, 0, t) + \ldots,$$

and again neglecting quadratic (and higher) terms, we obtain $v(x, 0, t) = \partial \eta / \partial t$. By virtue of eqn (3.14) this becomes

$$\partial \phi / \partial y = \partial \eta / \partial t \qquad \text{on } y = 0. \tag{3.21}$$

A similar treatment of the pressure condition, eqn (3.20), gives

$$\frac{\partial \phi}{\partial t} + g\eta = 0 \qquad \text{on } y = 0. \tag{3.22}$$

Dispersion relation

Let us now seek a sinusoidal travelling wave solution, so that the free surface is

$$\eta = A \cos(kx - \omega t), \tag{3.23}$$

where A is the amplitude of the surface displacement, ω is the frequency, and k is the wavenumber.

Inspection of eqns (3.21) and (3.22) suggests that the corresponding velocity potential $\phi(x, y, t)$ will be of the form

$$\phi = f(y)\sin(kx - \omega t).$$

Now ϕ satisfies Laplace's equation (3.15), i.e.

$$\frac{\partial^2 \phi}{\partial x^2} + \frac{\partial^2 \phi}{\partial y^2} = 0, \qquad (3.24)$$

so $f(y)$ must satisfy

$$f'' - k^2 f = 0,$$

the general solution of which is

$$f = Ce^{ky} + De^{-ky}.$$

We may take $k > 0$, without loss of generality. If the water is of infinite depth we must then choose $D = 0$ in order that the velocity be bounded as $y \to -\infty$. Thus

$$\phi = Ce^{ky} \sin(kx - \omega t).$$

Substituting this, together with eqn (3.23), into the free surface conditions (3.21) and (3.22), we obtain

$$Ck = A\omega, \qquad -C\omega + gA = 0,$$

so

$$\phi = \frac{A\omega}{k} e^{ky} \sin(kx - \omega t), \qquad (3.25)$$

and, most importantly,

$$\omega^2 = gk. \qquad (3.26)$$

This *dispersion relation* between ω and k is the key result. It takes the form $\omega^2 = g|k|$ if no restriction is placed on the sign of k.

Exact meaning of 'small amplitude'

The approximation of 'small amplitude' invites the practical question of small *compared with what*?

In eqn (3.18) we neglected the term $u\, \partial\eta/\partial x$ compared with the term v. Equations (3.14) and (3.25) show u and v to be of the same order of magnitude, $A\omega$. The approximation made in eqn (3.18) is thus essentially that the slope of the free surface is small, i.e. that the surface displacement is small compared with the wavelength of the waves.

Turning to eqn (3.20), we observe that $u^2 + v^2$ is of order $A^2\omega^2 = A^2gk$, and that this is negligible compared with $g\eta$ if Ak is small. Again, therefore, the condition is that A be small compared with the wavelength $\lambda = 2\pi/k$.

Particle paths

The velocity components are, from eqns (3.14) and (3.25):

$$u = A\omega e^{ky} \cos(kx - \omega t), \qquad v = A\omega e^{ky} \sin(kx - \omega t).$$

Assuming that any particle departs only a small amount (x', y') from its mean position $(\bar{x}, \bar{y})$ we may therefore find its position as a function of time by integrating

$$dx'/dt = A\omega e^{k\bar{y}} \cos(k\bar{x} - \omega t), \qquad dy'/dt = A\omega e^{k\bar{y}} \sin(k\bar{x} - \omega t);$$

whence

$$x' = -Ae^{k\bar{y}} \sin(k\bar{x} - \omega t), \qquad y' = Ae^{k\bar{y}} \cos(k\bar{x} - \omega t). \quad (3.27)$$

Particle paths are therefore circular, and the radius of the circles, $Ae^{k\bar{y}}$, decreases exponentially with depth, as do the fluid velocities themselves. Virtually all the energy of a surface water wave is contained within half a wavelength below the surface.

3.3. Dispersion: group velocity

Suppose we have a system that supports wave propagation, the dispersion relation being

$$\omega = \omega(k). \quad (3.28)$$

The *group velocity* is defined as

$$c_g = \frac{d\omega}{dk}, \quad (3.29)$$

and in a dispersive system c_g depends on k. This group velocity has several important properties:

(i) It is the velocity at which an isolated *wave packet* travels *as a whole* (Fig. 3.7(a)).

Waves

(ii) In the aftermath of some messy, localized, initial disturbance (such as dropping a stone into a pond) *it is the velocity at which you must travel if you wish to continually see waves of the same wavelength* $2\pi/k$ (Fig. 3.8).

(iii) It is the velocity at which *energy* is transported by waves of wavelength $2\pi/k$.

We now elaborate a little on each of these interpretations of group velocity.

Motion of a wave packet

Let us represent a general disturbance in the form of the Fourier integral

$$\eta(x, t) = \int_{-\infty}^{\infty} a(k) e^{i(kx - \omega t)} \, dk, \qquad (3.30)$$

it being understood that the real part of the right-hand side is to be taken. If the disturbance in question has the form of a single wave packet of almost constant wavenumber k_0, and if the amplitude of the wave packet varies slowly with x, so that the packet contains a large number of crests, then the amplitude distribution of the various Fourier components will be such that $|a(k)|$ is very small except when k is very near to k_0 (see Fig. 3.7 and Exercise 3.11). Now, when k is near to k_0,

$$\omega(k) \doteq \omega(k_0) + (k - k_0) c_g, \qquad (3.31)$$

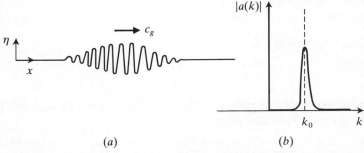

(a)   (b)

Fig. 3.7. A single, slowly varying wave packet and its spectrum.

where

$$c_g = \frac{d\omega}{dk}\bigg|_{k=k_0}. \qquad (3.32)$$

In the case of a slowly modulated wave packet, then, we should be able to replace $\omega(k)$ in eqn (3.30) by eqn (3.31), the idea being that for values of k not close to k_0, where eqn (3.31) fails, $|a(k)|$ will be so small that it makes no odds to the value of the integral. We thus find

$$\eta(x, t) \doteqdot e^{i[k_0 x - \omega(k_0)t]} \int_{-\infty}^{\infty} a(k) e^{i(k-k_0)(x-c_g t)} \, dk. \qquad (3.33)$$

Note that we have extracted a term representing a pure harmonic wave of wavenumber k_0, and that what remains is a function of x and t only through the particular combination $x - c_g t$. This shows, then, that the envelope of the wave in Fig. 3.7(a), and hence the packet as a whole, moves with the group velocity (3.32).

How might such a packet be generated? One might use a wavemaker which oscillates at a single frequency ω_0, first increasing its amplitude slowly from zero, up to some maximum value, and then bringing it slowly down to zero again. Most of the wave energy would then be concentrated in a narrow band of wavenumbers around $k = k_0$, where k_0 would be calculated in terms of ω_0 from the dispersion relation (3.28), which we presume to be known for the system in question.

Large-time response to a localized disturbance

Now imagine waves being set up in some far more natural, but messy, way—by some 2-D equivalent, perhaps, of a stone dropping into a pond. The early response of the system will be very complicated, but after a sufficiently long time the different Fourier components will have greatly dispersed. There will then be a slowly modulated wavetrain which is approximately sinusoidal everywhere, but with a *local* wavenumber $k(x, t)$ and frequency $\omega(x, t)$ which change gradually with x and t. In the case of surface waves on deep water, for example, we would expect that at any particular time the wavelength λ will gradually increase with x, as in Fig. 3.8, for long waves propagate faster

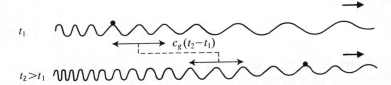

Fig. 3.8. A train of deep-water waves caused by an initial, localized disturbance. The double arrows show waves of some particular wavelength λ, and the dots mark a particular wavecrest.

(see eqns (3.26) and (3.2)). By the same token, at a later time, at any particular x, the local wavelength will be rather shorter than before.

Let us write a slowly varying wavetrain in the form

$$\eta(x, t) = A(x, t)e^{i\theta(x,t)}, \qquad (3.34)$$

where, again, it is understood that the real part of the right hand side is to be taken. The *phase function* $\theta(x, t)$ describes the oscillatory aspect of the wave, while $A(x, t)$ describes the gradual modulation of its amplitude with position and time. The local wavenumber k and frequency ω are defined by

$$k = \partial\theta/\partial x, \qquad \omega = -\partial\theta/\partial t. \qquad (3.35)$$

For a purely sinusoidal wave, in which θ is a linear function of x and t, k and ω are of course constants, and $\theta = kx - \omega t$. More generally, as in Fig. 3.8, k and ω will, like $A(x, t)$, be slowly varying functions of x and t. They will, however, be connected by the dispersion relation (3.28), because the wavetrain is approximately sinusoidal everywhere.

Now, it follows immediately from eqn (3.35) that

$$\frac{\partial k}{\partial t} + \frac{\partial \omega}{\partial x} = 0, \qquad (3.36)$$

and using eqn (3.28) this may be written

$$\frac{\partial k}{\partial t} + \frac{d\omega}{dk}\frac{\partial k}{\partial x} = 0,$$

i.e.

$$\frac{\partial k}{\partial t} + c_g(k)\frac{\partial k}{\partial x} = 0. \tag{3.37}$$

Just as the operator $\partial/\partial t + u\,\partial/\partial x$ signifies differentiation with respect to time as we move with speed u (see eqn (1.6)), so eqn (3.37) means that $k(x, t)$ *is constant for an observer moving with the group velocity* $c_g(k)$, as was claimed earlier. An alternative, more formal, way of deducing this is to treat eqn (3.37) as a first-order partial differential equation with implicit solution

$$k = f[x - c_g(k)t], \tag{3.38}$$

where f is an arbitrary function to be determined by the initial conditions (cf. eqn (3.114)). Plainly k remains constant if $x - c_g(k)t$ does.

Example: surface waves on deep water

Suppose there is some localized initial disturbance, so that waves are sent out in both directions, those propagating to the right having dispersion relation

$$\omega = (gk)^{\frac{1}{2}}, \tag{3.39}$$

with $k > 0$. Now, to continue to see waves of some particular wavenumber k at large times after the initial disturbance we must move with velocity

$$c_g = d\omega/dk = \tfrac{1}{2}(g/k)^{\frac{1}{2}}.$$

In other words, the wavetrain will be locally sinusoidal with wavenumber k in the neighbourhood of a distance

$$x = c_g t \tag{3.40}$$

from the initial disturbance region. Thus at any particular (large) x and t the local wavenumber is

$$k = gt^2/4x^2. \tag{3.41}$$

According to eqn (3.39) the local frequency will be

$$\omega = gt/2x$$

so

$$\partial\theta/\partial x = gt^2/4x^2, \qquad -\partial\theta/\partial t = gt/2x.$$

The local phase function is therefore

$$\theta(x, t) = -\frac{gt^2}{4x} + \varepsilon,$$

where ε is a constant, and by eqn (3.34) the free surface displacement is

$$\eta(x, t) = A(x, t)e^{-i(gt^2/4x - \varepsilon)}. \tag{3.42}$$

The amplitude $A(x, t)$ must, of course, depend on the details of how the waves are being generated (see Lamb 1932, p. 398, for an example).

Note from eqn (3.41) that the local wavelength λ is of order x^2/gt^2, and that the change in λ over a distance δx is therefore of order $x\,\delta x/gt^2$. The change in λ over a distance of one wavelength is therefore of order $x\lambda/gt^2$, and the fractional change over such a distance is of order x/gt^2. This must obviously be small if the wavetrain is to be locally sinusoidal, as assumed, so the key requirement for the preceding theory is

$$gt^2/x \gg 1. \tag{3.43}$$

At any particular x this expression tells us how long we must wait before the various Fourier components have dispersed sufficiently for the wavetrain to be locally sinusoidal.

Energy is transported at the group velocity

In one particular circumstance, that of a single, slowly modulated wave packet (Fig. 3.7(a)), this is rather obvious; the packet is where all the energy is, the packet moves as a whole with the group velocity, therefore so does the energy.

It is the case, however, that a perfectly sinusoidal wave of fixed wavenumber k and frequency ω also transports energy not at the crest speed ω/k but at the group velocity defined by $d\omega/dk$. This is a general result, valid for any dispersive wave system (Lighthill 1978, pp. 254–260), although we confine ourselves here to demonstrating it by means of an example (Exercise 3.12).

3.4. Surface tension effects: capillary waves

Imagine a line drawn parallel to the wave crests in the surface of the water. On the surface particles either side of that line there will be a *surface tension* force T, per unit length of line, directed tangentially to the surface. The vertical component of this force will therefore be $T\,\partial\eta/\partial s$, where s denotes distance along the surface, but for small wave amplitudes this will be approximately $T\,\partial\eta/\partial x$. A small portion of surface of length δx will experience surface tension forces at both its ends, so the net upward force on it will be

$$T\frac{\partial \eta}{\partial x}\bigg|_{x+\delta x} - T\frac{\partial \eta}{\partial x}\bigg|_{x} \doteq T\frac{\partial^2 \eta}{\partial x^2}\,\delta x.$$

This gives a net upward force per unit area of surface of $T\,\partial^2\eta/\partial x^2$, and this must be balanced by the difference between the atmospheric pressure p_0 and the pressure p in the fluid just below the surface, so that

$$p_0 - p = T\frac{\partial^2 \eta}{\partial x^2} \qquad \text{at } y = \eta(x, t).$$

Using eqn (3.19) and linearizing we obtain, in place of eqn (3.22):

$$\frac{\partial \phi}{\partial t} + g\eta - \frac{T}{\rho}\frac{\partial^2 \eta}{\partial x^2} = 0 \qquad \text{at } y = 0. \tag{3.44}$$

The analysis leading to eqn (3.26) may then be re-worked to obtain the new dispersion relation. Alternatively, we may note by inspection of eqn (3.44) that, as we are looking for solutions of the form $\eta = A\cos(kx - \omega t)$, the result must be obtainable by replacing g in the previous theory by $g + Tk^2/\rho$. Either way:

$$\omega^2 = gk + \frac{Tk^3}{\rho}, \tag{3.45}$$

$$c = \left(\frac{g}{k} + \frac{Tk}{\rho}\right)^{\frac{1}{2}} \tag{3.46}$$

and

$$c_g = \frac{g + 3Tk^2/\rho}{2(gk + Tk^3/\rho)^{\frac{1}{2}}}. \tag{3.47}$$

The importance of surface tension is measured by the parameter

$$\mathscr{S} = Tk^2/\rho g. \tag{3.48}$$

For surface waves on water, $\mathcal{S} = 1$ when the wavelength is about 1.7 cm. If the wavelength is much longer than this, the effects of surface tension are negligible. If, on the other hand, the wavelength is much smaller, so that $\mathcal{S}$ is large, the waves are essentially *capillary waves,* dominated by the effects of surface tension:

$$\omega^2 = Tk^3/\rho, \qquad c = (Tk/\rho)^{\frac{1}{2}}, \qquad c_g = \tfrac{3}{2}c. \qquad (3.49)$$

For capillary waves, short waves propagate fastest, and the group velocity exceeds the phase velocity, so crests move backward through a wave packet as it moves along as a whole.

Capillary waves are produced when raindrops fall on a pond, and as short waves travel faster the wavelength decreases with radius at any particular time, as in Fig. 3.9(*a*). This is in marked contrast to the disturbance produced by dropping a large stone into a pond (Fig. 3.9(*b*)), where the effects of gravity predominate, on account of the longer wavelengths involved, so that long waves travel faster (see the photographs in Crapper 1984, pp. 110–115).

The differences between capillary and gravity waves are again in evidence in the steady streaming past a 2-D obstacle (Fig. 3.10). Each wavecrest is at rest, but relative to still water it is travelling upstream with crest (or phase) speed $c = U$. Now, an important feature of eqn (3.46) is that it implies a minimum phase speed of

$$c^{\min} = (4gT/\rho)^{\frac{1}{4}} \qquad \text{for } k = (\rho g/T)^{\frac{1}{2}} \qquad (3.50)$$

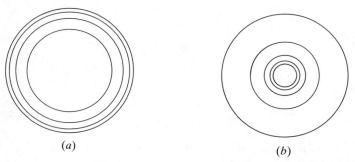

Fig. 3.9. Wave patterns produced by (*a*) a raindrop and (*b*) a large stone falling into a pond.

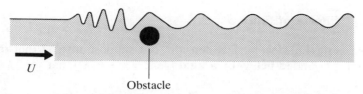

Fig. 3.10. Stationary waves generated by uniform flow, speed U, past a submerged obstacle.

(see Fig. 3.11). If $U < c^{min}$, then there are no steady waves generated by the obstacle, but if $U > c^{min}$ there are two values of k for which $c = U$. The smaller represents a gravity wave, the corresponding group velocity is less that c, and the energy of this relatively long-wavelength disturbance is thus carried downstream of the obstacle. The larger value of k represents a capillary wave, the corresponding group velocity is greater than c, and the energy of this relatively short-wavelength disturbance is therefore carried upstream of the obstacle, where it is rather quickly dissipated by viscous effects, on account of the short wavelength (Fig. 3.10). Nevertheless, these upstream waves may be readily observed, and by extending the above ideas to two dimensions one can account for the wave pattern set up by the fishing line in Fig. 3.2(*b*). (There are, again, excellent photographs of this on pp. 131–132 of Crapper 1984.)

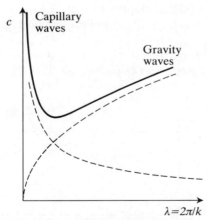

Fig. 3.11. The phase speed c for capillary-gravity waves.

3.5. Effects of finite depth

Let us now neglect surface tension, but suppose instead that the fluid is bounded below by a rigid plane $y = -h$, so that

$$\frac{\partial \phi}{\partial y} = 0 \quad \text{at } y = -h.$$

A simple re-working of the analysis in §3.2 (Exercise 3.1) shows that in place of eqn (3.26) we have the dispersion relation

$$\omega^2 = gk \tanh(kh). \tag{3.51}$$

The phase speed $c = \omega/k$ is thus given by

$$c^2 = \frac{g}{k} \tanh(kh), \tag{3.52}$$

and is sketched in Fig. 3.12. We note, in particular, that the phase speed cannot exceed $\sqrt{gh}$, and this is in fact true of the group velocity as well, so no small-amplitude disturbance travels faster than $\sqrt{gh}$ in water of uniform depth h.

If kh is large, i.e. $h \gg \lambda$, then $\tanh(kh) \doteq 1$ and

$$c^2 \doteq g/k, \tag{3.53}$$

as in the case of infinite depth. In fact, h has only to be greater than about $\tfrac{1}{3}\lambda$ for eqn (3.53) to provide a good approximation to the phase speed.

At the other extreme, however, of shallow water, i.e. $h \ll \lambda/2\pi$, we have $\tanh(kh) \doteq kh$, so the dispersion relation

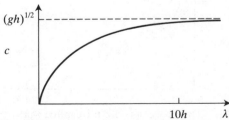

Fig. 3.12. Phase speed c of gravity waves in water of uniform depth h.

(3.52) becomes

$$c^2 \doteq gh. \qquad (3.54)$$

Notably, c is independent of k in this limit. Thus *gravity waves in shallow water are non-dispersive*, and a small-amplitude wave propagating in one direction will do so, whatever its form, without change of shape. This is highly significant, not least because it gives shallow-water waves something in common with sound waves, which we now discuss.

3.6. Sound waves

We now investigate, albeit briefly, the waves that result from the *compressibility* of a fluid. While the equation $\nabla \cdot \boldsymbol{u} = 0$ holds throughout most of this book, it does not hold here.

Euler's equation (1.12), which is a statement about the rate of change of *momentum* of fluid elements, holds as before:

$$\rho\left(\frac{\partial \boldsymbol{u}}{\partial t} + \boldsymbol{u} \cdot \nabla \boldsymbol{u}\right) = -\nabla p, \qquad (3.55)$$

although $\rho(\boldsymbol{x}, t)$ is of course now one of the variables, like p and $\boldsymbol{u}$. Individual fluid elements must still conserve their *mass*, so by Exercise 1.1 we have

$$\frac{\partial \rho}{\partial t} + \nabla \cdot (\rho \boldsymbol{u}) = 0. \qquad (3.56)$$

Finally, if we are dealing with a perfect gas, and if heat conduction within the gas is so slow as to be negligible, then

$$\frac{\mathrm{D}}{\mathrm{D}t}(p\rho^{-\gamma}) = 0, \qquad (3.57)$$

where γ denotes the 'ratio of specific heats', which is very nearly 1.4 for air at normal temperatures and pressures. This last equation expresses the fact that the *entropy*, which is proportional to $\log(p\rho^{-\gamma})$, is conserved by individual fluid elements.†

---

† Here we are taking a short-cut through the relevant thermodynamics (see Batchelor 1967, pp. 20–45). The equation of state for a perfect gas is $p = \rho R T$, where R is the gas constant and T the absolute temperature. Note that with p proportional to ρ^γ for a fluid element, T is proportional to $\rho^{\gamma-1}$, so a fluid element becomes hotter when it is compressed, as will be well known to anyone who has used a bicycle pump. Where does this thermal energy come from? Not by conduction from neighbouring elements, by hypothesis, but from work done by the pressure forces of the surrounding fluid in compressing the element in question.

Small-amplitude sound waves

Let the undisturbed state be one of rest, with constant pressure p_0 and constant density ρ_0. This state is a trivial exact solution of eqns (3.55)–(3.57).

Now let the system be slightly disturbed, so that

$$\boldsymbol{u} = \boldsymbol{u}_1, \qquad p = p_0 + p_1, \qquad \rho = \rho_0 + \rho_1. \qquad (3.58)$$

We intend to *linearize* the equations by neglecting quadratic and higher order terms in the perturbation variables $\boldsymbol{u}_1$, p_1, and ρ_1. As $p\rho^{-\gamma}$ stays constant for each fluid element, and was originally $p_0\rho_0^{-\gamma}$ for each such element, $p\rho^{-\gamma}$ is $p_0\rho_0^{-\gamma}$ everywhere. Thus

$$(p_0 + p_1)(\rho_0 + \rho_1)^{-\gamma} = p_0\rho_0^{-\gamma},$$

i.e.

$$\left(1 + \frac{p_1}{p_0}\right)\left(1 + \frac{\rho_1}{\rho_0}\right)^{-\gamma} = 1$$

and therefore

$$\left(1 + \frac{p_1}{p_0}\right)\left(1 - \frac{\gamma\rho_1}{\rho_0} + \ldots\right) = 1.$$

On neglecting higher order terms we obtain

$$p_1/p_0 = \gamma\rho_1/\rho_0,$$

and on defining

$$a_0 = (\gamma p_0/\rho_0)^{\frac{1}{2}} \qquad (3.59)$$

we may write the relation between p_1 and ρ_1 as

$$p_1 = a_0^2 \rho_1. \qquad (3.60)$$

Turning to eqn (3.55) we see that this equation reduces upon linearization to

$$\rho_0 \frac{\partial \boldsymbol{u}_1}{\partial t} = -\nabla p_1, \qquad (3.61)$$

while eqn (3.56) likewise becomes

$$\frac{\partial \rho_1}{\partial t} + \rho_0 \nabla \cdot \boldsymbol{u}_1 = 0. \qquad (3.62)$$

An effective way of dealing with these linearized equations, which have constant coefficients, is to take the divergence of eqn (3.61), and so obtain

$$\rho_0 \frac{\partial}{\partial t} \nabla \cdot \boldsymbol{u}_1 = -\nabla^2 p_1,$$

then use eqn (3.62) to eliminate $\nabla \cdot \boldsymbol{u}_1$, and finally use eqn (3.60) to eliminate ρ_1. The result is

$$\frac{\partial^2 p_1}{\partial t^2} = a_0^2 \nabla^2 p_1, \tag{3.63}$$

so that the perturbation pressure—and, in fact, any of the other perturbation variables—satisfies the classical 3-D wave equation.

For one-dimensional waves, with $p_1 = p_1(x, t)$, we have

$$\frac{\partial^2 p_1}{\partial t^2} = a_0^2 \frac{\partial^2 p_1}{\partial x^2}, \tag{3.64}$$

with general solution

$$p_1 = f(x - a_0 t) + g(x + a_0 t),$$

the first/second term corresponding to a wave propagating to the right/left with speed a_0 and *without change of shape*. Sound waves are therefore non-dispersive, and the speed of sound is a_0, given by eqn (3.59), which is approximately 340 m s^{-1} for air at 20°C and normal atmospheric pressure.

For spherically symmetric waves, with $p_1 = p_1(r, t)$, eqn (3.63) reduces to

$$\frac{\partial^2 p_1}{\partial t^2} = a_0^2 \frac{1}{r^2} \frac{\partial}{\partial r} \left(r^2 \frac{\partial p_1}{\partial r} \right), \tag{3.65}$$

which would be more difficult but for the happy circumstance that the substitution $p_1 = r^{-1} h(r, t)$ leads to

$$\frac{\partial^2 h}{\partial t^2} = a_0^2 \frac{\partial^2 h}{\partial r^2},$$

which is the same as eqn (3.64). It follows that

$$p_1 = \frac{1}{r} [F(r - a_0 t) + G(r + a_0 t)] \tag{3.66}$$

is the general solution to eqn (3.65). If the fluid domain extends to infinity, and the source of the sound is in some neighbourhood of the origin, we shall want to impose a *radiation condition* that there be no wave coming in from infinity. The solution will then be of the form

$$p_1 = \frac{1}{r} F(r - a_0 t),$$

corresponding to an outward-propagating wave. The decrease of amplitude with distance from the source, which is a purely geometrical effect, is of course a matter of common experience.

3.7. Supersonic flow past a thin aerofoil

Consider now steady 2-D compressible flow past an aerofoil, as in Fig. 3.3, and suppose that the aerofoil only causes a slight disturbance to the uniform flow U, with correspondingly small changes to the pressure p_0 and density ρ_0. We propose to write

$$\boldsymbol{u} = [U + u_1, v_1, 0], \qquad p = p_0 + p_1, \qquad \rho = \rho_0 + \rho_1 \quad (3.67)$$

in the steady versions of (3.55)–(3.57) and then neglect quadratic and higher order terms in the variables u_1, v_1, p_1, and ρ_1. Turning first to the steady version of eqn (3.57), we see that $p\rho^{-\gamma}$ is constant along a streamline. Provided that all the streamlines of the flow come from $x = -\infty$ it follows that $p\rho^{-\gamma}$ is everywhere equal to its upstream value $p_0 \rho_0^{-\gamma}$. The same few steps leading to eqn (3.60) then apply here also, so that

$$p_1 = a_0^2 \rho_1. \tag{3.68}$$

Turning to the steady version of eqn (3.55), the x-component of the left-hand side is

$$(\rho_0 + \rho_1)\left[(U + u_1)\frac{\partial}{\partial x} + v_1 \frac{\partial}{\partial y}\right](U + u_1),$$

and this is equal to

$$(\rho_0 + \rho_1)\left[(U + u_1)\frac{\partial u_1}{\partial x} + v_1 \frac{\partial u_1}{\partial y}\right]$$

as U is a constant. On expanding this expression we note that the only term linear in the perturbation variables is $\rho_0 U \, \partial u_1/\partial x$, so

$$\rho_0 U \frac{\partial u_1}{\partial x} = -\frac{\partial p_1}{\partial x}.$$

Similarly,

$$\rho_0 U \frac{\partial v_1}{\partial x} = -\frac{\partial p_1}{\partial y}.$$

Cross-differentiation gives

$$\rho_0 U \frac{\partial}{\partial x}\left(\frac{\partial v_1}{\partial x} - \frac{\partial u_1}{\partial y}\right) = 0,$$

so the vorticity is independent of x. But it is zero far upstream and far downstream, where the flow is uniform, so it is zero everywhere. We can therefore introduce a velocity potential $\phi(x, y)$ such that

$$u_1 = \frac{\partial \phi}{\partial x}, \qquad v_1 = \frac{\partial \phi}{\partial y}, \qquad (3.69)$$

and it then follows that

$$p_1 = -\rho_0 U \frac{\partial \phi}{\partial x}. \qquad (3.70)$$

We now turn to the steady version of eqn (3.56), i.e.

$$\rho \nabla \cdot \boldsymbol{u} + \boldsymbol{u} \cdot \nabla \rho = 0,$$

which is

$$(\rho_0 + \rho_1)\nabla \cdot \boldsymbol{u}_1 + (U + u_1)\frac{\partial \rho_1}{\partial x} + v_1 \frac{\partial \rho_1}{\partial y} = 0,$$

as U and ρ_0 are constants. On linearization this equation becomes

$$\rho_0\left(\frac{\partial u_1}{\partial x} + \frac{\partial v_1}{\partial y}\right) + U \frac{\partial \rho_1}{\partial x} = 0.$$

If we substitute for u_1 and v_1 from eqn (3.69) and for ρ_1 from eqn (3.68), the result is

$$\rho_0\left(\frac{\partial^2\phi}{\partial x^2}+\frac{\partial^2\phi}{\partial y^2}\right)+\frac{U}{a_0^2}\frac{\partial p_1}{\partial x}=0.$$

Finally, if we substitute for p_1 from eqn (3.70), we obtain

$$(1-M^2)\frac{\partial^2\phi}{\partial x^2}+\frac{\partial^2\phi}{\partial y^2}=0, \quad (3.71)$$

as claimed in eqn (3.9), where

$$M=U/a_0 \quad (3.72)$$

is the *Mach number* for the flow.

Consider now, for simplicity, flow past a thin symmetric aerofoil at zero incidence to the stream, as in Fig. 3.3. Let its upper surface be $y=f(x)$, for $0<x<L$. The flow there must be tangential to the surface, so

$$\frac{v_1}{U+u_1}=f'(x) \quad \text{on } y=f(x), \quad 0<x<L.$$

Equation (3.71) will describe the disturbance (u_1, v_1) to the uniform flow U only if it is small compared with U, so what we mean by a 'thin' aerofoil in the present context is

$$|f'(x)|\ll 1; \quad (3.73)$$

this is a key requirement of the theory. As u_1, v_1, and $|f'(x)|$ are all small, the boundary condition above reduces, after quadratically small terms have been neglected, to

$$v_1[x, f(x)]=Uf'(x), \quad 0<x<L.$$

When the left-hand side is expanded in a Taylor series,

$$v_1[x, 0+]+f(x)\frac{\partial v_1}{\partial y}[x, 0+]\ldots=Uf'(x), \quad 0<x<L,$$

the boundary condition reduces further to

$$\partial\phi/\partial y=Uf'(x) \quad \text{on } y=0+, \quad 0<x<L, \quad (3.74)$$

because $f(x)$ and $\partial v_1/\partial y$ are both small. There is, of course, a similar condition for the underside of the aerofoil on $y=0-$.

$M^2 > 1$: the streamline pattern

For supersonic flow, eqn (3.71) has precisely the form of the 1-D wave equation (3.64), so the general solution is

$$\phi = F(x - By) + G(x + By), \qquad (3.75)$$

where

$$B = (M^2 - 1)^{\frac{1}{2}}. \qquad (3.76)$$

The lines $x - By =$ constant, $x + By =$ constant are the *Mach lines* of the flow, and ϕ will be zero upstream of the leading Mach lines AC and AE in Fig. 3.3(b), for reasons explained in §3.1. We ensure the continuity of ϕ across AC by specifying

$$0 = F(0) + G(2x),$$

so G must be a constant. There is no loss of generality in taking G to be zero, and downstream of AC we then have

$$\phi = F(x - By), \quad \text{with } F(0) = 0.$$

To satisfy the boundary condition (3.74) we require

$$-BF'(x) = Uf'(x), \qquad 0 < x < L,$$

so

$$F(x) = -\frac{U}{B}f(x) + c, \qquad 0 < x < L,$$

where c is a constant. But with $f(0) = 0$, as in Fig. 3.3, we must choose $c = 0$, in order that $F(0) = 0$. Thus

$$\phi = -\frac{U}{B}f(x - By) \quad \text{for } y > 0, \quad 0 < x - By < L. \quad (3.77)$$

Similarly,

$$\phi = -\frac{U}{B}f(x + By) \quad \text{for } y < 0, \quad 0 < x + By < L.$$

Along the Mach lines BD and BF in Fig. 3.3(b), ϕ is therefore zero again, because $f(L) = 0$, and downstream of them the flow is undisturbed.

One outstanding feature of eqn (3.77) is that it implies

$$v_1 = Uf'(x - By) \quad \text{for } y > 0, \quad 0 < x - By < L, \quad (3.78)$$

so that in the region between the Mach lines AC and BD the slope of the streamlines (v_1/U in the linearized theory) is $f'(x - By)$, which is constant along lines $x - By =$ constant. Thus, in between the leading and trailing Mach lines *each streamline has exactly the same shape as that of the wing*, as in Fig. 3.3(b).

3.8. Internal gravity waves

Let us return now to the motion of an incompressible fluid. Even in the absence of a free surface, buoyancy forces can give rise to *internal gravity waves* if different fluid elements contain different concentrations of salt (say) and so have different densities. This happens in the oceans, and can be arranged quite easily in the laboratory. We suppose that the stratification of the fluid is such that the density $\rho_0(y)$ decreases with height y in some prescribed way when the fluid is at rest, the more salty water being further down, so that $\rho_0'(y) < 0$. The corresponding hydrostatic pressure distribution $p_0(y)$ will be given by

$$0 = -\frac{dp_0}{dy} - \rho_0 g. \quad (3.79)$$

Suppose now that the fluid is slightly disturbed from this state. Each element conserves its mass, its volume, and hence its density, so the governing equations are

$$\rho \left[\frac{\partial \boldsymbol{u}}{\partial t} + (\boldsymbol{u} \cdot \nabla) \boldsymbol{u} \right] = -\nabla p + \rho \boldsymbol{g},$$

$$\nabla \cdot \boldsymbol{u} = 0, \quad \frac{\partial \rho}{\partial t} + (\boldsymbol{u} \cdot \nabla) \rho = 0 \quad (3.80)$$

(see Exercise 1.1). Suppose, in addition, that the motion is two-dimensional and of small amplitude. We may then write

$$\boldsymbol{u} = [u_1(x, y, t), v_1(x, y, t), 0],$$

$$p = p_0(y) + p_1(x, y, t), \quad \rho = \rho_0(y) + \rho_1(x, y, t),$$

and on substituting in eqn (3.80) and neglecting quadratically small terms in u_1, v_1, p_1, and ρ_1 we obtain

$$\rho_0 \frac{\partial u_1}{\partial t} = -\frac{\partial p_1}{\partial x}, \qquad \rho_0 \frac{\partial v_1}{\partial t} = -\frac{\partial p_1}{\partial y} - \rho_1 g,$$
$$\frac{\partial u_1}{\partial x} + \frac{\partial v_1}{\partial y} = 0, \qquad \frac{\partial \rho_1}{\partial t} + v_1 \frac{d\rho_0}{dy} = 0. \tag{3.81}$$

Now, some of the coefficients of these linear equations, namely ρ_0 and $d\rho_0/dy$, are functions of y. In looking at one particular Fourier mode we must therefore write

$$v_1 = \hat{v}_1(y) e^{i(kx - \omega t)}, \tag{3.82}$$

together with similar expressions for u_1, p_1, and ρ_1 (the real part being understood). Substitution in eqn (3.81) leads to

$$\rho_0 \omega \hat{u}_1 = k \hat{p}_1, \qquad \rho_0 i \omega \hat{v}_1 = \hat{p}_1' + \hat{\rho}_1 g,$$
$$ik\hat{u}_1 + \hat{v}_1' = 0, \qquad -i\omega \hat{\rho}_1 + \hat{v}_1 \frac{d\rho_0}{dy} = 0, \tag{3.83}$$

where a dash denotes differentiation with respect to y. On eliminating all variables except $\hat{v}_1(y)$ we obtain

$$\hat{v}_1'' + \frac{\rho_0'}{\rho_0} \hat{v}_1' + k^2 \left(\frac{N^2}{\omega^2} - 1 \right) \hat{v}_1 = 0, \tag{3.84}$$

where the *buoyancy frequency* N is defined by

$$N^2 = -\frac{g}{\rho_0} \frac{d\rho_0}{dy}. \tag{3.85}$$

For simplicity, suppose now that ρ_0 decreases exponentially with height, so that $\rho_0 \propto e^{-y/H}$. The buoyancy frequency N and the coefficients in eqn (3.84) are then constant, and we may easily solve the equation to obtain wave-like solutions for v_1 of the form

$$v_1 \propto e^{y/2H} e^{i(kx + ly - \omega t)}, \tag{3.86}$$

where

$$\omega^2 = \frac{N^2 k^2}{\left(k^2 + l^2 + \dfrac{1}{4H^2} \right)}. \tag{3.87}$$

Waves

Using eqn (3.86) and the last of eqns (3.83) we see that

$$\rho_1 \propto e^{-y/2H} e^{i(kx+ly-\omega t)}; \tag{3.88}$$

it is usually these density variations that are most conveniently observed in laboratory experiments, by the same 'Schlieren' technique that is used to visualize shock waves in supersonic flow (see Fig. 3.21).

The 2-D propagation of these internal gravity waves is, of course, *anisotropic*; in the dispersion relation (3.87) ω depends on k and l in quite different ways. This leads to some rather surprising results, as we now see.

It is frequently the case in practice that the wavelength $\lambda = 2\pi/(k^2 + l^2)^{\frac{1}{2}}$ is small compared with the scale height H of the basic density distribution, and eqn (3.87) is then approximately

$$\omega^2 = \frac{N^2 k^2}{(k^2 + l^2)}. \tag{3.89}$$

Now consider not a single sinusoidal wave but a slowly modulated 2-D wave packet, as in Fig. 3.13. A natural extension of the argument following eqn (3.30) shows that the packet as a whole propagates with the group velocity

$$\mathbf{c}_g = \left(\frac{\partial \omega}{\partial k}, \frac{\partial \omega}{\partial l}\right), \tag{3.90}$$

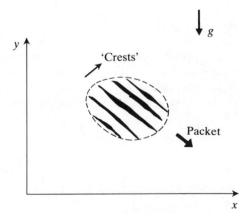

Fig. 3.13. Propagation of a 2-D packet of internal gravity waves; the 'crests' denote lines of constant phase $kx + ly - \omega t$.

and by differentiating eqn (3.89) we find

$$c_g = \frac{\omega l}{(k^2 + l^2)k}(l, -k). \qquad (3.91)$$

This result is surprising. Suppose, for instance, that ω, k, and l are all positive in eqn (3.86). The lines of constant phase, $kx + ly - \omega t = $ constant, are orientated as in Fig. 3.13, and as time proceeds these 'crests' move in the direction (k, l), i.e. to the right and upward. Plainly, however, the group velocity (3.91) is perpendicular to this direction and, moreover, has a vertically downward component. The packet as a whole, then, moves in the direction of the large arrow, while the individual 'crests' move sideways through it. While such behaviour may seem extraordinary, it is a clear prediction of the theory and has been well confirmed by experiment.

3.9. Finite-amplitude waves in shallow water

The rest of this chapter is concerned with non-linear waves, i.e. waves of finite amplitude.

We return first to 2-D surface waves, taking a flat bottom at $y = 0$ and a free surface $y = h(x, t)$. We shall *not* assume that the amplitude of the waves is small compared with the depth—so linearized theory does not apply (see Exercise 3.1)—but we shall assume that if h_0 is some typical value of $h(x, t)$ and L is a typical horizontal length scale of the wave, then

$$h_0 \ll L. \qquad (3.92)$$

This is the basis of the so-called *shallow-water approximation*.

Now, the full 2-D equations are

$$\frac{\partial u}{\partial t} + u\frac{\partial u}{\partial x} + v\frac{\partial u}{\partial y} = -\frac{1}{\rho}\frac{\partial p}{\partial x}, \qquad (3.93)$$

$$\frac{\partial v}{\partial t} + u\frac{\partial v}{\partial x} + v\frac{\partial v}{\partial y} = -\frac{1}{\rho}\frac{\partial p}{\partial y} - g, \qquad (3.94)$$

$$\frac{\partial u}{\partial x} + \frac{\partial v}{\partial y} = 0. \qquad (3.95)$$

In the shallow-water approximation we neglect the vertical component of acceleration Dv/Dt in comparison with g in eqn

90 Waves

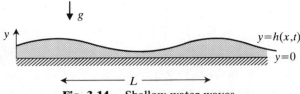

Fig. 3.14. Shallow-water waves.

(3.94); we verify this simplification a posteriori. The vertical pressure distribution is therefore essentially hydrostatic:

$$\partial p/\partial y = -\rho g.$$

On integrating and applying the condition $p = p_0$ at $y = h(x, t)$ we obtain

$$p = p_0 - \rho g[y - h(x, t)].$$

Substituting this into eqn (3.93) gives

$$\frac{\mathrm{D}u}{\mathrm{D}t} = -g\frac{\partial h}{\partial x}.$$

This equation implies that the rate of change of u for any particular fluid element is independent of y. Thus, if u is independent of y initially, it will remain independent of y for all t. The above equation then simplifies to

$$\frac{\partial u}{\partial t} + u\frac{\partial u}{\partial x} = -g\frac{\partial h}{\partial x}, \tag{3.96}$$

where u and h are functions of x and t only.

A second equation linking these quantities may be obtained by first integrating eqn (3.95) with respect to y:

$$v = -\frac{\partial u}{\partial x}y + f(x, t),$$

and then imposing the condition $v = 0$ at $y = 0$, which gives $f(x, t) = 0$. We then invoke the kinematic condition (3.18) at the free surface:

$$v = \frac{\partial h}{\partial t} + u\frac{\partial h}{\partial x} \quad \text{at } y = h(x, t),$$

whence

$$\frac{\partial h}{\partial t} + u\frac{\partial h}{\partial x} + h\frac{\partial u}{\partial x} = 0. \tag{3.97}$$

Before casting the *shallow-water equations* (3.96) and (3.97) into a more revealing form we should justify the neglect of the term Dv/Dt in eqn (3.94). Comparing the second and third terms in eqn (3.96) we infer that typical values of u are of order $(gh_0)^{\frac{1}{2}}$, and on comparing the first term with either of the others we then infer that $L/(gh_0)^{\frac{1}{2}}$ is a typical time scale on which events occur in this theory. These estimates are plainly in keeping with all the terms of eqn (3.97) being of the same order. Turning to eqn (3.95), we infer that v is small compared with u, of order $(gh_0)^{\frac{1}{2}}h_0/L$. In eqn (3.94), then, we find that all three terms that make up Dv/Dt are of order gh_0^2/L^2. They are therefore small compared with g—as assumed—by virtue of the shallow-water hypothesis (3.92).

It is now helpful to introduce the new variable

$$c(x, t) = (gh)^{\frac{1}{2}} \tag{3.98}$$

in place of $h(x, t)$. Then eqns (3.96) and (3.97) become

$$\frac{\partial u}{\partial t} + u\frac{\partial u}{\partial x} + 2c\frac{\partial c}{\partial x} = 0,$$

$$\frac{\partial}{\partial t}(2c) + u\frac{\partial}{\partial x}(2c) + c\frac{\partial u}{\partial x} = 0.$$

Adding and subtracting these equations gives

$$\left[\frac{\partial}{\partial t} + (u + c)\frac{\partial}{\partial x}\right](u + 2c) = 0, \tag{3.99}$$

$$\left[\frac{\partial}{\partial t} + (u - c)\frac{\partial}{\partial x}\right](u - 2c) = 0. \tag{3.100}$$

These partial differential equations lend themselves to treatment by the *method of characteristics*. Consider first eqn (3.99), and let us define parametrically a characteristic curve $x = x(s)$, $t = t(s)$ in the x–t plane, starting at some point (x_0, t_0), and such that

$$dt/ds = 1, \quad dx/ds = u + c. \tag{3.101}$$

We make this particular choice so that we may write eqn (3.99) in the form

$$\left(\frac{dt}{ds}\frac{\partial}{\partial t} + \frac{dx}{ds}\frac{\partial}{\partial x}\right)(u + 2c) = 0,$$

whence, by the chain rule,

$$\frac{d}{ds}(u + 2c) = 0. \tag{3.102}$$

Equation (3.99) thus reduces to the simple statement that $u + 2c$ is a constant along characteristic curves defined by eqn (3.101). A similar argument can be applied to eqn (3.100), and the outcome is that

$u \pm 2c$ is constant along 'positive'/'negative' characteristic curves defined by $dx/dt = u \pm c$. (3.103)

We do not know in advance what these characteristic curves will look like, of course, for we do not yet know how u and c depend on x and t; that is what we are trying to find.

The flow caused by a dam break

Suppose that water of uniform depth h_0 is contained at rest in $x < 0$ by a dam at $x = 0$, and that at $t = 0$ this dam suddenly breaks. We wish to use shallow-water theory to find the subsequent flow and, in particular, the subsequent shape of the free surface.

Consider first the region $x < 0$, $t < 0$. The fluid is at rest, so $u = 0$, and c is a constant, $c_0 = (gh_0)^{\frac{1}{2}}$. According to eqn (3.103) the characteristics are therefore straight lines $dx/dt = \pm c_0$, as indicated in Fig. 3.15(c).

Now consider a point P in $x < 0$, $t > 0$ such that it is the intersection of a positive and a negative characteristic emanating from the region $x < 0$, $t < 0$. We cannot assume in advance that these two characteristics remain straight as they emerge from $t < 0$; we simply observe that $u + 2c = 2c_0$ along the positive characteristic and $u - 2c = -2c_0$ along the negative one, by eqn (3.103). Thus $u = 0$ and $c = c_0$ at their point of intersection P. Throughout the region composed of such points, then, $u = 0$ and

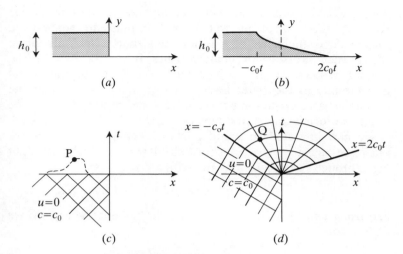

Fig. 3.15. The dam-break problem.

$c = c_0$, and the characteristics throughout that region are therefore straight with slope $dx/dt = \pm c_0$. The boundary of this 'undisturbed' region is therefore $x = -c_0 t$, as indicated in Fig. 3.15(b,d).

Now, in some region immediately to the right of $x = -c_0 t$ any point Q will lie on some positive characteristic emanating from the undisturbed region, and along any such characteristic $u + 2c$ is $2c_0$, so

$$u + 2c = 2c_0 \qquad (3.104)$$

throughout this region. From eqn (3.103) we have $u - 2c = k$ along the negative characteristic through Q, where k is a constant for that particular negative characteristic. Combining this with eqn (3.104) we see that u and c are both constant along the negative characteristic, which is therefore a straight line with slope $dx/dt = u - c$, according to eqn (3.103). Now, solving eqn (3.104) and $u - 2c = k$ will give $u - c$ as a function of k, and the various negative characteristics in this region will therefore be straight lines of differing slope. Crossing of like characteristics almost always implies an inconsistency; if the negative characteristics with $k = k_1$ and $k = k_2$ were to cross then

$u - 2c$ would need to be both k_1 and k_2 at that particular point in the x–t plane, yet u and c must in general be single-valued functions of position and time. The only place where such crossing of characteristics is appropriate in the present problem is at $x = 0$, $t = 0$, on account of the initial discontinuity in the water level at $x = 0$. We conclude that the negative characteristics must be straight lines passing through the origin, so forming a centred fan, as in Fig. 3.15(d).

We may now deduce u and c in the region occupied by this fan. The negative characteristics $\mathrm{d}x/\mathrm{d}t = u - c$, being straight lines through the origin, may be written

$$u - c = x/t,$$

and using eqn (3.104), which is valid throughout the region, we deduce that

$$c = \tfrac{1}{3}\left(2c_0 - \frac{x}{t}\right), \qquad u = \tfrac{2}{3}\left(c_0 + \frac{x}{t}\right). \qquad (3.105, 3.106)$$

Now, $c = (gh)^{\frac{1}{2}}$, so c cannot be negative, and the limit of the fan is therefore $x = 2c_0 t$. The shape of the free surface is thus given, as in Fig. 3.15(b), by

$$h(x, t) = h_0, \qquad x < -c_0 t,$$

$$h(x, t) = \frac{1}{9g}\left[2(gh_0)^{\frac{1}{2}} - \frac{x}{t}\right]^2, \qquad -c_0 t < x < 2c_0 t. \qquad (3.107)$$

$$h(x, t) = 0, \qquad x > 2c_0 t.$$

Non-linear wave distortion

We may view what is going on in Fig. 3.15 as a smoothing out of the initial discontinuity in $h(x, t)$, but the physical mechanism achieving this smoothing is certainly not the dispersion of §3.3; it is, instead, a finite-amplitude mechanism, and we may obtain further insight into it as follows.

In the region $-c_0 t < x < 2c_0 t$ we have $u + 2c = 2c_0$, so eqn (3.99) is satisfied. If we substitute for u in eqn (3.100) we obtain

$$\frac{\partial c}{\partial t} + (2c_0 - 3c)\frac{\partial c}{\partial x} = 0.$$

It is convenient to write

$$z = 3c - 2c_0 \tag{3.108}$$

so that

$$\frac{\partial z}{\partial t} - z\frac{\partial z}{\partial x} = 0. \tag{3.109}$$

This equation is sufficiently interesting that it is worth leaving our particular problem for a moment and considering eqn (3.109) on its own merits. Its general solution is

$$z = F(x + zt), \tag{3.110}$$

where $F(\xi)$ is an arbitrary differentiable function of $\xi = x + zt$, although this is only an implicit solution for $z(x, t)$, as $z(x, t)$ occurs in the argument of the function F. We may verify that eqn (3.110) satisfies eqn (3.109) by differentiating:

$$\frac{\partial z}{\partial t} = F'(\xi)\left(\frac{\partial z}{\partial t}t + z\right),$$

i.e.

$$\frac{\partial z}{\partial t} = \frac{zF'(\xi)}{1 - tF'(\xi)}. \tag{3.111}$$

Similarly,

$$\frac{\partial z}{\partial x} = \frac{F'(\xi)}{1 - tF'(\xi)}, \tag{3.112}$$

and in this way eqn (3.109) is plainly satisfied.

Now let us consider, at $t = 0$, some particular value of z. It will occur at some particular value of x. If, at a time t later, we want to find that same value of z again, we will clearly find it at a value of x that leaves the argument of F unchanged, i.e. at a value of x less than the original by an amount zt. We may say, then, that any particular value of z propagates to the left with speed z. The crux of the matter here is that larger values of z propagate faster.

Consider, then, what is happening in Fig. 3.15(*b*). A moment after the dam breaks $h(x, t)$ is a strongly decreasing function of x. Now, larger values of z propagate to the left faster; by eqn (3.108) the same is true of c, and hence of $h(x, t)$. This, then, is the reason that the free surface smooths out as time proceeds in this particular case.

96 Waves

But suppose now that we have a function $z(x, t)$ which is governed instead by

$$\frac{\partial z}{\partial t} + z \frac{\partial z}{\partial x} = 0. \tag{3.113}$$

The general solution is then

$$z = f(x - zt), \tag{3.114}$$

and larger values of z propagate to the right faster than smaller ones. A wave profile such as that in Fig. 3.16(a) accordingly *steepens* as time goes on, as in Fig. 3.16(b). Indeed, there will evidently come a time when the slope $\partial z/\partial x$ becomes infinite at some particular x, and beyond that time the solution will become multivalued. From eqn (3.114) we find that

$$\frac{\partial z}{\partial x} = \frac{f'(X)}{1 + tf'(X)}, \qquad X = x - zt, \tag{3.115}$$

so $\partial z/\partial x$ first becomes infinite at a time

$$t_c = \min_{(\text{over } X)} \{-1/f'(X)\}. \tag{3.116}$$

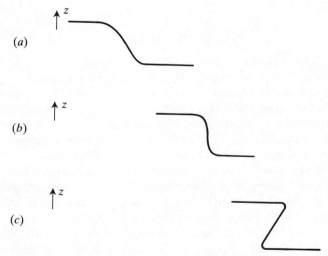

Fig. 3.16. Non-linear wave steepening.

Waves 97

This critical time is determined, not surprisingly, by the steepest negative slope of the initial profile. Another instructive way of viewing eqn (3.116) is in terms of the first crossing of characteristics (Exercise 3.19).

The formation of a bore

Suppose that fluid of uniform depth h_0 is contained at rest in $x > 0$ by a vertical plate, and that at $t = 0$ the plate is started into motion in the positive x-direction with speed $U = \alpha t$, where α is a constant. We may again use the method of characteristics, and an identical argument to that in the dam break problem leads to the conclusion that there is no disturbance at all ahead of the point $x = c_0 t$ (see Fig. 3.18).

Our attention switches to the region $\tfrac{1}{2}\alpha t^2 < x < c_0 t$. (The implied restriction $t < 2c_0/\alpha$ will not, in fact, concern us; the solution breaks down well before this.) Now, some neighbourhood of $x = c_0 t$ will be penetrated by negative characteristics

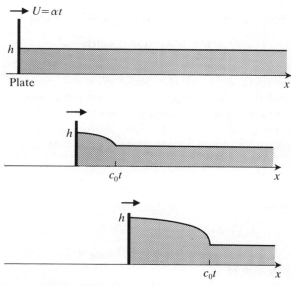

Fig. 3.17. The formation of a bore.

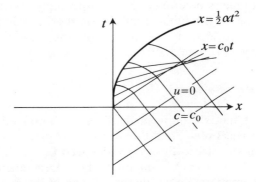

Fig. 3.18. The characteristic diagram.

from the undisturbed region, and throughout that neighbourhood

$$u - 2c = -2c_0, \tag{3.117}$$

by eqn (3.103). Substituting into eqn (3.99) we find

$$\frac{\partial u}{\partial t} + \left(\frac{3u}{2} + c_0\right)\frac{\partial u}{\partial x} = 0. \tag{3.118}$$

This is similar to eqn (3.113) and has general solution

$$u = \mathcal{F}[x - (\tfrac{3}{2}u + c_0)t]. \tag{3.119}$$

We expect u, at any particular time, to decrease from a maximum value of U at the plate to a minimum of zero at $x = c_0 t$, and it is fairly evident, then, that conditions prevail for the kind of non-linear steepening shown in Fig. 3.16.

In any event, we may determine the function $\mathcal{F}$ by applying the boundary condition at the plate, for $u = \alpha t$ at $x = \tfrac{1}{2}\alpha t^2$, so that

$$\alpha t = \mathcal{F}(-\alpha t^2 - c_0 t), \qquad t \geq 0. \tag{3.120}$$

Writing $\xi = -\alpha t^2 - c_0 t$ we deduce that

$$\mathcal{F}(\xi) = \tfrac{1}{2}[-c_0 + (c_0^2 - 4\alpha\xi)^{\frac{1}{2}}], \qquad \xi \leq 0,$$

the positive sign of the square root being taken because $\mathcal{F} \geq 0$, by virtue of eqn (3.120). So

$$2u = -c_0 + [c_0^2 - 4\alpha\{x - (\tfrac{3}{2}u + c_0)t\}]^{\frac{1}{2}},$$

and rearranging this as an explicit expression for $u(x,t)$ we obtain

$$2u = -(c_0 - \tfrac{3}{2}\alpha t) + [(c_0 - \tfrac{3}{2}\alpha t)^2 - 4\alpha(x - c_0 t)]^{\frac{1}{2}}. \quad (3.121)$$

Using eqn (3.117) we find that

$$c = c_0 - \tfrac{1}{4}(c_0 - \tfrac{3}{2}\alpha t) + \tfrac{1}{4}[(c_0 - \tfrac{3}{2}\alpha t)^2 - 4\alpha(x - c_0 t)]^{\frac{1}{2}}. \quad (3.122)$$

In this way c decreases with x, at any given time, from a value of $c_0 + \tfrac{1}{2}\alpha t$ at the plate to its undisturbed value c_0 at $x = c_0 t$.

Finally,

$$\frac{\partial c}{\partial x} = -\frac{\alpha}{2}[(c_0 - \tfrac{3}{2}\alpha t)^2 - 4\alpha(x - c_0 t)]^{-\frac{1}{2}},$$

and this first becomes infinite at $x = c_0 t$ at a time

$$t_c = \frac{2c_0}{3\alpha}. \quad (3.123)$$

At this particular time c is $\tfrac{4}{3}c_0$ at $x = \tfrac{1}{2}\alpha t^2$, so the height at the plate is $\tfrac{16}{9}h_0$. We have tried to indicate all these conclusions in Fig. 3.17.

The above treatment was slightly different to that used in the dam break problem in that we chose not to highlight the role played by the characteristics. One may, instead, use eqn (3.117) to show that the positive characteristic through any point $(\tfrac{1}{2}\alpha t_1^2, t_1)$ of the 'plate' curve in Fig. 3.18 has $u = \alpha t_1$ and $c = c_0 + \tfrac{1}{2}\alpha t_1$ everywhere along it, and is therefore the straight line

$$x - \tfrac{1}{2}\alpha t_1^2 = (c_0 + \tfrac{3}{2}\alpha t_1)(t - t_1).$$

By solving this equation for t_1 one may obtain eqns (3.121) and (3.122). Furthermore, the slope $dx/dt = c_0 + \tfrac{3}{2}\alpha t_1$ of these characteristics increases with t_1, as shown in Fig. 3.18, so they cross, and the first crossing of two such characteristics occurs at the time t_c given by eqn (3.123), the two characteristics in question coming from $t_1 = 0$ and $t_1 = 0+$. We leave all this as an exercise.

Beyond the time t_c given by (3.123), Fig. 3.16(c) indicates that $c(x,t)$ becomes a multivalued function of x. By virtue of eqn (3.117) this would mean $u(x,t)$ becoming a multivalued function of x also, which is physically impossible, so our model breaks

100 Waves

down at $t = t_c$. In reality, the wavefront for $t > t_c$ will take the form of a fairly sudden change in water level known as a *bore*. The term *hydraulic jump* is also used, particularly when the position of the change in water level is fixed relative to the observer.

3.10. Hydraulic jumps and shock waves

A hydraulic jump may be produced quite simply by first tilting a tray of shallow water, then quickly returning the tray to the horizontal. More easily still, one can turn on a kitchen tap. The water in the sink splays radially outwards in a thin layer, but at a certain radius, depending on the rate of flow, the height of this thin layer suddenly increases.

In the latter case the jump is stationary relative to the observer, and that is how we shall view it in the following analysis (Fig. 3.19). Across the jump:

(i) mass is conserved;

(ii) there is no loss of momentum, other than that caused by the difference in pressure on the two sides of the jump;

(iii) there is a loss of energy, owing to turbulence at the jump itself, which dissipates the lost energy as heat.

The first two of these statements imply

$$U_1 h_1 = U_2 h_2, \tag{3.124}$$

$$\tfrac{1}{2}g h_1^2 + h_1 U_1^2 = \tfrac{1}{2}g h_2^2 + h_2 U_2^2. \tag{3.125}$$

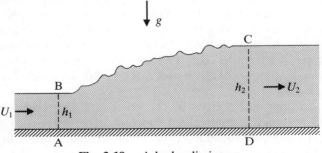

Fig. 3.19. A hydraulic jump.

[To see how eqn (3.125) comes about, consider the net force in the downstream direction on some dyed fluid that momentarily occupies some control region ABCDA; that net force is equal to the rate of change of momentum of the dyed fluid. Now, the momentum within the control region is not changing with time, so the rate of change of momentum of the dyed fluid is the rate at which momentum is being swept through CD minus the rate at which it is being swept through AB. Now, these cross-sections are taken either side of the jump, where the flow is uniform, so the pressure is hydrostatic. Taking atmospheric pressure as zero for simplicity, the net pressure force on AB is therefore

$$\int_0^{h_1} p \, dy = \int_0^{h_1} \rho g (h_1 - y) \, dy = \tfrac{1}{2} \rho g h_1^2.$$

A similar calculation gives $\tfrac{1}{2}\rho g h_2^2$ for the force in the upstream direction on CD, so the net force in the downstream direction is $\tfrac{1}{2}\rho g(h_1^2 - h_2^2)$. The rate at which mass is being swept through CD is $\rho h_2 U_2$, so the rate at which momentum is being swept through CD is $\rho h_2 U_2^2$. The rate at which it is being swept through AB is $\rho h_1 U_1^2$, and eqn (3.125) then follows.]

At this point it is helpful to introduce the *Froude number*

$$F = U/(gh)^{\frac{1}{2}}, \qquad (3.126)$$

i.e. the ratio of the flow speed to the speed of small-amplitude surface waves in shallow water. This dimensionless parameter has a role comparable with that of the Mach number in compressible flow, and similar terminology is in common use: a flow with $F < 1$ is termed *subcritical*, while a flow with $F > 1$ is termed *supercritical*.

Now, a little algebra with eqns (3.124) and (3.125) gives the following expressions for the Froude numbers upstream and downstream of the jump:

$$F_1^2 = U_1^2/gh_1 = (h_1 + h_2)h_2/2h_1^2, \qquad F_2^2 = U_2^2/gh_2 = (h_1 + h_2)h_1/2h_2^2. \qquad (3.127)$$

We finally use the third physical consideration, that of energy. The rate of loss of energy in the jump is

$$\frac{\rho g U_1}{4h_2}(h_2 - h_1)^3, \qquad (3.128)$$

(Exercise 3.20), and this must be positive, so

$$h_2 > h_1. \tag{3.129}$$

This accounts for the way the water level leaps up, not down, at a hydraulic jump. Further, it implies via eqn (3.127) that

$$F_1 > 1 \quad \text{and} \quad F_2 < 1, \tag{3.130}$$

so that a hydraulic jump changes a supercritical flow into a subcritical one.

Unsteady 1-D gas dynamics

We now describe some remarkable similarities between shallow-water theory and gas dynamics.

According to eqn (3.57), the quantity $p\rho^{-\gamma}$ remains constant for each element of the gas. Consequently, if the gas is initially at rest with constant pressure p_0 and density ρ_0, then $p\rho^{-\gamma}$ remains constant everywhere throughout the subsequent motion, and the flow is said to be *homentropic*. In 1-D homentropic flow, with $\boldsymbol{u} = [u(x, t), 0, 0]$, the governing equations reduce to

$$\left[\frac{\partial}{\partial t} + (u+a)\frac{\partial}{\partial x}\right]\left(u + \frac{2a}{\gamma - 1}\right) = 0, \tag{3.131}$$

$$\left[\frac{\partial}{\partial t} + (u-a)\frac{\partial}{\partial x}\right]\left(u - \frac{2a}{\gamma - 1}\right) = 0, \tag{3.132}$$

where the variable $a(x, t)$ is defined in terms of $p(x, t)$ and $\rho(x, t)$ by

$$a = (\gamma p/\rho)^{\frac{1}{2}} \tag{3.133}$$

(Exercise 3.22).

The similarity between these equations and the shallow-water equations (3.99) and (3.100) is immediately evident, and elementary solutions may again be found in certain special cases. The problem corresponding to that in Fig. 3.17, for instance, involves a piston being moved with speed $U = \alpha t$ into a long tube containing gas at rest with sound speed $a_0 = (\gamma p_0/\rho_0)^{\frac{1}{2}}$. A similar analysis shows that between the piston and the undisturbed gas there is a region of 'simple wave flow' in which

$$u - \frac{2a}{\gamma - 1} = -\frac{2a_0}{\gamma - 1} \qquad (3.134)$$

(cf. eqn (3.117)), so

$$\frac{\partial u}{\partial t} + [\tfrac{1}{2}(\gamma + 1)u + a_0]\frac{\partial u}{\partial x} = 0, \qquad (3.135)$$

by virtue of eqn (3.131). The general solution of eqn (3.135) is

$$u = F[x - \{\tfrac{1}{2}(\gamma + 1)u + a_0\}t], \qquad (3.136)$$

so larger values of u propagate faster in the positive x-direction. At early times the distribution of u with x looks something like Fig. 3.16(a), but at later times the waveform steepens up, eventually to the state in Fig. 3.16(b) where $\partial u/\partial x$ becomes infinite at a certain value of x. This happens at a time $t_c = 2a_0/(\gamma + 1)\alpha$ in the particular problem under consideration. Equation (3.135) breaks down at that time, so that we never reach the physically absurd situation in Fig. 3.16(c) where $u(x, t)$ is a multivalued function of x. Instead, a *shock*, i.e. a discontinuity in u, p, and ρ, forms at the point of breakdown and then propagates down the tube.

Normal shock waves

The analysis of a shock that is normal to the oncoming stream has much in common with that of a hydraulic jump. Again, it is simplest to adopt a frame of reference in which the shock is at rest (Fig. 3.20), and the various quantities on the two sides of the

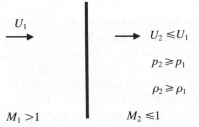

Fig. 3.20. Flow through a normal shock.

shock are then related by the *Rankine–Hugoniot equations*

$$\rho_1 U_1 = \rho_2 U_2, \qquad p_1 + \rho_1 U_1^2 = p_2 + \rho_2 U_2^2,$$

$$\tfrac{1}{2} U_1^2 + \frac{a_1^2}{\gamma - 1} = \tfrac{1}{2} U_2^2 + \frac{a_2^2}{\gamma - 1}, \qquad (3.137)$$

where $a_1^2 = \gamma p_1/\rho_1$ and $a_2^2 = \gamma p_2/\rho_2$. The first of these expresses conservation of mass, while the second is a statement that no momentum is lost in the shock, as may be established by an argument similar to that which gave eqn (3.125). The third equation says that no energy is lost in the shock (cf. Exercise 3.21).

There is a final physical statement to be made. According to the Second Law of Thermodynamics, entropy must not decrease across the shock, and as entropy is proportional to $\log(p\rho^{-\gamma})$ it follows that

$$p_2/p_1 \geq (\rho_2/\rho_1)^\gamma. \qquad (3.138)$$

After a great deal of algebra with eqns (3.137) and (3.138) one can deduce, in particular, the results displayed in Fig. 3.20:

(i) the flow into a normal shock is supersonic;

(ii) the flow out of a normal shock is subsonic;

(iii) both the pressure and density increase as the fluid passes through the shock.

Again, then, the analogy with the corresponding results (3.129) and (3.130) in shallow-water theory is quite strong.

High-speed flow past an aerofoil

We looked at the linearized theory for supersonic flow past a thin aerofoil in §3.7; we now look briefly at the more complicated flow fields that can arise when linearized theory is not valid (Fig. 3.21). *Oblique* shocks play an important role in deflecting the airstream. While the flow ahead of an oblique shock must be supersonic, the flow behind it may be subsonic or supersonic, as is evident from Fig. 3.21, and in this respect oblique shocks differ significantly from normal shocks.

It is as well to emphasize that with substantial variations of flow speed, pressure and density now present the Mach number

$$M = U/a \tag{3.139}$$

will vary significantly with position, it being understood that U here denotes the local flow speed and $a = (\gamma p/\rho)^{\frac{1}{2}}$ denotes the

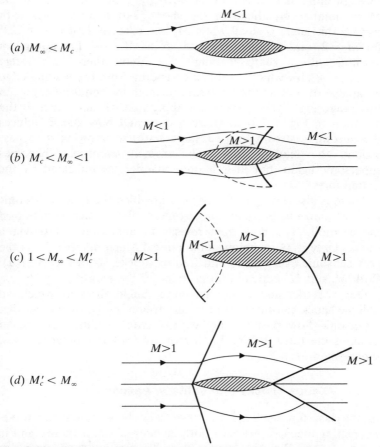

Fig. 3.21. Régimes of flow past a sharp-nosed aerofoil: (a) subsonic, (b) lower transonic, (c) upper transonic, and (d) supersonic.

local speed of small-amplitude sound waves. The free stream Mach number $M_\infty = U_\infty/a_\infty$ is, however, a prescribed quantity, and when this is sufficiently larger than 1 the flow is supersonic everywhere, as in Fig. 3.21(d). Oblique shocks extend from the sharp leading and trailing edges and provide sudden deflections of the airstream. In the limit of a thin aerofoil these shocks weaken and assume equal angles $\sin^{-1}(M_\infty^{-1})$ to the free stream; in this way we recover Fig. 3.3(b).

Oblique shocks, however orientated, cannot turn a stream through more than a certain angle $\theta_{\max}$, which depends on the Mach number M_1 ahead of the shock. For a perfect gas with $\gamma = 1.4$ this angle is small when M_1 just exceeds 1, it rises to 23° for $M_1 = 2$, and it tends to about 46° as $M_1 \to \infty$. For any given aerofoil with a sharp leading edge, then, there is a range $1 < M_\infty < M_c'$ for which no shock extending from the leading edge is capable of turning the airstream through the required angle. In this range Fig. 3.21(d) does not apply and we have instead the situation in Fig. 3.21(c), where a detached bow shock reduces the oncoming stream to a subsonic state, so allowing it to pass around the leading edge. The stream eventually becomes supersonic again, but in a smooth manner, as indicated by the dotted lines.

There is also a range $M_c < M_\infty < 1$ in which the flow is subsonic far upstream but becomes supersonic as the air accelerates past the aerofoil. The flow then reverts to a subsonic state via a shock. Only if M_∞ is below some critical Mach number M_c is the flow subsonic everywhere and free of shocks, as in Fig. 3.21(a). Both M_c and M_c' depend on the shape of the aerofoil.

Our remarks about the maximum angle through which an oblique shock can turn a supersonic stream imply, of course, that supersonic flow past a body with a *blunt* leading edge never assumes the form in Fig. 3.21(d); it always has a detached bow shock when $M_\infty > 1$.

3.11. Viscous shocks and solitary waves

In the last few sections we have seen how non-linear wave-steepening mechanisms exist both in shallow-water theory and in gas dynamics. We now discuss two different mechanisms which can exactly offset this wave steepening, so permitting finite amplitude waves to propagate without change of shape.

Weak viscous shocks

So far we have treated a shock as a genuine discontinuity. In practice a shock has a finite structure, although its thickness may be extremely small. One obvious limiting mechanism which comes into play as $\partial u/\partial x$ becomes large is viscosity, and it can be shown that for a weak viscous shock propagating in the positive x-direction into gas at rest the velocity $u(x, t)$ satisfies *Burgers' equation*

$$\frac{\partial u}{\partial t} + \{\tfrac{1}{2}(\gamma + 1)u + a_0\}\frac{\partial u}{\partial x} = \tfrac{2}{3}\nu\frac{\partial^2 u}{\partial x^2} \qquad (3.140)$$

(see, e.g., Ockendon and Tayler 1983, p. 88). Here ν may be written as μ/ρ_0, as changes in density are assumed to be small, and we are treating μ as a constant. When $\nu = 0$ the equation reduces to eqn (3.135).

The real value of Burgers' equation is as an evolution equation, which enables us to see how a finite shock structure emerges from some initial conditions as time goes on. For the time being, however, we simply seek a solution to eqn (3.140) in the form of a travelling wave

$$u = f(x - Vt), \qquad (3.141)$$

such that $u \to U_1$, say, as $x - Vt \to -\infty$ and $u \to 0$ as $x - Vt \to +\infty$. We then find (Exercise 3.24) that such a solution may be written

$$u(x, t) = \frac{U_1}{1 + \exp\{(x - Vt)/\Delta\}}, \qquad (3.142)$$

where the shock speed is

$$V = a_0 + \tfrac{1}{4}(\gamma + 1)U_1 \qquad (3.143)$$

and the shock thickness is

$$\Delta = \tfrac{8}{3}\frac{\nu}{(\gamma + 1)U_1}. \qquad (3.144)$$

In this way a rapid but smooth transition between $u = U_1$ and $u = 0$ is effected.

Note that the shock thickness is proportional to ν, so that when ν is small the shock wave is very thin.

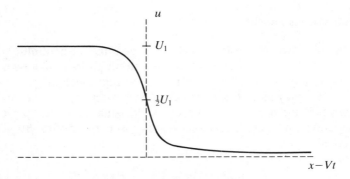

Fig. 3.22. A weak viscous shock.

Solitary waves in shallow water

Following on with the above theme, we now look for a non-linear wave of permanent form in shallow water, the difference being that the steepening effects of non-linearity will here be offset not by viscous diffusion but by weak *dispersion*.

Let h_0 denote the mean depth of the water, let η_0 be a typical magnitude of $\eta(x, t)$, the vertical displacement of the free surface, and let L denote a typical horizontal length scale of the wave. Then when the parameters

$$\varepsilon = \eta_0/h_0, \qquad \delta = h_0^2/L^2 \qquad (3.145)$$

are both small, and of the same order, the evolution of a wave travelling in the positive x-direction is governed approximately by the *Korteweg–de Vries equation*:

$$\frac{\partial \eta}{\partial t} + c_0 \frac{\partial \eta}{\partial x} + \frac{3c_0}{2h_0} \eta \frac{\partial \eta}{\partial x} + \tfrac{1}{6} c_0 h_0^2 \frac{\partial^3 \eta}{\partial x^3} = 0, \qquad (3.146)$$

where $c_0 = (gh_0)^{\frac{1}{2}}$. While we shall not derive this equation (see, e.g., Drazin and Johnson 1989, pp. 9–11; Ockendon and Tayler 1983, pp. 53–58), we may confirm quite easily that it is consistent with previous results in this chapter in two limiting cases, namely when either the third (non-linear) or fourth (dispersive) term is negligible (Exercise 3.25).

Our real interest, however, is in precisely the case when these two terms are of comparable magnitude. We want to see, in particular, whether the two physical effects can exactly offset each other so as to give a finite-amplitude wave of permanent shape. We accordingly seek a solution to eqn (3.146) of the form $\eta = f(x - Vt)$, where f and its derivatives are assumed to vanish as $x - Vt \to \pm\infty$. In this way we obtain

$$(c_0 - V)f' + \frac{3c_0}{2h_0}ff' + \tfrac{1}{6}c_0 h_0^2 f''' = 0,$$

and on integrating once and applying the boundary conditions we find

$$(c_0 - V)f + \frac{3c_0}{4h_0}f^2 + \tfrac{1}{6}c_0 h_0^2 f'' = 0.$$

We may then multiply by f', integrate again, and use the boundary conditions once more to obtain

$$\tfrac{1}{3}h_0^3 f'^2 = (a - f)f^2 \quad \text{where } a = 2h_0\left(\frac{V}{c_0} - 1\right).$$

On taking the square root of both sides we may then separate variables and integrate, most easily by the substitution $f = a \operatorname{sech}^2 p$. In this way we find that

$$\eta = a \operatorname{sech}^2[(3a/4h_0^3)^{\frac{1}{2}}(x - Vt)], \tag{3.147}$$

where

$$V = c_0\left(1 + \frac{a}{2h_0}\right). \tag{3.148}$$

The maximum height of the wave, a, is a free parameter in the theory, but eqn (3.146) only holds if both ε and δ in eqn (3.145) are small, so eqns (3.147) and (3.148) only hold if $a \ll h_0$. The

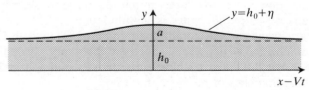

Fig. 3.23. A solitary wave.

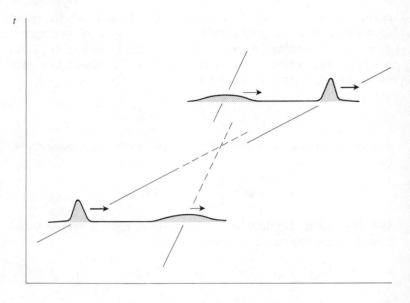

Fig. 3.24. Collision of two solitary waves.

wave has the form shown in Fig. 3.23, and propagates without change of shape at a speed just above the speed of infinitesimal waves c_0.

According to eqn (3.148), larger-amplitude solitary waves travel faster. In Fig. 3.24, then, one solitary wave catches up with the other. There is a period of complicated interaction, but eventually *both solitary waves emerge completely unscathed* (see Drazin and Johnson 1989, especially pp. 1–22 and their Fig. 4.3). There is nevertheless one crucial piece of evidence that a non-linear interaction must have taken place: in arriving at any particular position x the large-amplitude wave is slightly early and the small-amplitude wave is slightly late.

Solitary waves which retain their identity upon collision are called *solitons*. Twenty years ago their discovery caused something of a stir, and solitons have subsequently had a large impact on various branches of modern physics.

Exercises

3.1. Modify the analysis of §3.2 to establish the dispersion relation (3.52),

$$c^2 = \frac{g}{k}\tanh(kh),$$

for surface waves on water of uniform depth h.

In what precise sense must the surface displacement be small for the validity of the analysis? Find and sketch the particle paths.

3.2. Modify the analysis of §3.2 to establish the dispersion relation (3.6),

$$c^2 = \frac{g}{|k|}\left(\frac{\rho_1 - \rho_2}{\rho_1 + \rho_2}\right),$$

for waves on the interface between two fluids, the upper fluid being of density ρ_2 and the lower being of density $\rho_1 > \rho_2$.

3.3. Fluid of density ρ_2 lies on top of another of density $\rho_1 > \rho_2$, both fluids being confined between plane vertical boundaries at $x = 0$ and $x = a$ (see Fig. 3.25). The surface tension between the fluids is T. (We avoid consideration of capillary effects at the moving lines of contact between the fluid interface and the vertical boundaries.)

Derive the linearized boundary conditions to be satisfied at $y = 0$, and write down also the boundary conditions to be satisfied by the velocity potentials at $x = 0$ and $x = a$. Hence show that the normal modes of oscillation of the system have

$$\eta(x, t) = A_N \cos\frac{N\pi x}{a}\cos(\omega_N t + \varepsilon_N), \qquad N = 1, 2, \ldots,$$

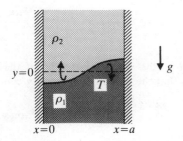

Fig. 3.25. Interfacial oscillations in a tank.

where

$$(\rho_1 + \rho_2)\omega_N^2 = \frac{N\pi}{a}\left[(\rho_1 - \rho_2)g + T\frac{N^2\pi^2}{a^2}\right].$$

Suppose that the fluids and the interface are initially at rest, with $\eta = \eta_0(x)$ at $t = 0$. Explain why, for $t \geq 0$,

$$\eta(x, t) = \sum_{N=1}^{\infty} A_N \cos\frac{N\pi x}{a} \cos \omega_N t,$$

where

$$A_N = \frac{2}{a}\int_0^a \eta_0(x)\cos\frac{N\pi x}{a}\,\mathrm{d}x.$$

3.4. *Rayleigh–Taylor instability.* Suppose that the upper fluid in Fig. 3.25 is the heavier, so that $\rho_2 > \rho_1$. Use the results of Exercise 3.3 to show that a small initial disturbance $\eta_0(x)$ will not remain small as time proceeds, so that the system is unstable, if

$$(\rho_2 - \rho_1)g > \frac{\pi^2}{a^2}T.$$

[Surface tension essentially stabilizes the system against short-wavelength disturbances, and the geometry puts an upper limit on the wavelength. For water overlying air, the above criterion amounts roughly to $a > 9$ mm, which explains why water can be retained in an inverted glass if the mouth of the glass is covered by a fine-meshed gauze.]

3.5. Water flows steadily with speed U over a corrugated bed $y = -h + \varepsilon \cos kx$, where $\varepsilon \ll h$, so that there is a time-independent disturbance $\eta(x)$ to the free surface, which would be at $y = 0$ but for the corrugations. By writing

$$u = U + \frac{\partial\phi}{\partial x}, \qquad v = \frac{\partial\phi}{\partial y},$$

where $\phi(x, y)$ denotes the velocity potential of the disturbance to the uniform flow, show that the linearized boundary conditions are

$$U\frac{\mathrm{d}\eta}{\mathrm{d}x} = \frac{\partial\phi}{\partial y}, \qquad U\frac{\partial\phi}{\partial x} + g\eta = 0 \qquad \text{on } y = 0,$$

$$\frac{\partial\phi}{\partial y} = -Uk\varepsilon \sin kx \qquad \text{on } y = -h,$$

and hence find $\eta(x)$. Deduce that crests on the free surface occur immediately above troughs on the bed if

$$U^2 < \frac{g}{k}\tanh(kh),$$

but that if this inequality is reversed the crests on the surface overlie the crests on the bed.

3.6. *Kelvin–Helmholtz instability.* One deep layer of inviscid fluid, density ρ_2, flows with uniform speed U over another deep layer of density $\rho_1 > \rho_2$ which is at rest. If the interface is

$$\eta(x, t) = A e^{i(kx - \omega t)},$$

where the real part of the right-hand side is understood, show that

$$(\rho_1 + \rho_2)\omega^2 - 2\rho_2 U k \omega + \rho_2 U^2 k^2 - |k|[k^2 T + (\rho_1 - \rho_2)g] = 0.$$

Deduce that the system is unstable if

$$U^2 > 2\left(\frac{1}{\rho_1} + \frac{1}{\rho_2}\right)[(\rho_1 - \rho_2)gT]^{\frac{1}{2}}.$$

[Note that both surface tension and gravity are needed to stop the instability. In the absence of one of them there will be instability for any U, however small, for disturbance wavelengths which are sufficiently long (in the case $g = 0$) or sufficiently short (in the case $T = 0$).]

3.7. When a stone is dropped into a deep pond, waves are eventually observed only beyond a central region of calm water which expands in radius with time (see Fig. 3.9). Furthermore, the wavelength just beyond this calm region is constant, about 4.5 cm.

Use 2-D plane wave theory, including both gravity and surface tension, to account broadly for these observations, and obtain an estimate for the speed at which the calm region expands.

$[g \doteqdot 9.81 \text{ m s}^{-2}; \rho = 10^3 \text{ kgm m}^{-3}, T \doteqdot 0.074 \text{ N m}^{-1}.]$

3.8. Surface waves generated by a mid-Atlantic storm arrive at the British coast with period 15 seconds. A day later the period of the waves arriving has dropped to 12.5 seconds. Roughly how far away did the storm occur?

3.9. A *ripple tank* is a device for simulating certain aspects of sound propagation (see Lighthill 1978, pp. 41–50, for some excellent photographs). It consists of a shallow layer of water in a glass-bottomed container, and is illuminated from below in such a way that images of the surface waves are thrown onto a screen. The waves have a *slight* tendency to disperse, on account of the small depth (see §3.5) and on account of surface tension (§3.4). Show that by choosing the depth h such that

$$h = (3T/\rho g)^{\frac{1}{2}} \doteq 0.5 \text{ cm},$$

these two dispersive agencies almost cancel each other, so that the wave speed is approximately 22 cm s^{-1} for all but the shortest wavelengths.

3.10. Show that when plane capillary waves have dispersed sufficiently to be locally sinusoidal,

$$\eta(x, t) \doteq A(x, t)\cos\left\{\frac{4\rho x^3}{27Tt^2} + \varepsilon\right\},$$

where ε is a constant and $A(x, t)$ is a slowly varying amplitude factor which will depend on the details of how the waves were generated. How does the local wavelength change with time as we move along with one particular crest?

Explain why, at any particular x, the above expression for $\eta(x, t)$ cannot be a good description of events if t is too large.

3.11. Consider the disturbance (3.30) at $t = 0$:

$$\eta(x, 0) = \int_{-\infty}^{\infty} a(k)e^{ikx}\, dk,$$

and suppose that

$$a(k) = a_0 e^{-\sigma(k-k_0)^2},$$

where a_0, k_0, and σ are constants. Show that

$$\eta(x, 0) = a_0(\pi/\sigma)^{\frac{1}{2}}e^{-x^2/4\sigma}e^{ik_0x},$$

the real part being understood. Sketch this *Gaussian wave packet* $\eta(x, 0)$ and show that if it contains a large number of crests then $a(k)$ is small except for values of k very close to k_0.

3.12. Calculate the total energy (per unit length in the z-direction) in one wavelength of a deep water wave:

$$E = \tfrac{1}{2}\rho \int_{x_0}^{x_0+\lambda} \int_{-\infty}^{\eta} u^2 + v^2\, dy\, dx + \int_{x_0}^{x_0+\lambda} \int_0^{\eta} \rho g y\, dy\, dx,$$

where the first term denotes the kinetic energy and the second the potential energy. Deduce that the average wave energy per unit length in the x-direction is $\bar{E} = \tfrac{1}{2}\rho g A^2$, where A is the amplitude of the free surface displacement (3.23).

Explain why at any fixed point the pressure perturbation caused by the presence of the waves is $p_1 = -\rho\, \partial\phi/\partial t$. Calculate

$$\int_{-\infty}^{\eta} p_1 u\, dy,$$

which is the rate at which those pressure perturbations do work on a section $x = $ constant in the fluid, and deduce that the average rate at which energy is transferred across a vertical cross-section is $\bar{F} = \frac{1}{4}\rho g A^2 c$, where $c = \omega/k$. Thus verify in this particular case that the energy propagation velocity, defined as $\bar{F}/\bar{E}$, is the same as the group velocity c_g, defined by eqn (3.29).

[Hint: the upper limit η may be replaced by zero to the level of approximation required, except in the expression for the potential energy.]

3.13. An inviscid, perfect gas is contained in a rigid sphere of radius L. Show that the natural frequencies of spherically symmetric oscillations of the gas are given by

$$\tan(\omega L/a_0) = \omega L/a_0.$$

3.14. Show that the drag on the upper surface of the thin aerofoil in §3.7 is approximately

$$\int_0^L p_1(x, 0) f'(x) \, dx,$$

and hence derive the expression (3.13).

3.15. Suppose that $H \gg (k^2 + l^2)^{-\frac{1}{2}}$ in eqn (3.86) so that the actual (i.e. real) vertical velocity is essentially

$$v_1 = A \cos(kx + ly - \omega t),$$

say, where A is a constant. Use eqn (3.81) to find corresponding expressions for u_1 and p_1.

The mean fluxes of energy in the x and y directions are $\overline{p_1 u_1}$ and $\overline{p_1 v_1}$ respectively, where an overbar denotes an average over one period (cf. Exercise 3.12). Show that

$$(\overline{p_1 u_1}, \overline{p_1 v_1}) = \frac{\rho_0 A^2}{2k^2} \frac{\omega l}{k} (l, -k),$$

and confirm that this is in the same direction and sense as the group velocity (3.91).

3.16. Consider a circular cylinder oscillating to and fro, with small amplitude and frequency ω, in a direction normal to its axis. If it is immersed in a compressible fluid, sound waves propagate outwards in all directions.

Suppose, instead, that it is immersed with its axis horizontal in a stratified fluid having buoyancy frequency N. Suppose too that its radius is small compared with H, the scale height of the basic density variations. It is then found that disturbances to the basic density field are

116 Waves

significant only in two planes extending from the cylinder, provided that $\omega < N$. Both planes make an angle α with the vertical, so forming a St Andrews cross (see Lighthill 1978, p. 314; and Tritton 1988, p. 212). Explain why this should be, and find an expression for α in terms of ω and N.

What happens when $\omega > N$?

3.17. *Waves in a rotating fluid.* Suppose an inviscid incompressible fluid is rotating uniformly with angular velocity $\mathbf{\Omega}$, and take Cartesian axes fixed in a frame rotating with that angular velocity. We show in §8.5 that the evolution of a *small* velocity field $\mathbf{u}_1$ relative to those rotating axes is governed by

$$\frac{\partial \mathbf{u}_1}{\partial t} + 2\mathbf{\Omega} \wedge \mathbf{u}_1 = -\frac{1}{\rho} \nabla p_1, \qquad \nabla \cdot \mathbf{u}_1 = 0,$$

where p_1 denotes a so-called 'reduced pressure'.

Write out these equations in Cartesian components, taking $\mathbf{\Omega} = (0, 0, \Omega)$, and by eliminating u_1, v_1, and w_1 show that

$$\left[\frac{\partial^2}{\partial t^2} \left(\frac{\partial^2}{\partial x^2} + \frac{\partial^2}{\partial y^2} + \frac{\partial^2}{\partial z^2} \right) + 4\Omega^2 \frac{\partial^2}{\partial z^2} \right] p_1 = 0.$$

Hence show that plane waves with

$$p_1 \propto e^{i(kx + ly + mz - \omega t)}$$

are possible provided that

$$\omega^2 = \frac{4\Omega^2 m^2}{k^2 + l^2 + m^2},$$

and deduce that the group velocity of a packet of such waves is perpendicular to the wavenumber vector $\mathbf{\kappa} = (k, l, m)$.

Show too that if r and z denote appropriate cylindrical polar coordinates then axisymmetric disturbances of the form

$$p_1 = \hat{p}_1(r, z) e^{i\omega t},$$

satisfy

$$\frac{\partial^2 \hat{p}_1}{\partial r^2} + \frac{1}{r} \frac{\partial \hat{p}_1}{\partial r} - \left(\frac{4\Omega^2}{\omega^2} - 1 \right) \frac{\partial^2 \hat{p}_1}{\partial z^2} = 0.$$

[I am grateful to C. Jones, N. Mottram, and N. Wright for pointing out a serious error in the original version of this problem.]

3.18. Water of depth h_0 lies at rest in the region $x < 0$, and there is a vertical plate at $x = 0$. At $t = 0$ the plate is moved into the region $x > 0$ with uniform speed $V < 2c_0$, where $c_0^2 = gh_0$.

Use an analysis similar to that for the dam break problem to show that:

(i) $c = c_0$ in $x < -c_0 t$;
(ii) $c = \frac{1}{3}(2c_0 - x/t)$ in $-c_0 t < x < (\frac{3}{2}V - c_0)t$;
(iii) $c = c_0 - \frac{1}{2}V$ in $(\frac{3}{2}V - c_0)t < x < Vt$;

so that the free surface takes the form indicated in Fig. 3.26.

[Hint: look at (i) and (iii) first, using very similar arguments to those in the text, and assume—and verify later—that the positive characteristics from region (i) reach the curve $x = Vt$ in the x–t plane, so that $u + 2c = 2c_0$ there.]

3.19. Consider the equation

$$\frac{\partial z}{\partial t} + z \frac{\partial z}{\partial x} = 0,$$

with initial condition $z = g(x)$ at $t = 0$. Show that the characteristic curve through the point $(x_0, 0)$ in the (x, t) plane is given by

$$x - g(x_0)t = x_0,$$

and sketch several such characteristics in the two cases (i) $g'(x) > 0$ for all x and (ii) $g'(x) < 0$ for all x.

In the latter case, show that characteristics from $(x_0, 0)$ and $(x_0 + \delta x_0, 0)$ intersect at a time

$$t_c = -1/g'(x_0),$$

so that the first such intersection takes place as in eqn (3.116).

3.20. Explain why, in 2-D flow, the flux of energy across a vertical section of height $h(x)$ is

$$\mathscr{F} = \int_0^h (p + \tfrac{1}{2}\rho u^2 + \rho g y) u \, dy,$$

and show that the flux of energy across such a section upstream of the

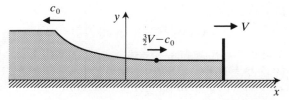

Fig. 3.26. A variant of the dam-break problem.

hydraulic jump in Fig. 3.19 is

$$\rho g U_1 h_1^2 + \tfrac{1}{2}\rho U_1^2 U_1 h_1.$$

Hence deduce that energy is lost in a hydraulic jump at a rate

$$\frac{\rho g U_1}{4h_2}(h_2 - h_1)^3.$$

3.21. Suppose a perfect gas is in steady, homentropic, and irrotational flow, so that $p\rho^{-\gamma}$ is a constant throughout the flow field and $\nabla \wedge \boldsymbol{u} = 0$. Prove that

$$\frac{\gamma p}{(\gamma - 1)\rho} + \tfrac{1}{2}\boldsymbol{u}^2 = \text{constant}.$$

3.22. A perfect gas is in unsteady, 1-D homentropic flow, so that $\boldsymbol{u} = [u(x, t), 0, 0]$ and $p\rho^{-\gamma}$ is a constant throughout the flow field. Show that the momentum equation and the mass conservation equation reduce to

$$\rho\left(\frac{\partial u}{\partial t} + u\frac{\partial u}{\partial x}\right) = -\frac{\partial p}{\partial x}, \qquad \frac{\partial \rho}{\partial t} + \frac{\partial}{\partial x}(\rho u) = 0,$$

and show also that these may be recast in the form

$$\frac{\partial u}{\partial t} + u\frac{\partial u}{\partial x} + \frac{2a}{\gamma - 1}\frac{\partial a}{\partial x} = 0,$$

$$\frac{2}{\gamma - 1}\left(\frac{\partial a}{\partial t} + u\frac{\partial a}{\partial x}\right) + a\frac{\partial u}{\partial x} = 0,$$

where $a^2 = \gamma p/\rho$. Hence obtain the evolution equations (3.131) and (3.132).

3.23. A perfect gas has been set into 1-D unsteady simple wave flow in such a way that

$$\frac{\partial u}{\partial t} + \{\tfrac{1}{2}(\gamma + 1)u + a_0\}\frac{\partial u}{\partial x} = 0,$$

(see eqn (3.135)), and

$$u(x, 0) = \tfrac{1}{2}U[1 - \tanh(x/L)],$$

where U and L are constants. Sketch $u(x, 0)$, sketch how $u(x, t)$ will evolve, and show that the solution breaks down at a time $4L/(\gamma + 1)U$.

3.24. Seek a travelling wave solution

$$u = f(\xi), \qquad \xi = x - Vt$$

to Burgers' equation (3.140), together with the conditions

$$f(-\infty) = U_1, \quad f(\infty) = 0,$$

as in Fig. 3.22. Hence establish the expression (3.143) for the shock speed, together with the equation

$$\tfrac{1}{4}(\gamma + 1)(f - U_1)f = \tfrac{2}{3}\nu f'.$$

Integrate this, and show that expressions (3.142) and (3.144) follow, if we choose our origin for ξ such that $f(0) = \tfrac{1}{2}U_1$.

3.25. (i) Show that if the third, non-linear, term in the Korteweg–de Vries equation (3.146) may be neglected then the equation admits a simple harmonic wave solution with $\eta = A \cos(kx - \omega t)$, provided that

$$\omega/k = c_0(1 - \tfrac{1}{6}k^2 h_0^2).$$

Show also that this is in keeping with the exact dispersion relation (3.51) for waves of infinitesimal amplitude, when that is expanded to second order for small kh_0.

(ii) Consider the shallow-water equations (3.99) and (3.100) in the case of a right-travelling simple wave flow into water of depth h_0 at rest, so that $u - 2c = -2c_0$. Show that

$$\frac{\partial c}{\partial t} + (3c - 2c_0)\frac{\partial c}{\partial x} = 0,$$

and show too that if $h = h_0 + \eta$, where $\eta \ll h_0$, this equation is in keeping with the Korteweg–de Vries equation (3.146) when the fourth, dispersive, term is neglected.

4 Classical aerofoil theory

4.1. Introduction

Let us begin by noting some of the key events in the early days of aerodynamics.

1894 F. W. Lanchester presents a paper, 'The soaring of birds and the possibilities of mechanical flight', to a meeting of the Birmingham Natural History and Philosophical Society. It contains the elements of the circulation theory of lift, but not in conventional terms.

1897 Lanchester submits a written version of his paper for publication by the Physical Society. It is rejected.

1901 The Wright brothers encounter failure with their first attempts at glider design. One of them is heard to mutter that 'nobody will fly for a thousand years'.

1902 Kutta publishes a short paper, 'Lifting forces in flowing fluids'. It contains the solution for 2-D irrotational flow past a circular arc, with circulation round the surface and a finite velocity at the trailing edge (Exercise 4.8). The connection between circulation and lift is recognized, though not in the form of the general theorem (1.35).

1903 17 December: The Wright brothers achieve their first powered flight. It lasts for 12 seconds, although they improve on this later the same day.

1904 Prandtl presents his paper on boundary layers to the Third International Congress of Mathematicians at Heidelberg (see §8.1).

1906 Joukowski publishes the lift theorem (1.35):

If an irrotational two-dimensional fluid current, having at infinity the velocity V_∞, surrounds any closed contour on which the circulation of velocity is Γ, the force of the aerodynamic pressure acts on this contour in a direction perpendicular to the velocity and has the value

$$L' = \rho_\infty V_\infty \Gamma.$$

The direction of this force is found by causing to rotate through a right angle the vector V_∞ around its origin, in an inverse direction to that of the circulation.

1907 Lanchester publishes his *Aerodynamics*, although some of the most important results in the book date from as early as 1892. He was certainly years ahead of everyone else in recognizing the inevitability, and the importance, of trailing vortices from the tip of a wing of finite length (§1.7).

A list like this is a concise way of presenting some of the facts, but it can be misleading, for the events within it were, at the time, almost wholly unconnected. Thus Lanchester, Kutta, and Joukowski came to their various conclusions about aerodynamics quite independently, and Wilbur Wright, had he known, would probably not have had much time for any of them. He and his brother relied greatly on their own experimental work on wind-tunnel flows past aerofoils of various shapes, but as late as 1909 he wrote to Lanchester:

... I note such differences of information, theory, and even ideals, as to make it quite out of the question to reach common ground ... , so I think it will save me much time if I follow my usual plan and let the truth make itself apparent in actual practice.

Our first aim in this chapter is to establish that for uniform irrotational flow past an aerofoil with a sharp trailing edge there is just one value of the circulation Γ for which the velocity is finite everywhere (Kutta–Joukowski condition). In particular, we seek to show that in the case of a thin symmetrical aerofoil of length L making an angle of attack α with the oncoming stream the value Γ is given by

$$\Gamma = -\pi UL \sin \alpha. \qquad (4.1)$$

We set about doing this by first solving the comparatively easy problem of irrotational flow past a circular cylinder, and then using the method of conformal mapping to infer the irrotational flow past 2-D objects of more wing-like cross-section.

We must add one important warning before we start. The present chapter is full of irrotational flows which involve slip at rigid boundaries. While any particular flow may well serve a quite different purpose, *it will represent correctly the motion of a viscous fluid at high Reynolds number only if the slip velocity can*

122 Classical aerofoil theory

be adjusted to zero successfully, by a viscous boundary layer, without separation. Rough guidelines on whether or not separation will occur have already been presented in §2.1.

4.2. Velocity potential and stream function

The velocity potential

The velocity potential ϕ is something that exists *only if* $\nabla \wedge \boldsymbol{u} = 0$; it is defined at any point P by

$$\phi = \int_O^P \boldsymbol{u} \cdot d\boldsymbol{x} \tag{4.2}$$

where O is some arbitrary fixed point. In a simply connected fluid region ϕ is independent of the path between O and P, and thus a single-valued function of position (Exercise 4.1.) Partial differentiation of eqn (4.2) gives

$$\boldsymbol{u} = \nabla \phi, \tag{4.3}$$

and the vector identity (A.2) at once confirms that this flow is irrotational, as desired.

This representation of an irrotational flow, eqn (4.3), is valid also in multiply connected fluid regions, but the integral in eqn (4.2) may then depend on the path from O to P, in which case ϕ will be a multivalued function of position. In this case, it is worth noting at once that the circulation round any closed curve C in the flow is given by

$$\Gamma = \oint_C \boldsymbol{u} \cdot d\boldsymbol{x} = \oint_C \nabla \phi \cdot d\boldsymbol{x} = [\phi]_C, \tag{4.4}$$

where the last expression denotes the change (if any) in ϕ after one circuit round C (see eqn (A.12)).

Let us take some examples. The uniform flow $\boldsymbol{u} = (U, 0, 0)$ has velocity potential $\phi = Ux$ (plus an insignificant arbitrary constant, which has no effect on the flow (4.3)). The stagnation point flow of Exercise 1.7:

$$u = \alpha x, \quad v = -\alpha y, \quad w = 0$$

is irrotational, and writing

$$\partial \phi / \partial x = \alpha x, \quad \partial \phi / \partial y = -\alpha y, \quad \partial \phi / \partial z = 0$$

we may integrate to obtain
$$\phi = \tfrac{1}{2}\alpha(x^2 - y^2).$$
In both these cases ϕ is a single-valued function of position; there is therefore no circulation round any closed circuit lying in the flow domain.

Now take the line vortex flow (1.21):
$$\mathbf{u} = \frac{k}{r}\mathbf{e}_\theta,$$
which is an irrotational flow *except at the origin*, where it is not defined. To meet this difficulty, consider the flow domain to be $r \geq a$, which is not simply connected, for there are now some closed curves (i.e. those which enclose $r = a$) which cannot be shrunk to a point without leaving the flow domain. To find the velocity potential we integrate
$$\frac{\partial \phi}{\partial r} = 0, \qquad \frac{1}{r}\frac{\partial \phi}{\partial \theta} = \frac{k}{r}, \qquad \frac{\partial \phi}{\partial z} = 0,$$
and thus obtain
$$\phi = k\theta,$$
which is a multivalued function of position. As we go round any circuit not enclosing $r = a$ it is clear that θ, and hence ϕ, will return, at the end of that circuit, to its original value. There is therefore no circulation round such a circuit. But as we go round any closed curve which winds once round the cylinder $r = a$, θ increases by 2π, and the circulation round such a circuit will therefore be $\Gamma = 2\pi k$. Thus all circuits which wind once round the cylinder have the same circulation (cf. Exercise 1.6).

The stream function

This is a useful device for representing flows which are *incompressible and two-dimensional*. The essential idea is to write
$$u = \frac{\partial \psi}{\partial y}, \qquad v = -\frac{\partial \psi}{\partial x}, \qquad (4.5)$$
thus automatically satisfying the 2-D incompressibility condition
$$\frac{\partial u}{\partial x} + \frac{\partial v}{\partial y} = 0. \qquad (4.6)$$

That such a function $\psi(x, y, t)$ may be found can be shown by a similar argument to that used above (Exercise 4.1).

An important property of ψ follows immediately from eqn (4.5), for

$$(\boldsymbol{u} \cdot \nabla)\psi = u\frac{\partial \psi}{\partial x} + v\frac{\partial \psi}{\partial y} = \frac{\partial \psi}{\partial y}\frac{\partial \psi}{\partial x} - \frac{\partial \psi}{\partial x}\frac{\partial \psi}{\partial y} = 0, \quad (4.7)$$

so ψ *is constant along a streamline.* This gives an effective way of finding the streamlines for a 2-D incompressible flow; if we can just find $\psi(x, y, t)$ the equations for the streamlines can be written down immediately.

A useful way of viewing the representation (4.5) is as

$$\boldsymbol{u} = \nabla \wedge (\psi \boldsymbol{k}), \quad (4.8)$$

where $\boldsymbol{k}$ is the unit vector in the z-direction. It provides, in particular, a way of obtaining the plane polar counterparts to eqn (4.5). Regarding ψ instead as a function of r, θ, and t, we obtain at once

$$u_r = \frac{1}{r}\frac{\partial \psi}{\partial \theta}, \qquad u_\theta = -\frac{\partial \psi}{\partial r}, \quad (4.9)$$

and such a flow automatically satisfies the 2-D incompressibility condition in plane polar coordinates:

$$\frac{1}{r}\frac{\partial}{\partial r}(ru_r) + \frac{1}{r}\frac{\partial u_\theta}{\partial \theta} = 0 \quad (4.10)$$

(see eqn (A.35)).

4.3. The complex potential

Suppose now that we have a flow which is (i) two-dimensional, (ii) incompressible, and (iii) irrotational. Then the velocity field can be represented by both eqns (4.3) and (4.5), so that

$$u = \frac{\partial \phi}{\partial x} = \frac{\partial \psi}{\partial y}, \qquad v = \frac{\partial \phi}{\partial y} = -\frac{\partial \psi}{\partial x}. \quad (4.11)$$

The second of the equations in each pair constitute the well known Cauchy–Riemann equations of complex variable theory, and provided that the partial derivatives in eqn (4.11) are continuous it follows that

$$w = \phi + i\psi \tag{4.12}$$

is an analytic function of the complex variable $z = x + iy$ (Priestley 1985, pp. 16, 184). We call $w(z)$ the *complex potential*.

One of the most important properties of a 2-D incompressible, irrotational flow is that its velocity potential and stream function both satisfy Laplace's equation, so

$$\frac{\partial^2 \phi}{\partial x^2} + \frac{\partial^2 \phi}{\partial y^2} = 0 \tag{4.13}$$

and

$$\frac{\partial^2 \psi}{\partial x^2} + \frac{\partial^2 \psi}{\partial y^2} = 0, \tag{4.14}$$

as may be seen directly from eqn (4.11).

The velocity components u and v are directly related to dw/dz, which is most conveniently calculated as follows:

$$\frac{dw}{dz} = \frac{\partial \phi}{\partial x} + i \frac{\partial \psi}{\partial x} = u - iv. \tag{4.15}$$

(Note the negative sign.) The flow speed at any point is therefore

$$q = (u^2 + v^2)^{\frac{1}{2}} = \left| \frac{dw}{dz} \right|. \tag{4.16}$$

We now consider a number of examples.

Uniform flow at an angle α to the x-axis

Here

$$u = U \cos \alpha, \qquad v = U \sin \alpha,$$

so $dw/dz = Ue^{-i\alpha}$, and therefore

$$w = Uze^{-i\alpha}. \tag{4.17}$$

Line vortex

We may write this flow as

$$\boldsymbol{u} = \frac{\Gamma}{2\pi r} \boldsymbol{e}_\theta, \tag{4.18}$$

where Γ is the circulation round any simple circuit enclosing the vortex, and we already know from the previous section that

$$\phi = \Gamma\theta/2\pi. \qquad (4.19)$$

Using eqn (4.9) we may also write

$$\frac{1}{r}\frac{\partial\psi}{\partial\theta} = 0, \qquad -\frac{\partial\psi}{\partial r} = \frac{\Gamma}{2\pi r},$$

whence

$$\psi = -\frac{\Gamma}{2\pi}\log r.$$

Thus

$$\phi + i\psi = \frac{\Gamma}{2\pi}(\theta - i\log r) = -\frac{i\Gamma}{2\pi}(\log r + i\theta),$$

and the complex potential for a line vortex at the origin is therefore

$$w = -\frac{i\Gamma}{2\pi}\log z. \qquad (4.20)$$

By the same token, the complex potential for a line vortex at $z = z_0$ is

$$w = -\frac{i\Gamma}{2\pi}\log(z - z_0). \qquad (4.21)$$

2-D irrotational flow near a stagnation point

If the complex potential $w(z)$ is analytic in some region it will possess a Taylor series expansion in the neighbourhood of any point z_0 in that region (Priestley 1985, p. 69), i.e.

$$w(z) = w(z_0) + (z - z_0)w'(z_0) + \tfrac{1}{2}(z - z_0)^2 w''(z_0) + \ldots.$$

Now, the first term is an inconsequential constant which makes no difference to dw/dz, and if $z = z_0$ is a *stagnation point* for the flow, then $w'(z_0) = 0$, by virtue of eqn (4.15). Unless $w''(z_0)$ also happens to be zero, it follows that the flow in the immediate neighbourhood of the stagnation point will be determined by the

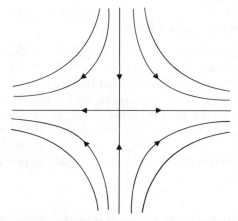

Fig. 4.1. 2-D irrotational flow near a stagnation point.

quadratic term in the above expression. Now, $w''(z_0)$ will typically be complex, $\alpha e^{i\beta}$, say, but by first shifting our coordinates:

$$z - z_0 = z_1,$$

so that the stagnation point is at $z_1 = 0$, and then rotating them so that

$$z_1 e^{i\beta/2} = z_2,$$

we may write

$$w = \text{constant} + \tfrac{1}{2}\alpha z_2^2 + \ldots.$$

Dropping the inconsequential constant, we see that relative to suitably located and orientated coordinates the complex potential in the neighbourhood of a stagnation point is

$$w = \tfrac{1}{2}\alpha z^2, \tag{4.22}$$

where α is real, the corresponding flow being

$$u = \alpha x, \qquad v = -\alpha y \tag{4.23}$$

(cf. Exercise 1.7). The stream function is

$$\psi = \alpha xy, \tag{4.24}$$

so the streamlines are rectangular hyperbolae, as in Fig. 4.1.

4.4. The method of images

Suppose there is a line vortex of strength Γ at a distance d from a rigid plane wall $x = 0$, as in Fig. 4.2(a). A clever trick for obtaining the flow is to imagine that the region $x \leq 0$ is also filled with fluid and that there is an equal and opposite vortex, i.e. of strength $-\Gamma$, at the mirror-image point, as in Fig. 4.2(b). The reason for doing this is that the x-components of velocity of the two vortices obviously cancel on $x = 0$, so there is no normal velocity component there. Thus the complex potential

$$w = -\frac{i\Gamma}{2\pi}\log(z-d) + \frac{i\Gamma}{2\pi}\log(z+d) \qquad (4.25)$$

serves not only for the flow problem in Fig. 4.2(b) but, in $x \geq 0$, for the flow in the presence of a wall in Fig. 4.2(a). This is a simple example of the *method of images*, which is all about getting flows that satisfy boundary conditions.

Let us examine the flow in Fig. 4.2 a little more carefully. The stream function ψ is obtained by writing

$$\phi + i\psi = -\frac{i\Gamma}{2\pi}\log\left(\frac{z-d}{z+d}\right), \qquad (4.26)$$

and the streamlines are therefore

$$\left|\frac{z-d}{z+d}\right| = \text{constant}. \qquad (4.27)$$

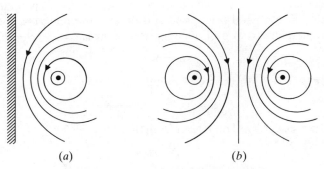

(a) (b)
Fig. 4.2. Flows due to line vortices.

Classical aerofoil theory

These are circles, the so-called coaxal circles of elementary geometry. Each circle cuts the circle $|z| = d$ orthogonally, and if the centre of any circle is distant c_1 and c_2 from the two vortices, then $c_1 c_2 = a^2$, where a is its radius.

It is a simple matter, then, to write down the flow inside a circular cylinder $|z| = a$ due to a line vortex at $z = c < a$: it will be as if the cylinder were not present and there were, instead, an equal and opposite line vortex at $z = a^2/c$. The complex potential for the flow in Fig. 4.3 is therefore

$$w = -\frac{i\Gamma}{2\pi}\log(z-c) + \frac{i\Gamma}{2\pi}\log\left(z - \frac{a^2}{c}\right). \qquad (4.28)$$

While it is not a matter of major concern at present, eqns (4.25) and (4.28) are, in fact, only *instantaneous* complex potentials corresponding to the momentary positions of the vortices; the vortices, and the whole streamline patterns associated with them, in fact move in a manner to be described in §5.6.

Milne-Thomson's circle theorem

Suppose we have a flow with complex potential $w = f(z)$, where all the singularities of $f(z)$ lie in $|z| > a$. Then

$$w = f(z) + \overline{f(a^2/\bar{z})}, \qquad (4.29)$$

where an overbar denotes complex conjugate, is the complex potential of a flow with (i) the same singularities as $f(z)$ in $|z| > a$ and (ii) $|z| = a$ as a streamline.

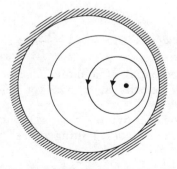

Fig. 4.3. Flow due to a line vortex inside a circular cylinder.

130 *Classical aerofoil theory*

The last property makes the circle theorem a sort of automated method of images for circular boundaries. To prove it, note first that as all the singularities of $f(z)$ are in $|z|>a$, all those of $f(a^2/\bar{z})$ are in $|a^2/\bar{z}|>a$, i.e. in $|z|<a$. Second, on the circle itself we have $z\bar{z}=a^2$, so

$$w = f(z) + \overline{f(z)} \qquad \text{on } |z|=a. \tag{4.30}$$

Thus w is real on $|z|=a$, so $\psi=0$ there, so $|z|=a$ is a streamline.

An elementary application of the circle theorem follows in the next section.

4.5. Irrotational flow past a circular cylinder

Consider irrotational flow, uniform with speed U at infinity, past a fixed circular cylinder $|z|=a$. If the stream is parallel to the x-axis the complex potential for the undisturbed flow is $f(z) = Uz$, which has a singularity only at infinity. Applying the circle theorem we find

$$f(a^2/\bar{z}) = Ua^2/\bar{z}, \qquad \overline{f(a^2/\bar{z})} = Ua^2/z,$$

so

$$w(z) = U\left(z + \frac{a^2}{z}\right) \tag{4.31}$$

is the complex potential of an irrotational flow, uniform at infinity, having $|z|=a$ as a streamline.

It is not the only irrotational flow satisfying these conditions; we may plainly superimpose a line vortex flow of arbitrary strength Γ to give

$$w(z) = U\left(z + \frac{a^2}{z}\right) - \frac{i\Gamma}{2\pi}\log z \tag{4.32}$$

as the complex potential of a more general irrotational flow having no normal velocity at $|z|=a$, yet being uniform, with speed U, at infinity.

Nevertheless, consider first the case (4.31) in which there is no circulation round the cylinder. Putting $z = re^{i\theta}$ we find that

$$\phi = U\left(r + \frac{a^2}{r}\right)\cos\theta \tag{4.33}$$

and
$$\psi = U\left(r - \frac{a^2}{r}\right)\sin\theta, \qquad (4.34)$$

whence†

$$u_r = U\left(1 - \frac{a^2}{r^2}\right)\cos\theta, \qquad u_\theta = -U\left(1 + \frac{a^2}{r^2}\right)\sin\theta. \qquad (4.35)$$

The flow is symmetric fore and aft of the cylinder, and some of the streamlines are sketched in Fig. 4.4(a).

There is evidently *slip* on the cylinder—according to this irrotational flow theory, at any rate—for

$$u_\theta = -2U\sin\theta \qquad \text{at } r = a. \qquad (4.36)$$

In discussing this it is convenient to use instead $u_s = -u_\theta$, which is positive, and $s = (\pi - \theta)a$, which is the distance along the top of the cylinder from the forward stagnation point. Thus

$$u_s = 2U\sin\frac{s}{a}, \qquad (4.37)$$

and

$$\frac{du_s}{ds} = \frac{2U}{a}\cos\frac{s}{a}.$$

The slip velocity therefore rises from zero at the front stagnation point to a maximum of $2U$ at $\theta = \pi/2$; it then decreases again to zero at the rear stagnation point.

When there is circulation Γ round the cylinder, as in eqn (4.32), the velocity components are

$$u_r = U\left(1 - \frac{a^2}{r^2}\right)\cos\theta, \qquad u_\theta = -U\left(1 + \frac{a^2}{r^2}\right)\sin\theta + \frac{\Gamma}{2\pi r}. \qquad (4.38)$$

Anticipating the applications to aerofoil theory that lie ahead, we have taken Γ to be negative in Fig. 4.4, so that the superimposed circulatory flow is clockwise. The character of the streamline

---

† We do not, of course, need the full apparatus of complex variable theory and circle theorem to establish this particular result; there is a much simpler way (Exercise 4.4).

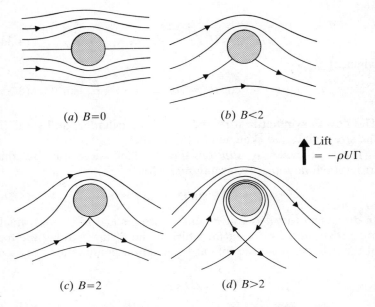

Fig. 4.4. Irrotational flows past a circular cylinder.

pattern depends crucially on the parameter
$$B = -\Gamma/2\pi U a, \qquad (4.39)$$
which is then positive.

One notable feature of the flow that changes with B is the location of the stagnation points. When $B < 2$ there are two of them, both located on the cylinder $r = a$, at $\sin\theta = -\tfrac{1}{2}B$. They therefore move round as B is increased and coalesce when $B = 2$ at $\theta = 3\pi/2$. When $B > 2$ there is only one stagnation point, and it lies off the cylinder at
$$\frac{r}{a} = \frac{B}{2} + \left(\frac{B^2}{4} - 1\right)^{\frac{1}{2}}, \qquad \theta = \frac{3\pi}{2}. \qquad (4.40)$$

This stagnation point thus moves further and further away from the cylinder as B increases, and the region of closed streamlines adjacent to the cylinder becomes steadily larger.

The net force on the cylinder may be calculated from the pressure distribution on $r = a$. As the cylinder is a streamline,

Classical aerofoil theory

and the motion is steady, Bernoulli's theorem gives
$$p + \tfrac{1}{2}\rho u^2 = \text{constant} \quad \text{on } r = a,$$
whence
$$\frac{p}{\rho} = \text{constant} - 2U^2 \sin^2\theta + \frac{U\Gamma}{\pi a} \sin\theta \quad \text{on } r = a.$$

This pressure distribution is symmetric fore and aft of the cylinder (i.e. unchanged by the transformation $\theta \Rightarrow \pi - \theta$), so any net force must be perpendicular to the oncoming stream. The force on a small element $a\,d\theta$ of the cylinder is $pa\,d\theta$ (per unit length in the z-direction). The y-component of this force is $-pa \sin\theta\,d\theta$, and there is therefore a net force on the cylinder of

$$\rho \int_0^{2\pi} \left(2U^2 \sin^2\theta - \frac{U\Gamma}{\pi a}\sin\theta\right) a \sin\theta\,d\theta = -\rho U\Gamma \quad (4.41)$$

in the y-direction, in keeping with the far more general Kutta–Joukowski Lift Theorem of §4.11.

There is positive 'lift', then, if $\Gamma < 0$, and it is easy to see why this should be so, as we have already observed in §1.6. On top of the cylinder in Fig. 4.4 the circulatory flow reinforces the oncoming stream (if $\Gamma < 0$), leading to high speeds and low pressures. Beneath the cylinder the circulatory flow opposes the oncoming stream, leading to low speeds—as evinced by the stagnation points—and high pressures.

Before proceeding further we should emphasize again that we are currently using the irrotational flows in Fig. 4.4 purely as a mathematical device for the calculation of irrotational flows past a thin aerofoil. We are deferring, in particular, all question of whether the flows of Fig. 4.4 are *themselves* observable for a real (i.e. viscous) fluid, whether at high Reynolds number or otherwise (see §§5.7 and 8.6, cf. §7.7).

For what follows it is convenient, in fact, to take the oncoming stream at an angle α to the x-axis. The complex potential of the undisturbed flow is $Uze^{-i\alpha}$, by virtue of eqn (4.17). Applying the circle theorem and superposing a line vortex flow of strength Γ then gives

$$w(z) = U\left(ze^{-i\alpha} + \frac{a^2}{z}e^{i\alpha}\right) - \frac{i\Gamma}{2\pi}\log z \quad (4.42)$$

as our starting point, and this corresponds to the flows of Fig. 4.4 turned anticlockwise through an angle α.

4.6. Conformal mapping

Let $w(z)$ be the complex potential of some 2-D irrotational flow in the z-plane, with $w = \phi + i\psi$. Suppose now that we choose

$$Z = f(z) \tag{4.43}$$

as some analytic function of z, with an inverse

$$z = F(Z) \tag{4.44}$$

which is an analytic function of Z. Then

$$W(Z) = w\{F(Z)\} \tag{4.45}$$

is an analytic function of Z. Now write

$$Z = X + iY \tag{4.46}$$

and split $W(Z)$ into its real and imaginary parts:

$$W(Z) = \Phi(X, Y) + i\Psi(X, Y). \tag{4.47}$$

As W is an analytic function of Z, Φ and Ψ satisfy the Cauchy–Riemann equations, and it follows that the two functions

$$u_*(X, Y) = \partial\Phi/\partial X = \partial\Psi/\partial Y, \quad v_*(X, Y) = \partial\Phi/\partial Y = -\partial\Psi/\partial X, \tag{4.48}$$

represent the velocity components of an irrotational, incompressible flow in the Z-plane.

Further, because $W(Z)$ and $w(z)$ take the same value at corresponding points of the two planes (i.e. points related by eqns (4.43) or (4.44)) it follows that Ψ and ψ are the same at corresponding points. Thus streamlines are mapped into streamlines. In particular, a fixed rigid boundary in the z-plane, which is necessarily a streamline, gets mapped into a streamline in the Z-plane, which could accordingly be viewed as a rigid boundary for the flow in the Z-plane. The key question, then, is: Given flow past a circular cylinder in the z-plane (see eqn (4.42)), can we choose the mapping (4.43) so as to obtain in the Z-plane uniform flow past a more wing-like shape?

Classical aerofoil theory

What happens to the circulation round a closed circuit is important in this connection. Evidently Φ and ϕ are the same at corresponding points of the two planes, and it follows that if we go once round some closed circuit of the z-plane and obtain some consequent change in ϕ, we will obtain the same change in Φ on going once round the corresponding circuit in the Z-plane. Appealing to eqn (4.4), then, we see that the circulations round two such corresponding circuits must be the same.

What happens to the flow at infinity is also of importance. Plainly

$$\frac{dW}{dZ} = \frac{dw/dz}{dZ/dz}, \qquad (4.49)$$

so

$$u_* - iv_* = (u - iv)/f'(z). \qquad (4.50)$$

If we want to map uniform flow past some object into the same uniform flow past another object we must therefore choose $f(z)$ such that $f'(z) \to 1$ as $|z| \to \infty$.

One last general observation concerns a strictly local property of conformal mapping which gives the method its name. Take some point z_0 in the z-plane, with a corresponding point Z_0 in the Z-plane, and let $f^{(n)}(z_0)$ be the first non-vanishing derivative of the function $f(z)$ at z_0. Typically, n will be 1, but there will be occasions in what follows when $f'(z_0) = 0$ but $f''(z_0) \neq 0$, in which case $n = 2$. Let δz denote a small element in the z-plane, originating at $z = z_0$, and let δZ denote the corresponding element in the Z-plane, originating at $Z = Z_0$. By expanding $f(z)$ in a Taylor series we find that

$$\delta Z = \frac{(\delta z)^n}{n!} f^{(n)}(z_0) + O(\delta z)^{n+1}.$$

To first order in small quantities, then,

$$\arg(\delta Z) = n \arg(\delta z) + \arg\{f^{(n)}(z_0)\},$$

and it follows that if δz_1 and δz_2 denote two small elements in the z-plane, both originating at z_0, then

$$\arg(\delta Z_2) - \arg(\delta Z_1) = n[\arg(\delta z_2) - \arg(\delta z_1)]. \qquad (4.51)$$

Thus when two short intersecting elements in the z-plane are mapped into two short intersecting elements in the Z-plane, the angle between them is multiplied by n. Usually, $n = 1$, and such angles are preserved. The shape of a small figure in the z-plane (e.g. a small parallelogram) is then preserved by the mapping—hence the name 'conformal'.

A very effective transformation for our purposes is the Joukowski transformation,

$$Z = z + \frac{c^2}{z}, \qquad (4.52)$$

and we shall exploit the fact that $f'(\pm c) = 0$ but $f''(\pm c) \neq 0$, so that angles between two short line elements which intersect at either $z = c$ or $z = -c$ are doubled by the transformation. The inverse of eqn (4.52) is

$$z = \tfrac{1}{2}Z + (\tfrac{1}{4}Z^2 - c^2)^{\frac{1}{2}}, \qquad (4.53)$$

although we have to take steps to pin down the meaning of this, for there are branch points at $Z = \pm 2c$. In all that follows we shall (i) cut the Z-plane along the real axis between $Z = -2c$ and $Z = 2c$, which stops eqn (4.53) from being multivalued, and (ii) interpret $(\tfrac{1}{4}Z^2 - c^2)^{\frac{1}{2}}$ as meaning that branch of the function which behaves like $\tfrac{1}{2}Z$ (as opposed to $-\tfrac{1}{2}Z$) as $|Z| \to \infty$, which ensures that $z \sim Z$ when $|Z|$ is large.

4.7. Irrotational flow past an elliptical cylinder

Consider the effect of the Joukowski transformation (4.52) on the circle $z = a\mathrm{e}^{\mathrm{i}\theta}$, where $0 \leq c \leq a$. Plainly

$$X + \mathrm{i}Y = \left(a + \frac{c^2}{a}\right)\cos\theta + \mathrm{i}\left(a - \frac{c^2}{a}\right)\sin\theta,$$

so the circle is mapped into the ellipse

$$\frac{X^2}{(a + c^2/a)^2} + \frac{Y^2}{(a - c^2/a)^2} = 1 \qquad (4.54)$$

in the Z-plane (see Fig. 4.5).

Substituting eqn (4.53) into eqn (4.42) we thus obtain

$$W(Z) = U\mathrm{e}^{-\mathrm{i}\alpha}[\tfrac{1}{2}Z + (\tfrac{1}{4}Z^2 - c^2)^{\frac{1}{2}}] + U\mathrm{e}^{\mathrm{i}\alpha}\frac{a^2}{c^2}[\tfrac{1}{2}Z - (\tfrac{1}{4}Z^2 - c^2)^{\frac{1}{2}}]$$

$$- \frac{\mathrm{i}\Gamma}{2\pi}\log[\tfrac{1}{2}Z + (\tfrac{1}{4}Z^2 - c^2)^{\frac{1}{2}}] \qquad (4.55)$$

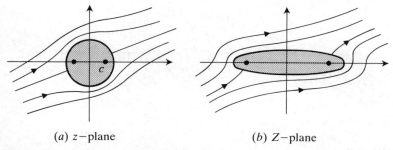

(a) z−plane   (b) Z−plane

Fig. 4.5. Flow past an elliptical cylinder by conformal mapping; no circulation.

as the complex potential for uniform flow at an angle α past the ellipse (4.54), with circulation Γ. It is an elementary, but messy, exercise to write $Z = X + iY$ and then extract the imaginary part of $W(Z)$, namely $\Psi(X, Y)$. The streamlines are sketched in Fig. 4.5(b) for the case $\Gamma = 0$.

4.8. Irrotational flow past a finite flat plate

If we choose $c = a$, so that

$$Z = z + \frac{a^2}{z}, \qquad (4.56)$$

the ellipse (4.54) collapses to a flat plate of length $4a$. Consider the velocity components u_* and v_* in the Z-plane:

$$u_* - iv_* = \frac{dW}{dZ} = \frac{dw/dz}{dZ/dz} = \left(Ue^{-i\alpha} - Ue^{i\alpha}\frac{a^2}{z^2} - \frac{i\Gamma}{2\pi z}\right) \bigg/ \left(1 - \frac{a^2}{z^2}\right). \qquad (4.57)$$

Using eqn (4.53) we can write them in terms of Z, but the comparative simplicity of eqn (4.57) can be more helpful for many purposes.

In particular, the flow speed is in general infinite at the ends of the plate ($Z = \pm 2a$), as these points correspond to the points $z = \pm a$. The status of these sharp edges as singular points in the flow is confirmed by a glance at the streamline pattern for the case $\Gamma = 0$ in Fig. 4.6(a).

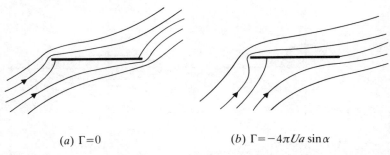

(a) $\Gamma=0$ (b) $\Gamma=-4\pi Ua\sin\alpha$

Fig. 4.6. Irrotational flow past a finite flat plate.

Notably, however, the singularity at the trailing edge $Z = 2a$ (i.e. $z = a$) may be removed *if the circulation Γ is chosen so that the numerator in eqn* (4.57) *vanishes at the trailing edge.* Thus if

$$Ue^{-i\alpha} - Ue^{i\alpha} - \frac{i\Gamma}{2\pi a} = 0,$$

i.e. if

$$\Gamma = -4\pi Ua \sin\alpha, \tag{4.58}$$

then by writing $z = a + \varepsilon$ in both the numerator and denominator of eqn (4.57) and taking the limit as $\varepsilon \to 0$ we find

$$u_* \to U\cos\alpha, \quad v_* \to 0 \quad \text{as } Z \to 2a,$$

so that the flow leaves the trailing edge smoothly and parallel to the plate, as in Fig. 4.6(b). The sense of the circulation is clockwise (for $\alpha > 0$), and this is why we chose to represent the effects of a clockwise circulation in Fig. 4.4.

Of course, the presence of this circulation still leaves a singularity in the velocity field at the leading edge in Fig. 4.6(b).

4.9. Flow past a symmetric aerofoil

In view of Figs 4.5 and 4.6 it will come as no surprise that if we use the mapping (4.56) on a circle in the z-plane which passes through $z = a$ but which encloses $z = -a$, we obtain an aerofoil with a rounded nose but a sharp trailing edge, as in Fig. 4.7(b). If the centre of the circle is on the real axis in the z-plane, at $z = -\lambda$, say, the aerofoil is symmetric and given in terms of the

Classical aerofoil theory 139

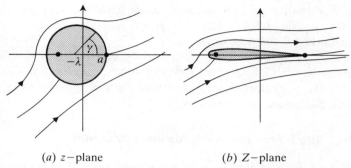

(a) z-plane   (b) Z-plane

Fig. 4.7. Flow past a symmetric Joukowski aerofoil by conformal mapping.

parameter γ by

$$Z = -\lambda + (a + \lambda)e^{i\gamma} + \frac{a^2}{-\lambda + (a + \lambda)e^{i\gamma}}. \quad (4.59)$$

Its shape and thickness depend on λ.

The complex potential $W(Z)$ corresponding to uniform flow past this aerofoil at angle of attack α is obtained by first modifying eqn (4.42) to take account of the new radius and location of the cylinder in the z-plane:

$$w(z) = U\left[(z + \lambda)e^{-i\alpha} + \frac{(a + \lambda)^2}{(z + \lambda)}e^{i\alpha}\right] - \frac{i\Gamma}{2\pi}\log(z + \lambda),$$

and then substituting $z = \frac{1}{2}Z + (\frac{1}{4}Z^2 - a^2)^{\frac{1}{2}}$.

The counterpart to eqn (4.57) is

$$\frac{dW}{dZ} = \left\{U\left[e^{-i\alpha} - \left(\frac{a+\lambda}{z+\lambda}\right)^2 e^{i\alpha}\right] - \frac{i\Gamma}{2\pi(z+\lambda)}\right\} \Big/ \left(1 - \frac{a^2}{z^2}\right), \quad (4.60)$$

but now it is only the vanishing of the denominator at $z = a$ ($Z = 2a$) that causes concern, for $z = -a$ corresponds to a point in the Z-plane which is inside the aerofoil. The value of Γ which makes the numerator in eqn (4.60) zero at the trailing edge ($z = a$) is

$$\Gamma = -4\pi U(a + \lambda)\sin\alpha. \quad (4.61)$$

140 Classical aerofoil theory

The flow is then smooth and free of singularities everywhere, as shown in Fig. 4.7(b), and this is an example of the *Kutta–Joukowski condition* at work.

When $\lambda \ll a$ the aerofoil described by eqn (4.59) is thin and symmetric, with length approximately $4a$ and maximum thickness $3\sqrt{3}\lambda$. By neglecting λ in comparison with a in eqn (4.61) we obtain the classic expression (4.1).

4.10. The forces involved: Blasius's theorem

Let there be a steady flow with complex potential $w(z)$ about some fixed body which has as its boundary the closed contour C, as in Fig. 4.8. If F_x and F_y are the components of the net force on the body, then

$$F_x - \mathrm{i}F_y = \tfrac{1}{2}\mathrm{i}\rho \oint_C \left(\frac{\mathrm{d}w}{\mathrm{d}z}\right)^2 \mathrm{d}z. \tag{4.62}$$

This is *Blasius's theorem*.

To prove it, let s denote arc length along C, and let θ denote the angle made with the x-axis by the tangent to C. Then the force on a small element δs of the boundary is $(-\sin\theta, \cos\theta)p\,\delta s$, so

$$\delta F_x - \mathrm{i}\,\delta F_y = -p(\sin\theta + \mathrm{i}\cos\theta)\,\delta s = -p\mathrm{i}\mathrm{e}^{-\mathrm{i}\theta}\,\delta s.$$

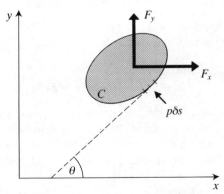

Fig. 4.8. Definition sketch for proof of Blasius's theorem.

Now, C is a streamline for the flow, so
$$u = q\cos\theta, \quad v = q\sin\theta \quad \text{on } C,$$
where $q = (u^2 + v^2)^{\frac{1}{2}}$, so
$$\frac{dw}{dz} = u - iv = qe^{-i\theta} \text{ on } C.$$

Using Bernoulli's equation we may write
$$\delta F_x - i\,\delta F_y = (\tfrac{1}{2}\rho q^2 - k)ie^{-i\theta}\,\delta s,$$
where k is a constant, and substituting for q we find
$$\delta F_x - i\,\delta F_y = \tfrac{1}{2}i\rho\left(\frac{dw}{dz}\right)^2 e^{i\theta}\,\delta s - ki(\delta x - i\,\delta y).$$

Now, $e^{i\theta}\,\delta s = \delta z$. On integrating round the closed contour C the final term disappears and we obtain eqn (4.62).

In a similar way we may establish a formula for $\mathcal{N}$, the moment about the origin of the forces on the body:
$$\mathcal{N} = \text{Real part of}\left[-\tfrac{1}{2}\rho \oint_C z\left(\frac{dw}{dz}\right)^2 dz\right] \quad (4.63)$$

(see Exercise 4.5).

We now consider two examples.

Uniform flow past a circular cylinder

We have, of course, already calculated the net force in this case by direct integration of the pressure distribution in §4.5. Nevertheless, the complex potential is, in the case $\alpha = 0$:
$$w = U\left(z + \frac{a^2}{z}\right) - \frac{i\Gamma}{2\pi}\log z,$$
so applying Blasius's theorem:
$$F_x - iF_y = \tfrac{1}{2}i\rho \oint_C \left[U\left(1 - \frac{a^2}{z^2}\right) - \frac{i\Gamma}{2\pi z}\right]^2 dz.$$

When the integrand is expanded only the z^{-1} term gives a contribution to the integral. The coefficient of that term is

142 *Classical aerofoil theory*

$-iU\Gamma/\pi$, so a simple application of the residue calculus gives
$$F_x - iF_y = \tfrac{1}{2}i\rho \cdot 2\pi i \cdot \left(-\frac{iU\Gamma}{\pi}\right) = i\rho U\Gamma.$$
Thus
$$F_x = 0, \qquad F_y = -\rho U\Gamma, \tag{4.64}$$
as found previously.

Uniform flow past an elliptical cylinder

Consider for simplicity the case when there is no circulation, as in Fig. 4.5(*b*). By the Kutta–Joukowski Lift Theorem (§4.11) there will be no net force on the ellipse, but there will in general be a torque about the origin given by eqn (4.63), i.e.
$$\text{Real part of } \left[-\tfrac{1}{2}\rho \oint_{\text{ellipse}} Z\left(\frac{dW}{dZ}\right)^2 dZ \right].$$

Now, the expression (4.55) for W in terms of $Z = z + c^2/z$ is quite complicated, even in the case $\Gamma = 0$. It is more sensible, then, to write
$$\frac{dW}{dZ} = \frac{dw}{dz}\frac{dz}{dZ}$$
and change the variable of integration from Z to z, so calculating
$$\text{Real part of } \left[-\tfrac{1}{2}\rho \oint_{|z|=a} Z\left(\frac{dw}{dz}\right)^2 \frac{dz}{dZ}\, dz \right].$$

Now, when $\Gamma = 0$,
$$w = U\left(ze^{-i\alpha} + \frac{a^2}{z}e^{i\alpha}\right),$$
so the torque on the ellipse is the real part of
$$-\tfrac{1}{2}\rho U^2 \oint_{|z|=a} \left(z + \frac{c^2}{z}\right)\left(e^{-i\alpha} - \frac{a^2}{z^2}e^{i\alpha}\right)^2 \left(1 - \frac{c^2}{z^2}\right)^{-1} dz.$$

The integrand has poles at $-c$, 0, and c, all within the contour (as $0 < c < a$). Expanding the whole integrand in a Laurent series valid for $|z| > c$, and therefore valid on the integration contour,

we obtain
$$\left(z+\frac{c^2}{z}\right)\left(e^{-2i\alpha}-\frac{2a^2}{z^2}+\frac{a^4}{z^4}e^{2i\alpha}\right)\left(1+\frac{c^2}{z^2}+\frac{c^4}{z^4}+\ldots\right).$$

The coefficient of z^{-1} is
$$c^2 e^{-2i\alpha} - 2a^2 + c^2 e^{-2i\alpha},$$
and the torque on the ellipse is therefore the real part of
$$-\tfrac{1}{2}\rho U^2 \cdot 2\pi i \cdot (2c^2 e^{-2i\alpha} - 2a^2),$$
i.e.
$$\mathcal{N} = -2\pi\rho U^2 c^2 \sin 2\alpha. \tag{4.65}$$

For the flow in Fig. 4.5(b) the torque is negative, i.e. clockwise. More generally, it is such as to tend to align the ellipse so that it is broadside-on to the stream.

4.11. The Kutta–Joukowski Lift Theorem

Consider steady flow past a two-dimensional body, the cross-section of which is some simple closed curve C, as in Fig. 4.9. Let the flow be uniform at infinity, with speed U in the x-direction, and let the circulation round the body be Γ. Then
$$F_x = 0, \quad F_y = -\rho U \Gamma. \tag{4.66}$$

To prove this theorem, first choose the origin O so that it lies inside the body. Then, assuming the flow to be free of singularities, dw/dz will be an analytic function of z in the flow domain and can be expanded in a Laurent series valid for $R < |z| < \infty$, where R is the radius of the smallest circle centred on O which encloses the body. Furthermore, the form of this series must be
$$\frac{dw}{dz} = U + \frac{a_1}{z} + \frac{a_2}{z^2} + \ldots \tag{4.67}$$
because the flow is uniform, speed U, at infinity.

Now, we stated Blasius's theorem in the form of an integral (4.62) taken round the contour C of the body, but if the flow is free of singularities we may, by a cross-cut argument and use of Cauchy's theorem, take the integral equally well round any simple closed contour C' which surrounds the body. In

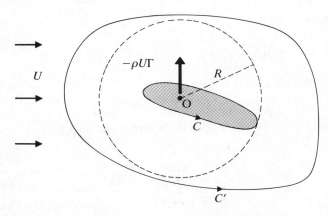

Fig. 4.9. Definition sketch for proof of the Kutta–Joukowski Lift Theorem.

particular, if we take it round a contour C', such as that in Fig. 4.9, which lies wholly in the region $|z| > R$, we may use eqn (4.67) to write

$$F_x - iF_y = \tfrac{1}{2}i\rho \oint_{C'} \left(U + \frac{a_1}{z} + \frac{a_2}{z^2} + \ldots \right)^2 dz.$$

On expanding the integrand only the z^{-1} term contributes to the integral, and with residue $2Ua_1$ at $z = 0$ this gives

$$F_x - iF_y = \tfrac{1}{2}i\rho \cdot 2\pi i \cdot 2Ua_1 = -2\pi\rho Ua_1. \qquad (4.68)$$

To find a_1, use eqn (4.67) to write

$$2\pi i a_1 = \oint_{C'} \frac{dw}{dz} dz,$$

where C' lies wholly in $|z| > R$. We may then appeal again to Cauchy's theorem and a cross-cut argument to justify taking the integral round C instead of C', as dw/dz is analytic in the whole of the flow region. Thus

$$2\pi i a_1 = \oint_{C} \frac{dw}{dz} dz = [w]_C = [\phi + i\psi]_C.$$

But C is a streamline, so the change in ψ after one journey round C is zero. The change in ϕ, on the other hand, is simply Γ, the circulation round the body (see eqn (4.4)). Thus

$$2\pi i a_1 = \Gamma, \quad (4.69)$$

and substituting this in eqn (4.68) establishes the theorem, eqn (4.66).

4.12. Lift: the deflection of the airstream

Notwithstanding the importance of circulation, the Kutta–Joukowski condition, and the theorem of §4.11, an aerofoil obtains lift essentially by imparting downward momentum to the oncoming airstream. In the case of a single aerofoil in an infinite expanse of fluid this elementary truth is disguised, perhaps, by the way that the deflection of the airstream tends to zero at infinity. But in uniform flow past an infinite array of aerofoils, as in Fig. 4.10, there is a finite deflection of the airstream at infinity, so that the downward momentum flux is more readily apparent. Moreover, the deflection is related in a most instructive way to both the circulation and the lift. For this reason, it is worth exploring, and to do this we first need a reformulation of the equation of motion.

The steady momentum equation in integral form

For steady flow, and in the absence of body forces, Euler's equation (1.12) reduces to

$$\rho(\mathbf{u} \cdot \nabla)\mathbf{u} = -\nabla p,$$

and using a suffix notation and the summation convention this may be written

$$\rho u_j \frac{\partial u_i}{\partial x_j} = -\frac{\partial p}{\partial x_i}.$$

Let us integrate this over some fixed region V which is enclosed by a fixed surface S, so that fluid is flowing in through some parts of S and out at others. Then the left-hand side becomes

$$\int_V \rho u_j \frac{\partial u_i}{\partial x_j} \, dV = \int_V \rho \frac{\partial}{\partial x_j}(u_j u_i) \, dV = \int_S \rho u_j u_i n_j \, dS$$

$$= \int_S \rho(\mathbf{u} \cdot \mathbf{n}) u_i \, dS,$$

146 Classical aerofoil theory

the first equation holding because $\partial u_j / \partial x_j = \nabla \cdot \boldsymbol{u} = 0$, and the second holding by virtue of eqn (A.18). Thus

$$\int_S \rho(\boldsymbol{u} \cdot \boldsymbol{n}) u_i \, \mathrm{d}S = -\int_V \frac{\partial p}{\partial x_i} \, \mathrm{d}V = -\int_S p n_i \, \mathrm{d}S,$$

where we have used eqn (A.15). In vector terms, then,

$$-\int_S p\boldsymbol{n} \, \mathrm{d}S = \int_S \rho \boldsymbol{u}(\boldsymbol{u} \cdot \boldsymbol{n}) \, \mathrm{d}S. \tag{4.70}$$

Now, $\rho \boldsymbol{u}$ is the momentum per unit volume of a fluid element, and $(\boldsymbol{u} \cdot \boldsymbol{n}) \, \delta S$ is the volume rate at which fluid is leaving a small portion δS of the surface S, so the right-hand side represents the rate at which momentum is getting carried out of S. The equation states, then, that the total force on S is equal to the rate at which momentum is carried out of S.

Flow past a stack of aerofoils

Let the (identical) aerofoils be a distance d apart, as in Fig. 4.10. Consider the flow in and out of the control surface ABCDA, where AB and DC are portions of identical streamlines a distance d apart, AD being far upstream, where the velocity is $(U, 0)$, and BC being far downstream, where we assume the velocity to be

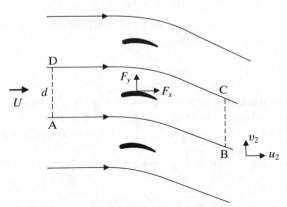

Fig. 4.10. Flow past a stack of aerofoils.

uniform again, but equal to (u_2, v_2). Now, because the fluid is incompressible the volume flux across AD must be equal to that across BC, so $Ud = u_2 d$, and therefore

$$u_2 = U. \qquad (4.71)$$

We now apply the result (4.70) to the fixed region S which lies within ABCDA but excludes the aerofoil. If the lift on the aerofoil is F_y there is a vertical component of force $-F_y$ on S. (There is no other y-component to the first term in eqn (4.70), for those at BC and DA are zero and those at AB and CD cancel, because at any given x the pressures on AB and CD will be the same, as the flow repeats periodically in the y-direction.) There is no flux of momentum across either AB or CD, for they are streamlines, and there is no flux of vertical momentum across AD. Vertical momentum is, however, flowing out of BC at a rate $\rho v_2 Ud$ (per unit length in the z-direction). Equating this to the force exerted on S by the aerofoil, we have

$$F_y = -\rho U v_2 d. \qquad (4.72)$$

In this way we see clearly how the lift is related to the deflection of the airstream; a downward deflection ($v_2 < 0$) corresponds to positive lift. Moreover, it is clear, too, how the circulation is related to this deflection, and hence to the lift itself, for the circulation round ABCDA is

$$\Gamma = v_2 d, \qquad (4.73)$$

as the contribution from DA is zero and those from AB and CD cancel. Thus

$$F_y = -\rho U \Gamma, \qquad (4.74)$$

so that the Kutta–Joukowski result for a single aerofoil in fact holds in this rather different situation also.

4.13. D'Alembert's paradox

Consider the steady flow of an ideal fluid around a 3-D body which is placed in a long straight channel of uniform cross-section (Fig. 4.11). Let us apply eqn (4.70) to the fixed region bounded by the obstacle, two fixed cross-sections S_1 and S_2, and the channel walls. The net force in the downstream direction on the

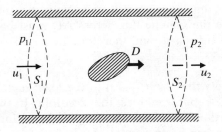

Fig. 4.11. Definition sketch for D'Alembert's paradox.

boundary of this region is

$$\int_{S_1} p_1 \, dS - \int_{S_2} p_2 \, dS - D,$$

where D is the drag exerted by the fluid on the obstacle. According to eqn (4.70), this net force is equal to the downstream component of the flux of momentum out of the region, which is

$$\rho \int_{S_2} u_2^2 \, dS - \rho \int_{S_1} u_1^2 \, dS,$$

where u_1 and u_2 are the velocity components parallel to the channel walls at S_1 and S_2. Thus

$$D = \int_{S_1} (p_1 + \rho u_1^2) \, dS - \int_{S_2} (p_2 + \rho u_2^2) \, dS. \qquad (4.75)$$

Now let us assume that the flow is uniform with speed U_0 far upstream, so that the pressure is a constant, p_0, there. Let us assume that conditions far downstream are similarly uniform; then considerations of mass flow show that the speed must again be U_0 far downstream, as the cross-sectional area of the channel has not changed. Applying the Bernoulli streamline theorem (1.16) to a streamline that runs along the channel walls from $x = -\infty$ to $x = +\infty$ we find that the uniform pressure far downstream must again be p_0.

If, then, we let the cross-sections S_1 and S_2 in Fig. 4.11 recede to infinity in the upstream and downstream directions, we see that the two competing integrals in eqn (4.75) tend to the same

limit, and we therefore deduce that
$$D = 0. \qquad (4.76)$$

This is one of several ways of presenting *D'Alembert's paradox*, namely that steady, uniform flow of an ideal fluid past a fixed body gives no drag on the body.

Another instructive way of viewing this result is as follows. Consider a finite rigid body which has as its boundary a simple closed surface S, and suppose that it is immersed in an infinite expanse of ideal fluid, the entire system being initially at rest. Suppose that the body now moves with speed $U(t)$ in the negative x-direction. The resulting flow is necessarily irrotational (§5.2), and it is, at any instant, unique (Exercise 5.24), determined entirely by the instantaneous normal component of velocity at the surface of the body. Indeed, at any instant the kinetic energy $T(t)$ of the fluid is proportional to the square of $U(t)$, the constant of proportionality being simply a function of the shape and size of the body (see, e.g., Exercise 5.27). Now, if D is the drag exerted on the body (i.e. the force opposite to the direction of $U(t)$), then the rate at which the fluid does work on the body is $-DU$. Equivalently, the body does work on the fluid at a rate DU, and the only way this energy can appear, in the present circumstances,† is as the kinetic energy of the fluid. So

$$DU = dT/dt. \qquad (4.77)$$

There is therefore a drag on the body during the starting process, because the body needs to do work to set up all the kinetic energy of the fluid. But suppose that after a certain time the translational velocity U is held constant. D is then zero, according to eqn (4.77), because the kinetic energy of the fluid remains constant (although it is redistributed, of course, in a rather trivial way, as the whole streamline pattern shifts to follow the body).

The above energy argument can be adapted quite easily for 2-D flow past a 2-D object, provided that there is no circulation; if there is circulation round the object the kinetic energy T is typically infinite, and the argument based on eqn (4.77) breaks

---

† Equation (4.77) does not hold for a viscous fluid, because this energy can then be dissipated (§6.5). Nor does it hold when water waves or sound waves are present, because they can radiate energy to infinity (see, e.g., §3.7).

down. The result nevertheless obtains; according to the Kutta–Joukowski Lift Theorem (4.66) the drag is zero, whether or not there is any circulation.

The result flies in the face of common experience; bodies moving through a fluid are usually subject to a substantial resistance, or drag. In Fig. 4.12 we see the drag on a circular cylinder plotted as a function of the Reynolds number, and it remains substantial even when R is changed from 10^2 to 10^7, which is equivalent to decreasing the viscosity by five orders of magnitude. But then, as the sketches indicate, the flow as a whole shows no sign of settling down to the form in Fig. 4.4(a) as $\nu \to 0$. This is because the mainstream flow speed would, in that event, decrease very substantially along the boundary at the rear of the cylinder, and there would therefore be a strong adverse pressure gradient. An attached boundary layer cannot cope with that (see §2.1), and separation of the boundary layer leads instead to a substantial *wake* behind the cylinder. This wake changes in character with increasing R, as in Fig. 4.12, but shows no sign of disappearing as $R \to \infty$.

D'Alembert described his result of zero drag as 'a singular paradox'. His original argument ($c.$ 1745) was in fact quite different to any of those above, and applied only to flow past

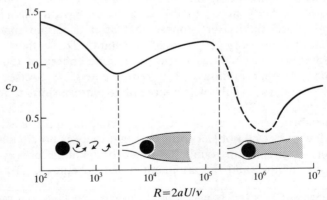

Fig. 4.12. Drag coefficient $c_D = D/\rho U^2 a$ for flow past a circular cylinder of radius a.

Classical aerofoil theory

bodies, such as a sphere, that have fore–aft symmetry (see Exercise 5.26). Such an appeal to symmetry is unnecessary, and Euler came across the full 'paradox' quite independently. His argument involved consideration of the balance of momentum, but it differed significantly from the first argument presented above, not least because the concept of internal pressure p was not secure at the time (see §6.1).

Lighthill (1986) argues that 'D'Alembert's paradox' might better be designated 'D'Alembert's theorem', for if only a body is designed so as to avoid the kind of boundary layer separation evident in Fig. 4.12, then very low drag forces may indeed be achieved. The key feature in this respect is a long, slowly tapering rear to the body—as with an aerofoil—for this typically implies a very *weak* adverse pressure gradient at the rear of the body, enabling the boundary layer to remain attached. For flow past such a 'streamlined' body c_D is typically $O(R^{-\frac{1}{2}})$ as $R \to \infty$ (see eqn (8.24)).

Exercises

4.1. (i) Show that in a simply connected region of irrotational fluid motion the integral (4.2) is independent of the path between O and P.

(ii) Show that in a simply connected region of two-dimensional, incompressible fluid motion the integral

$$\psi = \int_O^P u \, dy - v \, dx$$

is independent of the path between O and P, and hence serves as a definition of the stream function.

4.2. The velocity field

$$u_r = \frac{Q}{2\pi r}, \qquad u_\theta = 0,$$

where Q is a constant, is called a *line source* flow if $Q > 0$ and a *line sink* if $Q < 0$. Show that it is irrotational and that it satisfies $\nabla \cdot \boldsymbol{u} = 0$, save at $r = 0$, where it is not defined. Find the velocity potential and the stream function, and show that the complex potential is

$$w = \frac{Q}{2\pi} \log z.$$

152   *Classical aerofoil theory*

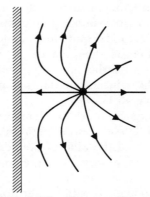

Fig. 4.13.   Irrotational flow due to a line source near a wall.

Observe that the stream function is a multivalued function of position. Why does this not contradict part (ii) of Exercise 4.1?

Fluid occupies the region $x \geq 0$, and there is a plane rigid boundary at $x = 0$. Find the complex potential for the flow due to a line source at $z = d > 0$, and show that the pressure at $x = 0$ decreases to a minimum at $|y| = d$ and thereafter increases with $|y|$.

[Any attempt to reproduce the flow of Fig. 4.13 at high Reynolds number would be fraught with difficulties. A viscous boundary layer would be present, to satisfy the no-slip condition, but for $|y| > d$ the substantial adverse pressure gradient along the boundary would make separation inevitable (see §2.1). More fundamentally still, there are considerable practical difficulties in producing a line source, as opposed to a line sink, at high Reynolds number. These are more easily seen by considering the corresponding 3-D problem; a point sink can be simulated quite well by sucking at a small tube inserted in the fluid, but blowing down such a tube produces not a point source but a highly directional and usually turbulent jet (see, e.g. Lighthill 1986, pp. 100–103). The streamline pattern in Fig. 4.13 may nevertheless be observed in a *Hele–Shaw cell* (§7.7), although viscous effects are then paramount throughout the whole flow, so the pressure distribution is not given by Bernoulli's equation.]

4.3.   An irrotational 2-D flow has stream function $\psi = A(x - c)y$, where A and c are constants. A circular cylinder of radius a is introduced, its centre being at the origin. Find the complex potential, and hence the stream function, of the resulting flow. Use Blasius's theorem (4.62) to calculate the force exerted on the cylinder.

Classical aerofoil theory

4.4. Show that the problem of irrotational flow past a circular cylinder may be formulated in terms of the velocity potential $\phi(r, \theta)$ as follows:

$$\frac{\partial^2 \phi}{\partial r^2} + \frac{1}{r}\frac{\partial \phi}{\partial r} + \frac{1}{r^2}\frac{\partial^2 \phi}{\partial \theta^2} = 0,$$

with

$$\phi \sim Ur \cos \theta \quad \text{as } r \to \infty, \qquad \partial \phi / \partial r = 0 \quad \text{on } r = a,$$

and obtain the solution (4.33) by using the method of separation of variables.

When there is circulation round the cylinder, derive eqn (4.40), and confirm that the stagnation points vary in position with the parameter B in the manner of Fig. 4.4.

4.5. Establish the expression (4.63) for the moment, $\mathcal{N}$, of forces on a body in irrotational flow, using an argument similar to that for Blasius's theorem.

4.6. By writing $z = a + \varepsilon$ in eqn (4.57) and taking the limit $\varepsilon \to 0$ check that the choice of circulation (4.58) does indeed lead to a finite velocity at the trailing edge.

4.7. According to eqns (4.1) and (4.66), the force on a thin symmetric aerofoil with a sharp trailing edge is

$$\mathcal{L} = \pi \rho U^2 L \sin \alpha$$

in a direction *perpendicular to the uniform stream*. This amounts to a component $\mathcal{L} \cos \alpha$ perpendicular to the aerofoil and a component $\mathcal{L} \sin \alpha$ parallel to the aerofoil, directed towards the leading edge. This latter component is, at first sight, rather curious; it might be thought that the net effect of a pressure distribution on a thin symmetric aerofoil should be almost normal to the aerofoil. That it is not is due to *leading edge suction*, i.e. a severe drop in pressure in the immediate vicinity of the rounded leading edge, this pressure drop being sufficient to make itself felt despite the small thickness of the wing on which it acts.

To see evidence of this, consider the extreme case of flow past a flat plate with circulation, as in Fig. 4.6(b) or Fig. 4.15. First, use eqns (4.56) and (4.57), on $z = ae^{i\theta}$, with Γ chosen according to eqn (4.58), to show that the flow speed on the plate is

$$U \left| \cos \alpha \pm \left(\frac{1-s}{1+s} \right)^{\frac{1}{2}} \sin \alpha \right|,$$

where the upper/lower sign corresponds to the upper/lower side of the plate, and s denotes $X/2a$, which therefore runs between -1 at the leading edge and $+1$ at the trailing edge.

154 Classical aerofoil theory

Show that the corresponding pressure distributions are

$$p(s) = p(1) - \tfrac{1}{2}\rho U^2 \left[\left(\frac{1-s}{1+s}\right)\sin^2\alpha \pm 2\left(\frac{1-s}{1+s}\right)^{\frac{1}{2}} \sin\alpha\cos\alpha \right],$$

(see Fig. 4.14). Note that there is a (negative) pressure singularity at the leading edge, whereas if the leading edge were rounded this pressure drop would be finite.

As far as the force component normal to the plate is concerned, note that the pressure difference across the plate is

$$p_D = 2\rho U^2 \left(\frac{1-s}{1+s}\right)^{\frac{1}{2}} \sin\alpha\cos\alpha.$$

This too has a singularity at the leading edge, but it is integrable. Show that

$$\int_{-2a}^{2a} p_D \, dX = \mathscr{L} \cos\alpha,$$

in keeping with the Kutta–Joukowski Lift Theorem.

Finally, show that eqn (4.65) holds even if there is circulation Γ round the ellipse, and then take the case $c = a$ to show that the torque on a flat plate about the origin is $-\mathscr{L}a \cos\alpha$, i.e. as if the whole lift force $\mathscr{L}$ were

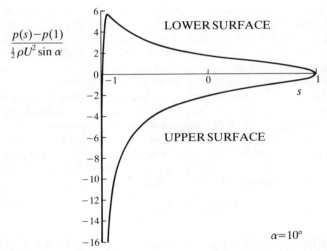

Fig. 4.14. Theoretical pressure distribution on a flat plate at a 10° angle of attack.

Classical aerofoil theory

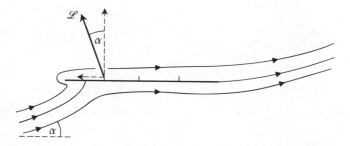

Fig. 4.15. The torque on a flat plate in uniform flow is as if the lift $\mathscr{L}$ were concentrated at a point one-quarter of the way along the plate from the leading edge.

applied at a point one-quarter of the way along the plate, as indicated in Fig. 4.15.

[The fact that this point is independent of α is of practical value, and makes for smooth control of an aircraft.]

4.8. Show that the Joukowski transformation $Z = z + a^2/z$ can be written in the form

$$\frac{Z-2a}{Z+2a} = \left(\frac{z-a}{z+a}\right)^2,$$

so that, in particular,

$$\arg(Z - 2a) - \arg(Z + 2a) = 2[\arg(z - a) - \arg(z + a)].$$

Consider the circle in the z-plane which passes through $z = -a$ and $z = a$ and has centre $ia \cot \beta$. Show that the above transformation takes it into a circular arc between $Z = -2a$ and $Z = 2a$, with subtended angle 2β (Fig. 4.16). Obtain an expression for the complex potential in the Z-plane, when the flow is uniform, speed U, and parallel to the real axis. Show that the velocity will be finite at both the leading and trailing edges if

$$\Gamma = -4\pi U a \cot \beta.$$

[This exceptional circumstance arises only when the undisturbed flow is parallel to the chord line of the arc.]

4.9. Provided that $f'(z_0) \neq 0$, points in the neighbourhood of $z = z_0$ are mapped by $Z = f(z)$, according to Taylor's theorem, in such a way that

$$Z - Z_0 = f'(z_0)(z - z_0) + O(z - z_0)^2,$$

156 Classical aerofoil theory

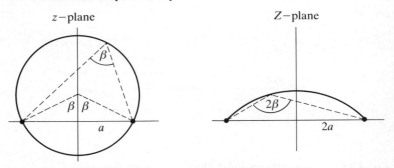

Fig. 4.16. Generation of a circular arc by a Joukowski transformation.

where $Z_0 = f(z_0)$. Use this to show that a line source of strength Q at $z = z_0$ is mapped into a line source of strength Q at $Z = Z_0$, provided that $f'(z_0) \neq 0$.

Fluid occupies the region between two plane rigid boundaries at $y = \pm b$, and there is a line source of strength Q at $z = 0$. Find the complex potential $w(z)$ for the flow

(i) by the method of images,

(ii) by using the mapping $Z = e^{\alpha z}$ with a suitably chosen $\alpha > 0$.

4.10. Use the momentum equation in its integral form (4.70) to show that there is a non-zero drag

$$F_x = \rho \Gamma^2 / 2d$$

on each of the aerofoils in Fig. 4.10.

Is this at odds with the Kutta–Joukowski Lift Theorem (4.66)?

5 Vortex motion

5.1. Kelvin's circulation theorem

THEOREM. *Let an inviscid, incompressible fluid of constant density be in motion in the presence of a conservative body force $g = -\nabla \chi$ per unit mass. Let $C(t)$ denote a closed circuit that consists of the same fluid particles as time proceeds (Fig. 5.1). Then the circulation*

$$\Gamma = \int_{C(t)} \boldsymbol{u} \cdot \mathrm{d}\boldsymbol{x} \tag{5.1}$$

round $C(t)$ is independent of time.

Proof. We appeal to the following lemma:

$$\frac{\mathrm{d}}{\mathrm{d}t} \int_{C(t)} \boldsymbol{u} \cdot \mathrm{d}\boldsymbol{x} = \int_{C(t)} \frac{\mathrm{D}\boldsymbol{u}}{\mathrm{D}t} \cdot \mathrm{d}\boldsymbol{x} \tag{5.2}$$

(Exercise 5.2). Then, by Euler's equation (1.12),

$$\frac{\mathrm{d}\Gamma}{\mathrm{d}t} = -\int_{C(t)} \nabla\left(\frac{p}{\rho} + \chi\right) \cdot \mathrm{d}\boldsymbol{x} = -\left[\frac{p}{\rho} + \chi\right]_C,$$

where the last term denotes the change in $p/\rho + \chi$ on going once round C (see eqn (A.12)). But this change is zero, as p, ρ, and χ are all single-valued functions of position. This proves the theorem.

Notes on the theorem

(a) C denotes a 'dyed' circuit, composed of the same fluid particles as time proceeds; the result is not true in general if C is a closed curve fixed in space.

(b) The conditions of incompressibility and constant density are not essential: Kelvin established his result subject to weaker restrictions (Exercise 5.4).

158 *Vortex motion*

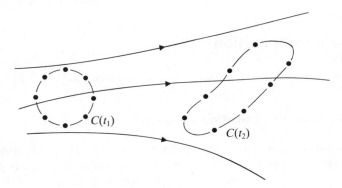

Fig. 5.1. Definition sketch for Kelvin's theorem, showing eight fluid particles along a 'dyed' circuit C at time t_1, and their positions at time t_2.

(c) The theorem does not require the fluid region to be simply connected, i.e. it does not require the dyed circuit C to be spannable by a surface S lying wholly in the fluid.

(d) The inviscid equations of motion enter the proof only in helping to evaluate a line integral round C, so if viscous forces happened to be important elsewhere in the flow, i.e. off the curve C, this would not affect the conclusion that Γ remains constant round C.

The generation of lift on an aerofoil

We mentioned in §1.1 how the shedding of a starting vortex is essential to the generation of lift on an aerofoil, and we now investigate why this should be so.

Consider the situation at a time t after the start. Vorticity and viscous forces will be confined to (i) a thin boundary layer on the aerofoil, (ii) a thin wake, and (iii) the rolled-up 'core' of the starting vortex, as indicated by the shading in Fig. 5.2. Consider now a dyed circuit abcda which is large enough to have been clear of all these regions since the start of the motion. As the original state was one of rest the circulation round that circuit was originally zero. By Kelvin's circulation theorem, then, the circulation round that circuit will still be zero at time t (see especially note (d) above). Thus if we sketch in a line aec—an

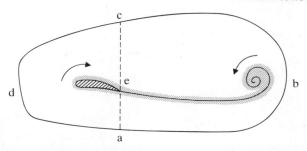

Fig. 5.2. The generation of circulation by means of vortex shedding.

instantaneous line in space at time t such that the curve aecda encloses the aerofoil but not the wake or the starting vortex—then the circulation round aecda must be equal and opposite to that round abcea.

What happens, then, as the aerofoil starts to move, is that positive vorticity is shed in the form of a starting vortex. By Stokes's theorem,

$$\int_S \boldsymbol{\omega} \cdot \boldsymbol{n} \, \mathrm{d}S = \int_C \boldsymbol{u} \cdot \mathrm{d}\boldsymbol{x},$$

this gives a positive circulation round abcea. This in turn implies, by the preceding argument, a negative circulation round aecda, and this circulation is very evident in some classic photographs taken by Prandtl and Tietjens (see, e.g., Batchelor 1967, Plate 13). The vortex shedding continues until the circulation round the aerofoil is sufficient to make the main, irrotational flow smooth at the trailing edge, as in Fig. 1.10(*b*), at which stage no further net vorticity is shed into the wake from the boundary layers on the upper and lower surfaces of the aerofoil. Thereafter the aerofoil retains its final 'Kutta–Joukowski' value of the circulation.

A novel mechanism of lift generation for hovering insects

An exotic variation on the above theme was discovered by Weis-Fogh (1973, 1975) in the hovering motions of the tiny chalcid wasp *Encarsia formosa* (wing chord ~0.2 mm). This insect claps its wings together, then flings them open about a

160 *Vortex motion*

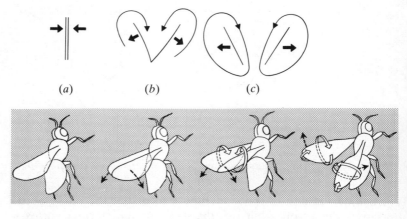

Fig. 5.3. The Weis-Fogh mechanism of lift generation. The first three sketches give a 2-D model of (*a*) the 'clap', (*b*) the 'fling', and (*c*) the parting of the wings. The remaining sketches (after Dalton 1977) show the mechanism in practice, and the final sketch indicates also the flow associated with the vortex (not shown) that extends, in a circular arc, between the wing tips (cf. Fig. 1.12).

horizontal line of contact, so that air has to rush in to fill the gap (Fig. 5.3(*b*)). Then it moves its wings apart, by which time each one has acquired during the 'fling' movement a circulation of the correct sign to give lift in the subsequent motion.

In practice, viscous effects are important, especially in causing large leading-edge separation vortices (see the excellent photographs in Spedding and Maxworthy 1986). Nevertheless, one remarkable feature of this novel lift generation mechanism is that it could work, in principle, in a strictly inviscid fluid (Lighthill 1973). In this sense it differs markedly from the conventional method for lift generation which we have just discussed, for that relies in an essential way on viscous effects for boundary layer formation, separation at the trailing edge, and consequent vortex shedding. In the Weis-Fogh mechanism the circulation round one wing essentially acts as the starting vortex for the other.

At first sight, perhaps, Kelvin's circulation theorem does not permit the situation in Fig. 5.3(*c*) for a strictly inviscid fluid: if one views the circuits there as dyed circuits then the circulations round them must have remained constant. Yet one cannot claim that those circulations are zero, even if the fluid were wholly at rest at stage (a), for neither dyed circuit at stage (c) was a *closed* circuit at stage (b), an unusual circumstance that arises only because the topology of the fluid domain has changed in the meantime.

The word 'meantime' gives, in fact, rather too leisurely an impression; *Encarsia formosa* goes through the sequence in Fig. 5.3 roughly 400 times a second.

5.2. The persistence of irrotational flow

Let an inviscid, incompressible fluid of constant density move in the presence of a conservative body force. Then if a portion of the fluid is initially in irrotational motion, that portion will always be in irrotational motion.

To prove this *Cauchy–Lagrange theorem* suppose that the vorticity $\omega = \nabla \wedge u$ were *not* identically zero throughout that portion of fluid at a later time. By virtue of Stokes's theorem:

$$\int_C u \cdot dx = \int_S \omega \cdot n \, dS,$$

and it would then be possible to select some small closed dyed circuit around which the circulation would be non-zero. But this would violate Kelvin's circulation theorem, because the circulation round such a circuit must initially have been zero, on account of Stokes's theorem and the fact that ω was initially zero. Our initial assumption must therefore be false. This completes the proof.

For 2-D flows the result is obvious from the vorticity equation (1.27); if ω is zero for a portion of the fluid at $t = 0$ then, according to eqn (1.27), ω remains zero for each fluid element constituting that portion for all time t. But in three dimensions the result is not obvious from eqn (1.25), and it is here that the theorem comes into its own. (Although it is of course quite evident that if ω is everywhere zero at $t = 0$ then $\omega = 0$ everywhere for all t is *one* solution of eqn (1.25).)

Irrotational flows are important, then, even in three dimensions. The velocity field can then be written as

$$u = \nabla \phi, \tag{5.3}$$

and ϕ will be a single-valued function of position when the flow region is simply connected (see §4.2). [In other circumstances—as, for example, with the irrotational part of the flow due to a vortex ring (Fig. 5.7(b))—ϕ may be multivalued.] As the fluid is incompressible, $\nabla \cdot \boldsymbol{u} = 0$, so ϕ satisfies *Laplace's equation*

$$\nabla^2 \phi = 0. \tag{5.4}$$

The general theory of irrotational flow is a classical and important part of fluid dynamics, and we explore something of it in Exercises 5.23–5.29. We should emphasize, however, that much of the present chapter is concerned with fluid motions in which the vorticity is *not* zero, in which case there is no such thing as a velocity potential ϕ and $\boldsymbol{u}$ *cannot* be written in the form (5.3).

5.3. The Helmholtz vortex theorems

A *vortex line* is, at any particular time t, a curve which has the same direction as the vorticity vector

$$\boldsymbol{\omega} = \nabla \wedge \boldsymbol{u} \tag{5.5}$$

at each point. Mathematically, then, a vortex line $x = x(s)$, $y = y(s)$, $z = z(s)$, is obtained by solving

$$\frac{\mathrm{d}x/\mathrm{d}s}{\omega_x} = \frac{\mathrm{d}y/\mathrm{d}s}{\omega_y} = \frac{\mathrm{d}z/\mathrm{d}s}{\omega_z}$$

at a particular time t.

The vortex lines which pass through some simple closed curve in space are said to form the boundary of a *vortex tube* (Fig. 5.4(a)).

Suppose now that we have an *inviscid, incompressible fluid of constant density moving in the presence of a conservative body force* (so that Kelvin's circulation theorem applies). Then

(1) *The fluid elements that lie on a vortex line at some instant continue to lie on a vortex line, i.e. vortex lines 'move with the fluid'.*

An immediate consequence of this is that vortex tubes move with the fluid in a like manner.

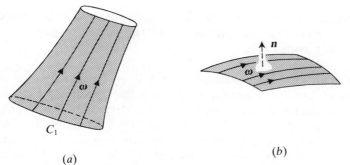

Fig. 5.4. (a) A vortex tube. (b) A vortex surface.

(2) *The quantity*

$$\Gamma = \int_S \boldsymbol{\omega} \cdot \boldsymbol{n} \, dS \tag{5.6}$$

is the same for all cross-sections S of a vortex tube. Furthermore, Γ is independent of time.

The quantity Γ is therefore a conserved property of the tube as a whole, called the *strength* of the tube.

Proof of (1). We first define a *vortex surface* as a surface such that $\boldsymbol{\omega}$ is tangent to the surface at every point (Fig. 5.4(b)). The proof proceeds by viewing the vortex line, in its initial configuration, as the intersection of two vortex surfaces. Mark the particles which occupy one of the vortex surfaces, at $t = 0$, with dye. Consider a closed circuit C made up of a particular set of dyed particles and spanned by a portion S_* of the vortex surface. At $t = 0$ the circulation round C is zero, for by Stokes's theorem

$$\int_C \boldsymbol{u} \cdot d\boldsymbol{x} = \int_{S_*} \boldsymbol{\omega} \cdot \boldsymbol{n} \, dS,$$

and $\boldsymbol{\omega} \cdot \boldsymbol{n}$ is zero on S_*. Now, as time proceeds the dyed sheet of fluid will deform, but the circulation round C will remain zero, by Kelvin's circulation theorem. This being so for all circuits such as C it follows, by using Stokes's theorem again, that $\boldsymbol{\omega} \cdot \boldsymbol{n}$ will remain zero at all points of the dyed sheet of fluid. That sheet therefore remains a vortex surface as time proceeds. The proof is completed by noting that the intersection of two such dyed sheets therefore remains the intersection of two vortex surfaces, i.e. it remains a vortex line.

Proof of (2). The statement that Γ is independent of the cross-section S has nothing to do with the equations of motion, but is simply a consequence of the fact that the vorticity $\omega = \nabla \wedge \boldsymbol{u}$ is divergence-free (Exercise 5.5). The statement that Γ is independent of time follows on considering a circuit, such as C_1 in Fig. 5.4(a), composed of fluid particles which lie on the wall of the vortex tube and encircle it. By Stokes's theorem, Γ is the circulation round C_1, and by Kelvin's circulation theorem this remains constant as time proceeds.

It is instructive to consider the particular case of a *thin* vortex tube in which ω is virtually constant across any particular cross-section. In that case Γ is essentially just the product $\omega \, \delta S$, where δS is the normal cross-section of the tube. But δS is also the normal cross-section of the fluid continually occupying the tube, and as the fluid must conserve its volume δS will vary inversely with the length l of a small section of the tube. Thus the vorticity ω varies in proportion to l; stretching of vortex tubes by the fluid motion intensifies the local vorticity.

In a tornado, for example, the strong thermal updraughts into the thunderclouds overhead produce intense stretching of vortex tubes, and hence the potentially devastating rotary motions observed. The funnel cloud serves, in fact, as a direct marker of the vortex tube, rather than the air occupying it, because it essentially marks regions of very low pressure (where the air rapidly expands and condenses), and these in turn are located in the core of the vortex, where all the vorticity is concentrated (see Exercise 1.3). Thus when the thunderclouds move on, and the funnel cloud tips over in the manner of Fig. 5.5, we have a vivid illustration of Helmholtz's first vortex theorem at work.

In contrast, it is the shortening of vortex tubes that is responsible for the gradual 'spin-down' of a stirred cup of tea (Fig. 5.6). The main body of the fluid is essentially inviscid and in rapid rotation, the centrifugal force being (almost) balanced by a radially inward pressure gradient. This pressure gradient also imposes itself throughout the thin viscous boundary layer on the

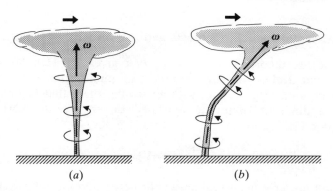

Fig. 5.5. The deformation of a tornado as the thunderclouds move overhead.

bottom of the cup, where it is stronger than required, for the fluid in the boundary layer rotates much less rapidly. That fluid therefore spirals inward (as evinced by the way in which tea leaves on the bottom of the cup congregate in the middle), and eventually turns up and out of the boundary layer, as in Fig. 5.6. In this way vortex tubes in the main body of the fluid become shorter and expand in cross-section, so that the vorticity decreases with time. It is by this subtle mixture of inviscid and viscous dynamics that the apparently innocuous spin-down of a stirred cup of tea is achieved (see §8.5).

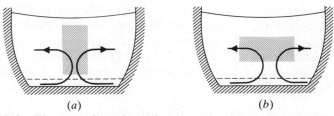

Fig. 5.6. The secondary circulation in a stirred cup of tea is driven by the bottom boundary layer (beneath the dotted line) and turns a tall, thin column of 'dyed' fluid into a short, fat one, so decreasing its angular velocity.

The Helmholtz vortex theorems and the vorticity equation

The vortex theorems above were first given by Helmholtz in 1858, but Kelvin did not obtain and publish his circulation theorem until 1867. It goes without saying, then, that Helmholtz took a different route; he appealed directly to the vorticity equation (1.25):

$$\frac{D\boldsymbol{\omega}}{Dt} = (\boldsymbol{\omega} \cdot \nabla)\boldsymbol{u}. \tag{5.7}$$

We will not give his actual argument here,† but consider instead the relationship between eqn (5.7) and the vortex theorems in some simple specific cases.

It is possible, for instance, to see by inspection of eqn (5.7) how stretching the fluid that lies along a vortex line leads to an intensification of the local vorticity field. Suppose, for example, that the vortex lines are almost in the z-direction, as in Fig. 5.5(a), so that $\boldsymbol{\omega} \doteqdot \omega \boldsymbol{k}$ and

$$\frac{D\boldsymbol{\omega}}{Dt} \doteqdot \omega \frac{\partial \boldsymbol{u}}{\partial z}. \tag{5.8}$$

The z-component of this equation gives

$$\frac{D\omega}{Dt} \doteqdot \omega \frac{\partial w}{\partial z},$$

and the vorticity of a particular fluid element therefore increases with time if $\partial w/\partial z > 0$, i.e. if the instantaneous vertical velocity increases with z. Such is the case, of course, if fluid elements are being stretched in the vertical direction, whereas if they were being carried up or down without any vertical stretching or squashing, w would be independent of z.

A particularly simple case is that of 2-D flow. Vortex tubes are aligned with the z-axis, and $w = 0$. There is no stretching of vortex tubes, and

$$\frac{D\omega}{Dt} = 0, \tag{5.9}$$

---

† It in fact contains a flaw, which may however be corrected (see, e.g. Lamb 1932, p. 206; Rosenhead 1963, pp. 122–123).

so that the vorticity ω of any particular fluid element is conserved.

A more revealing case in the present context is that of *axisymmetric* flow:

$$\boldsymbol{u} = u_R(R, z, t)\boldsymbol{e}_R + u_z(R, z, t)\boldsymbol{e}_z, \quad (5.10)$$

where (R, ϕ, z) denote cylindrical polar coordinates.† The velocity components are then independent of ϕ, the streamlines all lie in planes $\phi =$ constant, and the vorticity is $\boldsymbol{\omega} = \omega \boldsymbol{e}_\phi$, where

$$\omega = \frac{\partial u_R}{\partial z} - \frac{\partial u_z}{\partial R}. \quad (5.11)$$

In axisymmetric flow the vortex tubes are therefore ring-shaped, around the symmetry axis. According to the first vortex theorem they move with the fluid. In doing so they will, in general, expand and contract about the symmetry axis, and thus change in length. As the fluid is incompressible the cross-sectional area δS of a thin tube will be in inverse proportion to the length $2\pi R$ of the tube. But the second vortex theorem implies that $\omega \, \delta S$ will be a constant, so we conclude that ω will be proportional to the length of the tube $2\pi R$. We leave it as an instructive exercise (Exercise 5.7) to show that in the case of axisymmetric flow the vorticity equation (5.7) reduces to

$$\frac{\mathrm{D}}{\mathrm{D}t}\left(\frac{\omega}{R}\right) = 0, \quad (5.12)$$

which expresses just this result, that the vorticity of any particular fluid element changes in proportion to R as time proceeds.

When, in axisymmetric flow, an isolated vortex tube is surrounded by irrotational motion, we speak of it as a *vortex ring*. The familiar 'smoke-ring' is perhaps the most common example, and provides a vivid illustration of the Helmholtz vortex theorems, though the vortex core typically occupies only a fraction of the smoke ring as a whole (see Fig. 5.7).

---

† This is not our usual notation, as we are shortly to use spherical polar coordinates (r, θ, ϕ) for axisymmetric flow. It seemed best not to have the same symbol meaning two different things in the space of a few pages. Thus ϕ has the same meaning in the two cases, and $R = r \sin \theta$.

168 *Vortex motion*

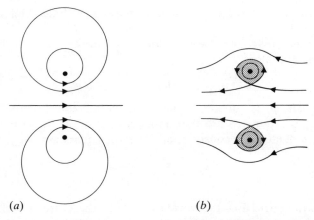

Fig. 5.7. Flow due to a vortex ring (*a*) relative to a fixed frame and (*b*) relative to a frame moving with the vortex core. Shading denotes smoke, in the case of a smoke ring, while the vortex core is indicated by the black dots.

5.4. Vortex rings

We showed in §5.1 how Kelvin's circulation theorem plays a key part in the mechanism by which an aircraft obtains lift at take-off. While this is one of the theorem's most elegant and significant applications, it is not of course what Kelvin had in mind in 1867. What he did have in mind is quite extraordinary, but clear enough from the following:

Jan. 22, 1867.

MY DEAR HELMHOLTZ—I have allowed too long a time to pass without thanking you for your kind letter .... Just now, ... *Wirbelbewegungen* have displaced everything else, since a few days ago Tait showed me in Edinburgh a magnificent way of producing them. Take one side (or the lid) off a box (any old packing-box will serve) and cut a large hole in the opposite side. Stop the open side loosely with a piece of cloth, and strike the middle of the cloth with your hand. If you leave anything smoking in the box, you will see a magnificent ring shot out by every blow. A piece of burning phosphorus gives very good smoke for the purpose; but I think nitric acid with pieces of zinc thrown into it, in the bottom of the box, and cloth wet with ammonia, or a large open dish of ammonia beside it, will answer better. The nitrite of ammonia makes fine white clouds in the air, which, I think, will be less pungent and disagreeable than the smoke from the phosphorus. We sometimes can make one ring shoot through another, illustrating perfectly your description; when one ring passes near another, each is much disturbed, and is seen to be in a state of violent vibration for a few seconds, till it settles again into its circular form. The accuracy of the circular form of the whole ring, and the fineness and roundness of the section, are beautifully seen. If you try it, you will easily make rings of a foot in diameter and an inch or so in section, and be able to follow them and see the constituent rotary motion. The vibrations make a beautiful subject for mathematical work. The solution for the longitudinal vibration of a straight vortex column comes out easily enough. The absolute permanence of the rotation, and the unchangeable relation you have proved between it and the portion of the fluid once acquiring such motion in a perfect fluid, shows that if there is a perfect fluid all through space, constituting the substance of all matter, a vortex-ring would be as permanent as the solid hard atoms assumed by Lucretius and his followers (and predecessors) to account for the permanent properties of bodies (as gold, lead, etc.) and the differences of their characters. Thus, if two vortex-rings were once created in a perfect fluid, passing through one another like links of a chain, they never could come into collision, or break one another, they would form an indestructible atom; every variety of combinations might exist. Thus a long chain of vortex-rings, or three rings, each running through each of the other, would give each very characteristic reactions upon other such kinetic atoms.

This atomic theory,† 40 years ahead of that of Niels Bohr, was no speculative sideline to Kelvin's hydrodynamic researches at the time; it was the main impetus behind them, and in the opening sentence of his 1867 paper he more or less says as much.

One hundred and twenty years later, vortex rings still exercise a certain fascination, although more modest and less dangerous ways of producing them are perhaps to be recommended. All that is needed is some arrangement for discharging smoke through a circular hole in a plane rigid boundary, where separation of the boundary layer can take place and be followed by the rolling up of the consequent vortex sheet (Fig. 5.9). Any simple apparatus which achieves this will suffice; I employ a syringe of the kind commonly used to squeeze icing on to cakes.

---

† Atiyah (1988) observes that one particular notion in this theory—that of using topology as a source of stability—may be said to have surfaced again in modern physics, albeit in a different guise.

170 *Vortex motion*

Fig. 5.8. Kelvin's sketches of knotted and linked vortex rings, the basis for his 'vortex atom' theory of matter.

A satisfactory procedure, having detached the nozzle itself, is as follows. Push the piston fully in, then puff cigar smoke through the circular hole while rapidly withdrawing the piston, so that the smoke is sucked into the syringe. As soon as the piston is fully withdrawn, put a hand over the hole to keep the smoke in. Allow a few moments for the motions inside to die down, and then generate vortex rings by holding the cylinder horizontally and giving the piston short, sharp taps. Each ring should travel a foot or so while maintaining its form, provided that the surrounding air is fairly still.

Helmholtz considered vortex rings in his 1858 paper, and after deducing that rings of smaller radius travel faster, went on:

> We can ... see how two ring-formed vortex filaments having the same axis would mutually affect each other, since each, in addition to its proper motion, has that of its elements of fluid as produced by the other ...

Fig. 5.9. Generation of a vortex ring by the discharge of fluid through a circular hole.

Vortex motion 171

If they have equal radii and equal and opposite angular velocities, they will approach each other and widen one another; so that finally, when they are very near each other, their velocity of approach becomes smaller and smaller, and their rate of widening faster and faster. If they are perfectly symmetrical, the velocity of fluid elements midway between them parallel to the axis is zero. Here, then, we might imagine a rigid plane to be inserted, which would not disturb the motion, and so obtain the case of a vortex-ring which encounters a fixed plane.

The last sentence is, of course, an interesting example of the method of images, while in saying earlier 'they will approach each other *and widen one another*' Helmholtz is applying his first vortex theorem.

He considers, too, the case when the vortex rings are travelling in the same direction. On the same basis he deduces:

... the foremost widens and travels more slowly, the pursuer shrinks and travels faster, till finally, if their velocities are not too different, it overtakes the first and penetrates it. Then the same game goes on in the opposite order, so that the rings pass through each other alternately.

Good photographs of this 'leap-frogging' phenomenon may be found in Yamada and Matsui (1978), in Oshima (1978) and on p. 46 of van Dyke (1982). In practice, of course, viscous effects act to stop such leap-frogging from continuing indefinitely; indeed they have profound effects, more generally, on the behaviour of real vortex rings (Maxworthy 1972).

Kelvin was of course well aware that real vortex rings do not, on account of viscous effects, wholly retain their identity in the manner indicated by Helmholtz's vortex theorems. One nevertheless wonders, given his hopes for the theory of vortex atoms, what he would have made of an experiment by Oshima and Asaka (1975) in which a red vortex ring and a yellow vortex ring (in water) collide at a certain angle. The rings merge, then break up again into two separate rings, each half yellow and half red. The way in which they do this is indicated in Fig. 5.10. In (*a*) the vortex rings are coming towards us, but they are also approaching one another. In (*b*) they collide, and after a distortion (*c*) of the resulting (single) vortex ring two separate rings are formed (*d*). These come towards us but move apart in a plane at right angles to the plane of approach. Oshima and Asaka provide excellent photographs of this collision process, and further photographs and analysis may be found in Fohl and Turner (1975).

172 *Vortex motion*

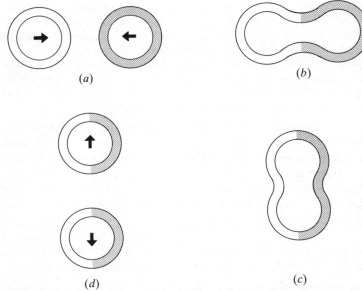

Fig. 5.10. The collision of two viscous vortex rings.

Even within the framework of strictly inviscid theory there are subtle aspects of vortex rings which have taken a long time to emerge. Kelvin himself expressed the view that 'the known phenomena of... smoke rings... convinces... us... that the steady configuration... is stable', and J. J. Thomson purported to demonstrate as much in his 1883 essay, *A treatise on vortex motion*. But Widnall and Tsai (1977) have carried out a more accurate calculation, and have shown that a vortex ring is in fact unstable, even according to ideal flow theory. The instability takes the form of bending waves around the perimeter, and these grow in amplitude as time proceeds (Fig. 5.11).

5.5. Axisymmetric flow

The uniform motion of a vortex ring—let alone its instability—presents theoretical difficulties, but there is one particular circumstance in which it is quite easy to calculate the self-induced motion of an isolated, axisymmetric patch of vorticity. Before doing this we introduce one or two concepts that are of more general value for axisymmetric flow.

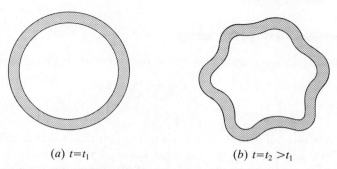

(a) $t=t_1$ (b) $t=t_2 > t_1$

Fig. 5.11. The instability of a vortex ring.

The Stokes stream function

For incompressible flow in two dimensions the stream function representation (4.8) ensures that $\nabla \cdot \boldsymbol{u} = 0$ is automatically satisfied. It is natural to enquire, then, whether for axisymmetric incompressible flow a representation of the form $\boldsymbol{u} = \nabla \wedge (\psi' \boldsymbol{e}_\phi)$ exists, ψ' being a function of R, z, and t only.

This is indeed the case, but a minor inconvenience is that ψ' turns out to be not constant along streamlines, but inversely proportional to R. We therefore write instead

$$\boldsymbol{u} = \nabla \wedge \left(\frac{\Psi}{R} \boldsymbol{e}_\phi\right), \qquad (5.13)$$

or, in spherical polars,

$$\boldsymbol{u} = \nabla \wedge \left(\frac{\Psi}{r \sin \theta} \boldsymbol{e}_\phi\right), \qquad (5.14)$$

whence

$$u_r = \frac{1}{r^2 \sin \theta} \frac{\partial \Psi}{\partial \theta}, \quad u_\theta = -\frac{1}{r \sin \theta} \frac{\partial \Psi}{\partial r}, \qquad (5.15)$$

Ψ being a function of r, θ, and t only. We may verify immediately that

$$(\boldsymbol{u} \cdot \nabla)\Psi = u_r \frac{\partial \Psi}{\partial r} + \frac{u_\theta}{r} \frac{\partial \Psi}{\partial \theta} = 0.$$

Thus the *Stokes stream function* Ψ, defined by eqn (5.15), is *constant along streamlines*.

Irrotational flow past a sphere

In steady axisymmetric flow the vorticity equation (5.12) reduces to

$$(\boldsymbol{u} \cdot \nabla)\left(\frac{\omega}{r \sin \theta}\right) = 0, \qquad (5.16)$$

so that $\omega/r \sin \theta$ is constant along streamlines. Consider, then, uniform inviscid flow past a rigid sphere $r = a$. If there are no closed streamlines in the flow, i.e. if all streamlines originate at infinity, where ω is zero, then ω is zero everywhere in $r > a$, so the flow is irrotational.

Now, the vorticity in axisymmetric flow is $\boldsymbol{\omega} = \omega \boldsymbol{e}_\phi$, where

$$\omega = \frac{1}{r}\frac{\partial}{\partial r}(r u_\theta) - \frac{1}{r}\frac{\partial u_r}{\partial \theta}, \qquad (5.17)$$

and this may be expressed in terms of the Stokes stream function as follows:

$$\omega = -\frac{1}{r \sin \theta}\left[\frac{\partial^2 \Psi}{\partial r^2} + \frac{\sin \theta}{r^2}\frac{\partial}{\partial \theta}\left(\frac{1}{\sin \theta}\frac{\partial \Psi}{\partial \theta}\right)\right]. \qquad (5.18)$$

Thus, for irrotational flow past a sphere, we wish to solve

$$\frac{\partial^2 \Psi}{\partial r^2} + \frac{\sin \theta}{r^2}\frac{\partial}{\partial \theta}\left(\frac{1}{\sin \theta}\frac{\partial \Psi}{\partial \theta}\right) = 0. \qquad (5.19)$$

in $r \geq a$, subject to $\Psi = 0$ on $r = a$ and

$$u_r \sim U \cos \theta, \quad u_\theta \sim -U \sin \theta \qquad \text{as } r \to \infty,$$

which, on using eqn (5.15), means

$$\Psi \sim \tfrac{1}{2}U r^2 \sin^2 \theta \qquad \text{as } r \to \infty. \qquad (5.20)$$

This last condition suggests trying a separable solution of the form $\Psi = f(r)\sin^2 \theta$, and this is indeed possible if

$$f'' - \frac{2f}{r^2} = 0,$$

i.e. if

$$f = Ar^2 + \frac{B}{r}.$$

Vortex motion 175

The boundary conditions then determine the arbitrary constants A and B, whence

$$\Psi = \tfrac{1}{2}U\left(r^2 - \frac{a^3}{r}\right)\sin^2\theta \quad \text{in } r \geq a. \tag{5.21}$$

The streamlines $\Psi = $ constant are sketched in Fig. 5.12(a). There is, inevitably, a velocity of slip

$$u_\theta = -\frac{1}{r\sin\theta}\frac{\partial\Psi}{\partial r} = -\tfrac{3}{2}U\sin\theta \quad \text{on } r = a, \tag{5.22}$$

and this implies, by Bernoulli's theorem, a severe adverse pressure gradient over the back of the sphere. In real, high Reynolds number flow past a sphere, no attached boundary layer can cope with this adverse pressure gradient, and separation of the boundary layer leads instead to a large wake (see §§2.1 and 8.6).

Hill's spherical vortex

Let us now suppose instead that the region $r < a$ is also filled with fluid. Remarkably, it is possible to find a closed-streamline inviscid flow in $r < a$ which matches on to eqn (5.21) in the sense that (i) Ψ is zero on $r = a$ and (ii) the tangential component of velocity u_θ matches with eqn (5.22) on $r = a$.

In this closed-streamline region (5.16) tells us only that $\omega/r\sin\theta$ is constant along each streamline; there is no reason to suppose it is the same constant along each one, let alone zero.

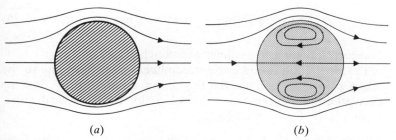

Fig. 5.12. (a) Irrotational flow past a sphere. (b) Hill's spherical vortex.

176 Vortex motion

The most we can claim, then, is that

$$\frac{\omega}{r \sin \theta} = c(\Psi) \quad \text{in } r \leq a,$$

where the function $c(\Psi)$ is at this stage unknown. Using eqn (5.18) this implies that

$$\frac{\partial^2 \Psi}{\partial r^2} + \frac{\sin \theta}{r^2} \frac{\partial}{\partial \theta} \left(\frac{1}{\sin \theta} \frac{\partial \Psi}{\partial \theta} \right) = -c(\Psi) r^2 \sin^2 \theta \quad (5.23)$$

is $r \leq a$, and $c(\Psi)$ is to be determined as part of the solution (if, indeed, such a solution exists).

Now, in order that u_θ matches with eqn (5.22) on $r = a$ we need

$$\frac{\partial \Psi}{\partial r} = \tfrac{3}{2} Ua \sin^2 \theta \quad \text{on } r = a, \quad (5.24)$$

and this suggests trying $\Psi = g(r) \sin^2 \theta$ in eqn (5.23). The left-hand side is then a function of r times $\sin^2 \theta$, and the form of the right-hand side then shows that $c(\Psi)$ will need to be a constant, c, if eqn (5.23) is to reduce to an ordinary differential equation for $g(r)$. The function $g(r)$ then emerges as

$$g(r) = Ar^2 + \frac{B}{r} - \tfrac{1}{10} c r^4.$$

We must choose $B = 0$ to keep $\boldsymbol{u}$ finite at $r = 0$, and A must then be chosen so that $\Psi = 0$ on $r = a$. Finally, eqn (5.24) implies that $c = -15U/2a^2$, so

$$\Psi = -\tfrac{3}{4} U r^2 \left(1 - \frac{r^2}{a^2} \right) \sin^2 \theta \quad \text{in } r \leq a. \quad (5.25)$$

The corresponding streamlines are sketched in Fig. 5.12(b).

The circulation round these streamlines varies from one to the other, of course, because the flow in $r \leq a$ has vorticity, but the circulation round the perimeter of a full hemispherical cross-section is, by Stokes's theorem,

$$\Gamma_{\max} = \int_0^\pi \int_0^a \omega r \, dr \, d\theta = c \int_0^\pi \int_0^a r^2 \sin \theta \, dr \, d\theta = -5Ua.$$

Equivalently, a Hill spherical vortex will travel through stationary fluid with uniform speed $\Gamma_{max}/5a$, distinguished from an ordinary smoke ring by the absence of a hole and by the way in which the vorticity is spread throughout the whole of the closed streamline region (cf. Fig. 5.7(b)).

5.6. Motion of a vortex pair

We now explore some aspects of 2-D vortex motion. Consider, for instance, the vortex pair of Fig. 5.13(a), and suppose that the core of each vortex, where all the vorticity is concentrated, is quite small. The fluid momentarily occupying one of the vortex cores will be swept downwards by the flow due to the other vortex, and by eqn (5.9) that fluid will retain its vorticity, so the vortex itself will be swept downwards. The two vortices therefore move down together, maintaining their relative positions. It is possible to observe this at airports by watching the trailing vortices from the wing-tips of departing aircraft (see Fig. 1.12(b)).

To make these ideas more specific we *treat each vortex as a line vortex which moves at the local flow velocity due to everything other than itself*. If the vortices are of strength Γ and $-\Gamma$, distance $2d$ apart, then each will induce a downward flow $\Gamma/4\pi d$ at the

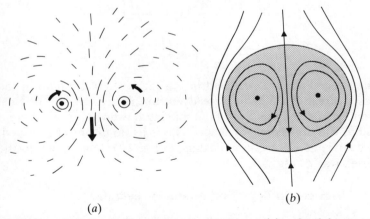

Fig. 5.13. Flow due to a vortex pair relative to (a) a fixed frame and (b) a frame moving with the vortices.

position momentarily occupied by the other, so the pair itself, and the whole instantaneous streamline pattern, will move downwards at this speed.

We may alternatively view the motion from a frame in which the vortices are fixed. This in turn is equivalent to superimposing a uniform upward flow with speed $\Gamma/4\pi d$, i.e. just that required to hold the vortices at rest. The complex potential for the resulting flow is clearly

$$w = -\frac{i\Gamma z}{4\pi d} - \frac{i\Gamma}{2\pi}\log(z-d) + \frac{i\Gamma}{2\pi}\log(z+d), \quad (5.26)$$

the first term representing the uniform upward flow, and the others representing the flows due to the two vortices. To confirm that the vortex at $z = d$ may indeed remain stationary in this situation we calculate the contribution to dw/dz at $z = d$ from everything but the vortex at $z = d$ itself. Thus if (U, V) denotes the translational velocity of the vortex at $z = d$ then

$$U - iV = \left[\frac{d}{dz}\left\{\frac{-i\Gamma z}{4\pi d} + \frac{i\Gamma}{2\pi}\log(z+d)\right\}\right]_{z=d}$$

$$= \left[\frac{-i\Gamma}{4\pi d} + \frac{i\Gamma}{2\pi(z+d)}\right]_{z=d} = 0. \quad (5.27)$$

The stream function for the flow (5.26) is

$$\psi = -\frac{\Gamma x}{4\pi d} - \frac{\Gamma}{2\pi}\log\left|\frac{z-d}{z+d}\right|$$

$$= -\frac{\Gamma}{4\pi}\left[\frac{x}{d} + \log\left\{\frac{(x-d)^2 + y^2}{(x+d)^2 + y^2}\right\}\right].$$

The streamlines are sketched in Fig. 5.13(b). If the fluid in the closed streamline region were dyed, an observer in the original frame would see this dyed fluid moving downward as a coherent entity, without change of shape. This is by no means unexpected, of course, as we are now dealing with a 2-D counterpart to the vortex ring of Fig. 5.7.

5.7. Vortices in flow past a circular cylinder

Let a circular cylinder of radius a be initially at rest in a fluid of kinematic viscosity ν. Suppose that it is suddenly translated with speed U perpendicular to its axis, and suppose too that the Reynolds number

$$R = \frac{2aU}{\nu} \qquad (5.28)$$

is somewhere in the region of 200 or so. With the simple home apparatus of §1.1 this might be achieved, for example, with the refill from a ballpoint pen (radius ~2 mm) and a towing speed U of about 5 cm s^{-1}.

The initial phase: almost irrotational flow

According to inviscid theory the response of the fluid to the motion of the cylinder will be determined by the vorticity equation (5.9):

$$\frac{D\omega}{Dt} = 0,$$

which says that the vorticity of each individual fluid element is conserved. Each has zero vorticity initially, as the fluid is at rest. Each element therefore continues to have zero vorticity and the subsequent flow is irrotational.

Consider now the real, viscous situation. During a very short initial phase, which is over by the time the cylinder has moved a distance comparable to its radius, the flow relative to the cylinder is indeed predominantly irrotational, as in Fig. 4.4(a). There is intense vorticity in the rapidly thickening boundary layer on the cylinder, but despite the large adverse pressure gradient at the rear of the cylinder there simply has not yet been time for separation to occur, and the vorticity in the boundary layer has not therefore found its way into the main flow.

During this initial phase irrotational flow theory plays a major role by determining the velocity at the edge of the boundary layer. This is important, for in impulsively started flows of this kind reversed flow in the boundary layer first occurs at the place where the velocity at the edge of the boundary layer decreases most rapidly with distance along the boundary. In the case of a circular cylinder, this place is the rear stagnation point, so this is where reversed flow first occurs (Fig. 5.14(a)).

Flow at a later stage: the von Kármán vortex street

Thereafter the flow diverges substantially from that predicted by irrotational flow theory. The two attached eddies behind the cylinder grow in size, as in Fig. 5.14(b). At a later time still the flow ceases to be symmetric about the centreline (Fig. 5.14(c)) and, even more strangely, it ceases to be steady relative to the cylinder, even though the flow at infinity (relative to the cylinder) is constant. Instead, the flow settles into an unsteady but highly structured form in which vortices are shed alternately from the two sides of the cylinder, so giving the remarkable *von Kármán vortex street* of Fig. 5.14(d, e).

Von Kármán's interest in the phenomenon stemmed from about 1911, when he was a graduate assistant in Prandtl's laboratory in Göttingen. He tells of those early days in his *Aerodynamics* (1954):

... Prandtl had a doctoral candidate, Karl Hiemenz, to whom he gave the task of constructing a water channel in which he could observe the separation of the flow behind a cylinder. The object was to check experimentally the separation point calculated by means of the boundary-layer theory. For this purpose, it was first necessary to know the pressure distribution around the cylinder in a steady flow. Much to his surprise, Hiemenz found that the flow in his channel oscillated violently.

When he reported this to Prandtl, the latter told him: 'Obviously your cylinder is not circular.'

However, even after very careful machining of the cylinder, the flow continued to oscillate. Then Hiemenz was told that possibly the channel was not symmetric, and he started to adjust it.

I was not concerned with this problem, but every morning when I came in the laboratory I asked him, 'Herr Hiemenz, is the flow steady now?'

He answered very sadly, 'It always oscillates.'

It must be said that this picture of events is valid for a certain range of Reynolds numbers only. Thus at $R = 2000$ the wake is essentially turbulent, with only traces of the periodic structure of Fig. 5.14(d, e). At $R = 30$, on the other hand, the wake develops into two symmetrically disposed vortices which remain attached as time proceeds, much as in Fig. 5.14(b). There are many

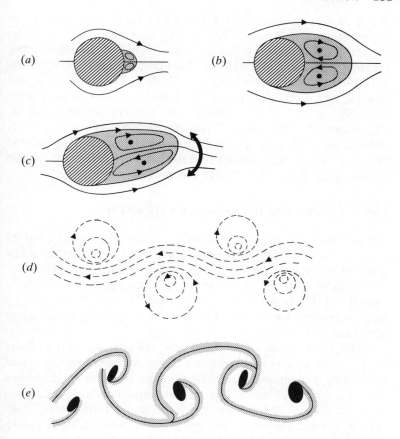

Fig. 5.14. Time-development of flow due to an impulsively moved circular cylinder. (*a*), (*b*), (*c*) Instantaneous streamlines *relative to axes moving with the cylinder* at three fairly early times. (*d*) The instantaneous streamlines, as implied by a streak photograph, *relative to fixed axes*, at a rather later time; the cylinder has moved out of the picture and left behind a trail of von Kármán vortices which follow it by moving to the left at a much slower speed than that of the cylinder. (*e*) At that same later time, typical dye traces, the dye essentially marking those fluid elements which were, at $t = 0$, close to the cylinder boundary; to a fair degree, then, the dye traces also mark regions of strong vorticity.

182 Vortex motion

excellent photographs in the literature of these attached vortices (Coutanceau and Bouard 1977; van Dyke 1982, pp. 28–30), the early evolution of the wake at rather higher Reynolds number (Prandtl and Tietjens 1934, pp. 279–280; Bouard and Coutanceau 1980; van Dyke 1982, pp. 36–37; Perry *et al.* 1982; Loc and Bouard 1985), the subsequent von Kármán vortex street (Goldstein 1938, p. 552; Rouse 1946, p. 241; Rosenhead 1963, opp. p. 105; Batchelor 1967, plate 2; van Dyke 1982, pp. 4–5, 56–57; Perry *et al.* 1982, opp. p. 90; Tritton 1988, pp. 25–26), and the turbulent wake that occurs instead at still higher Reynolds number (van Dyke 1982, p. 31; Tritton 1988, p. 30).

The von Kármán vortex street: a simple model

We now model a fully formed vortex street (Fig. 5.14(d, e)) by one set of line vortices of strength Γ at $z = na$, and another set of strength $-\Gamma$ at $z = (n + \frac{1}{2})a + ib$, with $n = 0, \pm 1, \pm 2 \ldots$ (see Fig. 5.15). As in §5.6 we assume that each line vortex moves at the local flow velocity due to everything other than itself, this being a crude substitute for having finite patches of vorticity which move according to eqn (5.9).

Consider any vortex. The local flow velocity due to the others in the same row is zero, because their contributions cancel in pairs. The y-components of velocity due to those in the other row also cancel in pairs, but the x-components reinforce each other to give a certain velocity V to the left (if $\Gamma > 0$). This velocity is common to all the vortices, so the whole array moves to the left at this speed, while maintaining its form.

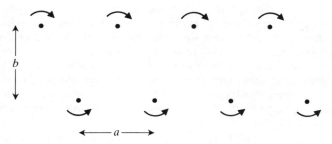

Fig. 5.15. Line vortex representation of a von Kármán vortex street.

To find V, let us calculate dw/dz at, say, $z = \tfrac{1}{2}a + ib$, where w is the complex potential due to the whole of the bottom row. The complex potential due to the member of that row at $z = na$ may be written as

$$-\frac{i\Gamma}{2\pi}\log(z - na),$$

but we note, as a preliminary, that an equally good representation of this particular flow is

$$-\frac{i\Gamma}{2\pi}\log\left(1 - \frac{z}{na}\right), \qquad (n \neq 0),$$

for the two differ only by an additive constant, which makes no difference to the resulting flow. The complex potential due to the whole of the bottom row can therefore be written

$$w = -\frac{i\Gamma}{2\pi}\sum_{-\infty}^{-1}\log\left(1 - \frac{z}{na}\right) - \frac{i\Gamma}{2\pi}\log z - \frac{i\Gamma}{2\pi}\sum_{1}^{\infty}\log\left(1 - \frac{z}{na}\right)$$

$$= -\frac{i\Gamma}{2\pi}\log\left[z\prod_{n=1}^{\infty}\left(1 - \frac{z^2}{n^2 a^2}\right)\right]$$

$$= -\frac{i\Gamma}{2\pi}\log\left(\sin\frac{\pi z}{a}\right) + \text{constant}, \qquad (5.29)$$

where we have used an identity drawn from complex variable theory (e.g. Carrier et al. 1966, p. 97). Thus

$$\frac{dw}{dz} = -\frac{i\Gamma}{2a}\cot\left(\frac{\pi z}{a}\right),$$

whence

$$\left.\frac{dw}{dz}\right|_{z=\tfrac{1}{2}a+ib} = \frac{i\Gamma}{2a}\tan\left(\frac{i\pi b}{a}\right) = -\frac{\Gamma}{2a}\tanh\left(\frac{\pi b}{a}\right).$$

The whole vortex street therefore moves to the left with speed

$$V = \frac{\Gamma}{2a}\tanh\left(\frac{\pi b}{a}\right). \qquad (5.30)$$

This accounts for why the von Kármán vortices in Fig. 5.14(d) give chase to the cylinder, and why, relative to the cylinder, they are not swept downstream at quite the free stream speed U.

5.8. Instability of vortex patterns

Von Kármán went on to consider what happens if the vortices in Fig. 5.15 are slightly displaced from their correct positions. He showed that such displacements do not remain small as time proceeds, so that the basic configuration is unstable, except in the case

$$\cosh\frac{\pi b}{a} = \sqrt{2}, \quad \text{i.e. } b/a \doteqdot 0.281, \tag{5.31}$$

when his analysis revealed no instability. The relevance or otherwise of this special value of b/a to real vortex streets has caused much consternation over the years, the issue being clouded by the subsequent discovery that the system is more weakly unstable even in the case (5.31).

Another classical problem involves the stability of n line vortices spaced regularly around the circumference of a circle of radius a. Now, it is obvious that two vortices of strength Γ, placed a distance $2a$ apart, will rotate about the mid-point of the line joining them with angular velocity $\Gamma/4\pi a^2$, because each induces a velocity $\Gamma/4\pi a$ perpendicular to that line at the position occupied by the other (Fig. 5.16(a)). More generally, it can be shown that n equal line vortices can maintain themselves in a circular array by rotating with angular velocity

$$\Omega = (n-1)\frac{\Gamma}{4\pi a^2}, \tag{5.32}$$

where Γ denotes the circulation around any one such vortex

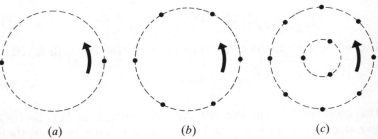

Fig. 5.16. Stable rotating configurations of 2, 6, and 11 line vortices.

(Exercise 5.14). The stability of this motion was first investigated in 1883 by J. J. Thomson, who later discovered the electron. He concluded that the motion was stable when $n < 7$ and unstable when $n \geq 7$, but the case $n = 7$ was subsequently shown to be neutrally stable by Havelock in 1931. It is chastening to find this apparently academic problem having very real application, nigh on a century after Thomson's analysis, to superfluid hydrodynamics.† In liquid helium, at temperatures extremely close to absolute zero, unusual line vortices are observed, each with a circulation Γ which is quantized and equal to $\hbar/m$, where $\hbar$ is Planck's constant and m is the mass of the ^{4}He II atom. These vortices can be observed rotating in various types of array (e.g. Fig. 5.16(c)), but, notably, only in the singly circular arrays of Fig. 5.16(a, b) if $n < 7$, as the stability results would suggest (see the photograph in Yarmchuk et al. (1979) and Table II of Campbell and Ziff (1979)).

We turn now to the evolution of finite patches of concentrated vorticity. An early example was provided by Kirchhoff in 1876, who showed that an elliptical patch of uniform vorticity ω will rotate with angular velocity

$$\Omega = \frac{ab}{(a+b)^2}\omega, \qquad (5.33)$$

where a and b denote the semi-axes of the elliptical region (see Lamb 1932, p. 232). Some years later, in 1893, Love showed that this simple motion is unstable if b/a is greater than 3 or less than $\frac{1}{3}$, and the subsequent evolution of such a vortex has been investigated by Dritschel 1986 (see especially his Figs 12–14).

A circular array of n finite patches of vorticity—a sort of smeared-out version of Fig. 5.16(a, b)—turns out to be unstable even when $n < 7$, if the patches are big enough, the critical size being larger for smaller values of n (Dritschel 1985, see especially his Fig. 2 and §7).

We remarked above that the classical von Kármán vortex street is stable for just one spacing ratio $b/a = 0.281$, at least according to linear theory (exemplified by Exercise 5.13). If the

---

† This field seems to provide a wealth of other exotic applications of classical, strictly inviscid, flow theory (see Roberts and Donnelly 1974, especially pp. 184–186, 196–199, 210–211; also Donnelly 1988).

186 *Vortex motion*

vortices have small but finite cross-sectional area A, there remains just one spacing ratio for which the street is stable on linear theory, this ratio being close to the von Kármán value and only weakly dependent on the small parameter A/a^2 (Meiron *et al.* 1984). This hard-earned result was somewhat unexpected (but see the survey of the whole problem in the introduction to Jimenez (1987)).

The evolution of a continuous 2-D distribution of vorticity

$$\omega = \frac{\partial v}{\partial x} - \frac{\partial u}{\partial y}$$

is of course governed by the vorticity equation (5.9)

$$\frac{\partial \omega}{\partial t} + u \frac{\partial \omega}{\partial x} + v \frac{\partial \omega}{\partial y} = 0, \qquad (5.34)$$

together with

$$\frac{\partial u}{\partial x} + \frac{\partial v}{\partial y} = 0. \qquad (5.35)$$

Now, eqn (5.34) implies that ω is conserved for an individual fluid element, and the incompressibility condition (5.35) implies that the element's cross-sectional area in the x–y plane is conserved, so

$$\int \omega \, dS = \text{constant}, \qquad (5.36)$$

the integral being taken over the whole plane of the flow. There are other relationships of this kind:

$$\int x\omega \, dS = \text{constant}, \quad \int y\omega \, dS = \text{constant} \qquad (5.37)$$

(see Batchelor (1967, p. 528), and see Exercise 5.15 for the equivalent result for line vortices), and such conserved quantities provide valuable constraints on how distributions of vorticity can evolve.

A particularly interesting case is that of *vortex merging*. Suppose that, at $t = 0$, two circular patches of uniform and equal vorticity, each of radius R, have centres a distance d apart. Then if d/R is greater than about 3.5 the (deformed) patches end up rotating about a common centre, much as do two line vortices of equal strength (Fig. 5.16(a)). But if d/R is less than 3.5 the vortex patches quickly merge, and to satisfy the conservation laws they do this by wrapping around each other with irrotationally moving fluid entrained between them like the jam in a Swiss roll (Aref (1983), and see also the outstanding photographs of a computer simulation of this process by Seren *et al.*, in Reed (1987)).

While two nearby like-signed vortices tend to merge in this way, two nearby patches of vorticity of opposite sign stand a chance of escaping from the vicinity of other such patches, essentially as a lone vortex pair. An interesting example of this occurs in the work of Cattaneo and Hughes (1988; see especially their Figs 6 and 8). This behaviour has also been observed in the truly remarkable soap-film experiments of Couder and Basdevant (1986). By towing a cylinder through a soap film they produce some extraordinary phenomena which are, presumably, lurking in the 2-D equations of motion, but which are usually obscured in more conventional experiments by an assortment of 3-D instabilities (see especially their Figs 3 and 7).

5.9. A steady viscous vortex maintained by a secondary flow

The Helmholtz vortex theorems are about the convection of vortex lines with the fluid and the intensification of vorticity when vortex lines are stretched. In a viscous fluid there is also diffusion of vorticity (see §§2.3–2.5), and the three processes correspond, respectively, to the second, third, and fourth terms in the vorticity equation (2.39):

$$\frac{\partial \boldsymbol{\omega}}{\partial t} + (\boldsymbol{u} \cdot \nabla)\boldsymbol{\omega} = (\boldsymbol{\omega} \cdot \nabla)\boldsymbol{u} + \nu \nabla^2 \boldsymbol{\omega}. \qquad (5.38)$$

There is one exact solution of the Navier–Stokes equations—known as the *Burgers vortex*—which involves all three processes. It is essentially the vortex of Fig. 2.12, but with the radially outward diffusion of vorticity countered by a secondary flow (Fig. 5.17) which (i) sweeps the vorticity back towards the axis and (ii) intensifies the vorticity by stretching fluid elements in the z-direction. The result is a steady, rather than decaying, vortex

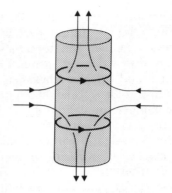

Fig. 5.17. The Burgers vortex.

of the form

$$u_r = -\tfrac{1}{2}\alpha r, \qquad u_z = \alpha z, \qquad u_\theta = \frac{\Gamma}{2\pi r}(1 - e^{-\alpha r^2/4\nu}) \quad (5.39)$$

(Exercise 5.19), where $\alpha > 0$ and Γ are constants. The velocity profile is sketched in Fig. 5.18.

The vorticity

$$\boldsymbol{\omega} = \frac{\alpha \Gamma}{4\pi\nu} e^{-\alpha r^2/4\nu} \boldsymbol{e}_z \quad (5.40)$$

is concentrated in a vortex core of radius of order $(\nu/\alpha)^{\frac{1}{2}}$, which is smaller for small viscosity fluids and for strong secondary flows, as one would expect.

The Burgers vortex provides an excellent example of a balance between convection, intensification and diffusion of vorticity, and it is easy to show that without diffusion ($\nu = 0$) the secondary flow makes the vortex stronger and stronger as time proceeds (Exercise 5.18).

The Burgers vortex is, unfortunately, untypical of real vortices in one important respect; the radius of the core is firmly linked to the strength of the secondary flow (via α), but the magnitude of the rotary flow is not—Γ and α are both free parameters in eqn (5.39). This is essentially because there are no rigid boundaries. For real vortices the presence of rigid boundaries plays a crucial part by coupling the magnitudes of the rotary and secondary flows (see §8.5.)

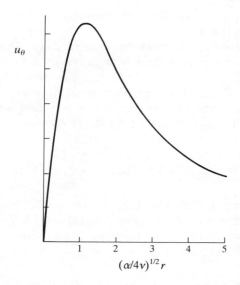

Fig. 5.18. The variation of u_θ with r in a Burgers vortex.

5.10. Viscous vortices: the Prandtl–Batchelor theorem

In the 2-D, steady motion of an inviscid fluid, the equation for the vorticity $\boldsymbol{\omega} = \omega \boldsymbol{k}$ reduces to

$$(\boldsymbol{u} \cdot \nabla)\omega = 0,$$

(see eqn (1.29)), so that ω is constant along any streamline. We may write, then,

$$\boldsymbol{\omega} = [0, 0, \omega(\psi)], \tag{5.41}$$

the stream function ψ being constant along streamlines, as implied by its definition:

$$u = \partial\psi/\partial y, \qquad v = -\partial\psi/\partial x$$

(see eqns (4.5) and (4.7)). The representation (5.41) emphasizes how ω may well be a different constant on different streamlines.

In some cases, as in Fig. 1.8, ω can be determined everywhere without much more ado. In the case of Fig. 1.8, all streamlines can be traced back far upstream, where the vorticity is zero. It follows immediately that the vorticity is zero everywhere.

What happens, however, if there is a region of *closed streamlines* in the flow?

As long as we consider wholly inviscid theory there is in fact nothing we can say about how ω might vary from one streamline to another in such a region. We are, however, as usual, only interested in inviscid theory insofar as it may describe the behaviour of a real fluid *in the limit* $\nu \to 0$, and for the steady flow of a fluid of *non-zero* (but constant) viscosity ν it is the case that

$$\int_C (\nabla \wedge \boldsymbol{\omega}) \cdot d\boldsymbol{x} = 0, \tag{5.42}$$

where C is any closed streamline (Exercise 5.20).

It is important to note that this integral constraint is exact, and holds for any non-zero ν, however small. Now, in the limit $\nu \to 0$, eqn (5.41) holds and

$$\nabla \wedge (0, 0, \omega) = \left(\frac{\partial \omega}{\partial y}, -\frac{\partial \omega}{\partial x}, 0\right) = \omega'(\psi)\left(\frac{\partial \psi}{\partial y}, -\frac{\partial \psi}{\partial x}, 0\right).$$

Combining this with eqn (5.42) we obtain

$$\omega'(\psi) \int_C \boldsymbol{u} \cdot d\boldsymbol{x} = 0, \tag{5.43}$$

the function $\omega'(\psi)$ being taken outside the integral because ψ is a constant on the streamline C. The line integral is of course the circulation round the closed streamline C, and will be zero only in exceptional cases; in a typical closed streamline region (such as either of the two eddies in Fig. 5.14(*b*)) the flow will be in the same sense all round C. So $\omega'(\psi)$ is zero, and this is the Prandtl–Batchelor theorem: in steady, 2-D viscous flow the vorticity is constant throughout any region of closed streamlines *in the limit* $\nu \to 0$.

The computations by Fornberg (1985) of steady flow past a circular cylinder provide a recent example of the theorem at work. These computations give two attached eddies in the wake of the cylinder, so that there are two regions of closed streamlines. Such flows are unstable at high Reynolds number, but they are nevertheless of some interest and importance. In particular, the vorticity in each closed streamline region becomes

Vortex motion 191

progressively more uniform as the Reynolds number R increases, and at $R = 600$ it looks very uniform indeed (see Fornberg's Fig. 9).

Exercises

5.1. Let a closed circuit C of fluid particles be given, at $t = 0$, by
$$x = (a \cos s, a \sin s, 0), \quad 0 \leq s < 2\pi,$$
so that each value of s between 0 and 2π corresponds to a particular fluid particle. Let $C(t)$ be given subsequently by
$$x = (a \cos s + a\alpha t \sin s, a \sin s, 0), \quad 0 \leq s < 2\pi.$$
Find the velocity $u(s, t)$ of each fluid particle, and show that the particles $s = 0$ and $s = \pi$ remain at rest. Find the acceleration of each fluid particle, show that
$$u = (\alpha y, 0, 0),$$
and sketch how the shape of $C(t)$ changes with time.
Now, by definition,
$$\Gamma = \int_{C(t)} u \cdot dx = \int_0^{2\pi} u \cdot \frac{\partial x}{\partial s} ds.$$
Calculate the last integral explicitly at time t, confirming that it is independent of t, in accord with Kelvin's circulation theorem.

5.2. Let $C(t)$ denote a closed circuit composed of the same fluid particles as time proceeds. Then
$$\frac{d}{dt} \int_{C(t)} u \cdot dx = \int_{C(t)} \frac{Du}{Dt} \cdot dx. \tag{5.44}$$
To prove this, let $x = x(s, t)$ be a parametric representation of $C(t)$, so that each fluid particle has, throughout the motion, a particular value of s lying between, say, 0 and 1. Then
$$\frac{d}{dt} \int_{C(t)} u \cdot dx = \frac{d}{dt} \int_0^1 u \cdot \frac{\partial x}{\partial s} ds = \int_0^1 \frac{\partial}{\partial t} \left(u \cdot \frac{\partial x}{\partial s} \right) ds,$$
where $\partial/\partial t$ denotes differentiation with respect to t *holding s constant*, s being the variable of integration and the limits on s being fixed. Continue the analysis to establish the result, eqn (5.2).

5.3. Let an ideal fluid be in 2-D motion. By virtue of eqn (5.9) the vorticity ω of any fluid element is conserved. The fluid element must

192 *Vortex motion*

also conserve its volume, and because it is not being stretched in the z-direction its cross-sectional area δS in the x–y plane must therefore be conserved. It follows that the integral

$$\int \omega \, dS$$

taken over a dyed cross-section S in the x–y plane, must be independent of time. By Stokes's theorem, or by Green's theorem in the plane (A.24), it follows that Γ, the circulation round the dyed circuit which forms the perimeter of S, must also be independent of time.

This is in some respects a nice way of seeing how Kelvin's circulation theorem comes about. It is, however, a wholly 2-D argument, and that theorem is certainly not restricted to 2-D flows. What is the other serious limitation to the above point of view?

5.4. Show that if we relax the assumptions of incompressibility and constant density in §5.1 then

$$\frac{d\Gamma}{dt} = \int_{C(t)} -\frac{1}{\rho} \nabla p \cdot d\mathbf{x} = \int_{C(t)} -\frac{1}{\rho} \frac{\partial p}{\partial s} \, ds.$$

A *barotropic* fluid is one for which the pressure p is a function only of the density ρ, so that $p = f(\rho)$. Kelvin's circulation theorem holds for such a fluid; apply Stokes's theorem to the first integral above to give one demonstration of this. What unnecessary assumption is involved in this argument? Construct an alternative proof based on the second integral above.

Use Exercise 1.5 to show that the vorticity equation for a *barotropic* fluid is

$$\frac{D}{Dt}\left(\frac{\boldsymbol{\omega}}{\rho}\right) = \frac{\boldsymbol{\omega}}{\rho} \cdot \nabla \mathbf{u}, \qquad (5.45)$$

and note that this is just the vorticity equation (5.7), but with $\boldsymbol{\omega}/\rho$ in place of $\boldsymbol{\omega}$.

As Kelvin's circulation theorem holds, eqn (5.6) is independent of time for a barotropic fluid. Modify the thin vortex-tube argument following the proof of (2) in §5.3 to show that for a barotropic fluid it is $\boldsymbol{\omega}/\rho$, rather than $\boldsymbol{\omega}$, that varies in proportion to l, the length of a small section of the tube.

5.5. Prove that the quantity (5.6) is, at any time, the same for all cross-sections of a vortex tube.

5.6. Show that if $\mathbf{a}(\mathbf{x}, t)$ is any suitably smooth vector field and

$$\mathscr{C}(t) = \int_{C(t)} \mathbf{a} \cdot d\mathbf{x},$$

where $C(t)$ is a circuit consisting of the same fluid particles as time proceeds, then

$$\frac{d\mathscr{C}}{dt} = \int_{C(t)} \left\{ \frac{\partial \boldsymbol{a}}{\partial t} + (\nabla \wedge \boldsymbol{a}) \wedge \boldsymbol{u} \right\} \cdot d\boldsymbol{x}.$$

5.7. Use eqn (5.10) to show that the vorticity equation (5.7) reduces, in the case of axisymmetric flow, to eqn (5.12):

$$\frac{D}{Dt}\left(\frac{\omega}{R}\right) = 0.$$

5.8. Inviscid fluid occupies the region $x \geq 0$, $y \geq 0$ bounded by two rigid boundaries $x = 0$, $y = 0$. Its motion results wholly from the presence of a line vortex, which itself moves according to the Helmholtz vortex theorems. Show that the path taken by the vortex is

$$\frac{1}{x^2} + \frac{1}{y^2} = \text{constant}.$$

When an aircraft takes off, the two vortices that trail from its wing-tips (§1.7) are observed to move downwards under each other's influence and then to move further apart as they approach the ground. Why is this?

5.9. *A Bernoulli theorem for unsteady irrotational flow.* Use the momentum equation for an ideal fluid in the form (1.14) to show that for an *irrotational* flow:

$$\frac{\partial \phi}{\partial t} + \frac{p}{\rho} + \tfrac{1}{2}\boldsymbol{u}^2 + \chi = F(t),$$

where ϕ is the velocity potential and $F(t)$ is a function of time alone. (Note, too, that $F(t)$ may be taken to be zero if desired, for its presence is equivalent to adding $\int_0^t F(t_1)\,dt_1$ to the velocity potential ϕ, which is of no consequence, the velocity field being $\boldsymbol{u} = \nabla \phi$.)

5.10. Inviscid fluid occupies the region $x \geq 0$, and there is a plane rigid boundary at $x = 0$. A line vortex of strength Γ is at (d, y_0). Explain why the instantaneous complex potential is

$$w = -\frac{i\Gamma}{2\pi}\log(z - d - iy_0) + \frac{i\Gamma}{2\pi}\log(z + d - iy_0),$$

and why the vortex moves downward, parallel to the boundary, in such a way that

$$dy_0/dt = -\Gamma/4\pi d.$$

Consider, for simplicity, the motion when $y_0 = 0$. Show that at that instant

$$v = -\frac{\Gamma d}{\pi(y^2 + d^2)} \quad \text{and} \quad \frac{\partial \phi}{\partial t} = -\frac{\Gamma^2}{4\pi^2(y^2 + d^2)} \quad \text{on } x = 0,$$

and hence use Exercise 5.9 to calculate the net force

$$\int_{-\infty}^{\infty} p \, dy$$

exerted on the wall $x = 0$.

What would the force on the wall be if the vortex were somehow fixed at $(d, 0)$?

[This raises questions about the forces involved on the fluid in the core of a vortex when it moves in accord with Helmholtz's theorems, and Lamb (1932, p. 222) makes some interesting observations on the matter.]

5.11. Consider a *symmetric* vortex street in which one set of line vortices of strength Γ is at $z = na$ and the other set, of strength $-\Gamma$, is at $z = na + ib$. Show that the whole array may, in principle, maintain its form by moving to the left with speed

$$V = \frac{\Gamma}{2a} \coth\left(\frac{\pi b}{a}\right).$$

[This configuration is, however, unstable according to linear theory for all values of the spacing ratio b/a; there is no exceptional value corresponding to eqn (5.31). It is still not entirely clear whether this is of any significance in connection with the observed asymmetry of real von Kármán vortex streets, particularly as symmetric streets have, it seems, been observed, albeit under somewhat artificial circumstances (see Taneda 1965; Figs 8a and 9).]

5.12. Suppose that there is, in $y \geq 0$, the irrotational flow

$$u = -\alpha x, \quad v = \alpha y,$$

where α is a positive constant, and let there be a plane rigid boundary at $y = 0$. Suppose, in addition, there are two line vortices, one of strength $-\Gamma$ at $z = z_1(t)$ and the other of strength Γ at $z = z_2(t)$, where $z = x + iy$. Write down the instantaneous complex potential for the whole flow by the method of images and, by letting the vortices move with the fluid according to Helmholtz's first vortex theorem, show that

$$\frac{d\bar{z}_1}{dt} = -\frac{i\Gamma}{2\pi}\left[\frac{1}{z_1 - z_2} - \frac{1}{z_1 - \bar{z}_2} + \frac{1}{z_1 - \bar{z}_1}\right] - \alpha z_1,$$

$$\frac{d\bar{z}_2}{dt} = \frac{i\Gamma}{2\pi}\left[\frac{1}{z_2 - z_1} - \frac{1}{z_2 - \bar{z}_1} + \frac{1}{z_2 - \bar{z}_2}\right] - \alpha z_2, \quad (5.46)$$

where an overbar denotes the complex conjugate.

Verify that the vortices may remain at rest at

$$z_1 = d(-1 + i), \qquad z_2 = d(1 + i),$$

where $d^2 = \Gamma/8\pi\alpha$ (see Fig. 5.19(b)).

[This system, when rotated clockwise through 90°, may be regarded as a simple model for the attached vortices in Fig. 5.14(a, b).]

5.13. Investigate the stability of the vortex configuration in Fig. 5.19(b) as follows. Introduce dimensionless variables

$$z_1' = z_1/d, \qquad z_2' = z_2/d, \qquad t' = 4\alpha t,$$

and rewrite eqn (5.46) accordingly. Then disturb the vortices slightly, so that

$$z_1' = -1 + i + \varepsilon_1(t), \qquad z_2' = 1 + i + \varepsilon_2(t),$$

where $\varepsilon_1(t)$ and $\varepsilon_2(t)$ are complex variables with moduli which are small compared to 1. Expand the right-hand sides of eqn (5.46) binomially for small $|\varepsilon_1|$ and $|\varepsilon_2|$, and retain only terms of first order in small quantities to obtain

$$4\dot{\bar{\varepsilon}}_1 = -i(\varepsilon_2 - \bar{\varepsilon}_1) + \tfrac{1}{2}(\varepsilon_1 - \bar{\varepsilon}_2) - \varepsilon_1,$$

$$4\dot{\bar{\varepsilon}}_2 = i(\varepsilon_1 - \bar{\varepsilon}_2) + \tfrac{1}{2}(\varepsilon_2 - \bar{\varepsilon}_1) - \varepsilon_2,$$

where the dot denotes differentiation with respect to t'.

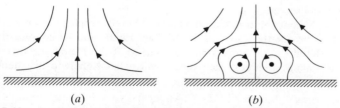

Fig. 5.19. Irrotational flow away from a stagnation point (a) without and (b) with 'attached' vortices.

196 Vortex motion

By introducing suitable new dependent variables in place of ε_1 and ε_2, or otherwise, solve these equations, and thus show that the vortex configuration is unstable, in that any small initial difference in the y-displacements of the two vortices will grow exponentially with time.

[An analysis of this kind with a cylindrical, rather than plane boundary was first carried out by Föppl in 1913, with similar result, and it was the basis for an early theory of how the asymmetry in the downstream positions of the two vortices in Fig. 5.14(c) might come about.]

5.14. Establish that an array of n line vortices of strength Γ, spaced equally around a circle of radius a, can rotate with angular velocity

$$\Omega = (n-1)\frac{\Gamma}{4\pi a^2}.$$

5.15. Let there be line vortices of strength Γ_k at $z = z_k$, where $k = 1, 2, \ldots, n$, each moving under the influence of all the others. Show that if the sth vortex has coordinates (x_s, y_s), then

$$\frac{dx_s}{dt} - i\frac{dy_s}{dt} = \frac{-i}{2\pi}\sum_{\substack{k=1 \\ k \neq s}}^{n}\frac{\Gamma_k}{z_s - z_k}.$$

Hence show that $\sum_{s=1}^{n}\Gamma_s x_s$ and $\sum_{s=1}^{n}\Gamma_s y_s$ are both constant.

5.16. The *helicity* of a blob of fluid is defined as

$$\int_V \boldsymbol{u}\cdot\boldsymbol{\omega}\,dV,$$

where the integral is taken over the volume of the blob. Using Reynolds's transport theorem (6.6a) we find that the rate of change of the helicity of a dyed blob of incompressible fluid is

$$\int_V \frac{D}{Dt}(\boldsymbol{u}\cdot\boldsymbol{\omega})\,dV.$$

Show that if $\boldsymbol{\omega}\cdot\boldsymbol{n} = 0$ on S, the boundary of V, then the helicity of the blob is conserved.

[The helicity of two closed vortex tubes is crucially dependent on whether or not they are *linked* (Moffatt 1969), and its conservation is then related to the immutability, by virtue of the Helmholtz theorems, of the linkage between such tubes, which led Kelvin to his theory of vortex atoms (see Fig. 5.8).]

5.17. *Ertel's theorem* (1942). Consider the vorticity equation in its form (1.24):

$$\frac{\partial\boldsymbol{\omega}}{\partial t} + \nabla\wedge(\boldsymbol{\omega}\wedge\boldsymbol{u}) = 0.$$

Vortex motion

Take the scalar product with $\nabla\lambda$, where $\lambda(\mathbf{x}, t)$ is any scalar function of position and time that we care to choose, and then use vector identities to show that

$$\frac{D}{Dt}(\boldsymbol{\omega} \cdot \nabla\lambda) = (\boldsymbol{\omega} \cdot \nabla)\frac{D\lambda}{Dt}.$$

Hence deduce that *if $\lambda(\mathbf{x}, t)$ is any scalar quantity which is conserved by individual fluid elements, then $\boldsymbol{\omega} \cdot \nabla\lambda$ is likewise conserved.*

[This is actually a special case of the theorem, which is not restricted to incompressible fluids of constant density.]

5.18. *An intensifying vortex.* Consider the flow

$$x = e^{-\frac{1}{2}\alpha t}\left[X\cos\left\{\frac{\Omega}{\alpha}(e^{\alpha t} - 1)\right\} - Y\sin\left\{\frac{\Omega}{\alpha}(e^{\alpha t} - 1)\right\}\right],$$

$$y = e^{-\frac{1}{2}\alpha t}\left[Y\cos\left\{\frac{\Omega}{\alpha}(e^{\alpha t} - 1)\right\} + X\sin\left\{\frac{\Omega}{\alpha}(e^{\alpha t} - 1)\right\}\right],$$

$$z = Ze^{\alpha t},$$

where (x, y, z) denotes the position at time t of the fluid particle that was, at $t = 0$, at (X, Y, Z) (see Exercise 1.7). Show that

$$\mathbf{u} = (-\tfrac{1}{2}\alpha x - \Omega y e^{\alpha t}, -\tfrac{1}{2}\alpha y + \Omega x e^{\alpha t}, \alpha z)$$

and

$$\boldsymbol{\omega} = (0, 0, 2\Omega e^{\alpha t}).$$

Verify that $\nabla \cdot \mathbf{u} = 0$. Show too that the inviscid vorticity equation (5.7) is satisfied, and note how it describes the rate of change of the vorticity in terms of the stretching of the vortex lines resulting from the increase of w with z.

Briefly describe the above flow.

5.19. *The Burgers vortex.* Seek an exact, steady solution to the Navier–Stokes equations of the form

$$\mathbf{u} = -\tfrac{1}{2}\alpha r \mathbf{e}_r + u_\theta(r)\mathbf{e}_\theta + \alpha z \mathbf{e}_z,$$

where α is a positive constant. Note that $\boldsymbol{\omega} = \omega \mathbf{e}_z$, where

$$\omega = \frac{1}{r}\frac{d}{dr}(ru_\theta).$$

Verify that $\nabla \cdot \mathbf{u} = 0$, and show that the equations of motion imply

$$-\tfrac{1}{2}\alpha r\omega = \nu\frac{d\omega}{dr}.$$

198   Vortex motion

Deduce that

$$u_\theta = \frac{\Gamma}{2\pi r}(1 - e^{-\alpha r^2/4\nu}),$$

where Γ is an arbitrary constant.

5.20. *Steady viscous flow with closed streamlines*. The steady momentum equation for an incompressible viscous fluid of constant density ρ is

$$(\boldsymbol{u} \cdot \nabla)\boldsymbol{u} = -\nabla(p/\rho) + \nu\nabla^2\boldsymbol{u}.$$

Rewrite the first and last terms by means of suitable vector identities, and then integrate both sides round a closed streamline C to show that

$$\nu \int_C (\nabla \wedge \boldsymbol{\omega}) \cdot d\boldsymbol{x} = 0,$$

where $\boldsymbol{\omega} = \nabla \wedge \boldsymbol{u}$.

5.21. *Cauchy's vorticity formula* (1815). Let a fluid particle be at position $\boldsymbol{X}$ at $t = 0$, and let the vorticity there be $\boldsymbol{\omega}_0$ at $t = 0$. Let the subsequent motion of the fluid particle be described by $\boldsymbol{x} = \boldsymbol{x}(\boldsymbol{X}, t)$ as, for example, in Exercise 5.18. (This description will have a unique inverse $\boldsymbol{X} = \boldsymbol{X}(\boldsymbol{x}, t)$.) Let the vorticity of the fluid at $\boldsymbol{x}$, the position of the particle at time t, be $\boldsymbol{\omega}$. Then Cauchy proved that $\boldsymbol{\omega}$ is related to $\boldsymbol{\omega}_0$ by

$$\omega_i = \omega_{0j}\frac{\partial x_i}{\partial X_j}, \qquad i = 1, 2, 3,$$

where $\boldsymbol{x} = (x_1, x_2, x_3)$, $\boldsymbol{X} = (X_1, X_2, X_3)$, and summation over $j = 1, 2, 3$ is understood, by virtue of the repeated suffix.

Confirm, first, that this formula holds in the particular case of Exercise 5.18, and then prove that it holds in general.

[One way is to use Ertel's theorem (Exercise 5.17) on three scalar quantities that are rather trivially constant following a particular fluid element; this gives $\boldsymbol{\omega}_0$ in terms of $\boldsymbol{\omega}$, which then has to be inverted.]

5.22. *Alternative proof of the laws of vortex motion*. Let $\boldsymbol{X} = \boldsymbol{X}(s)$ denote a line of dyed particles in the fluid, at $t = 0$, s denoting distance along the line at that time, and suppose that the line is also a vortex line. Use Cauchy's vorticity formula (Exercise 5.21) to show that the dyed particles continue to lie on a vortex line. Investigate, too, the magnitude of the vorticity, $|\boldsymbol{\omega}|$, in the neighbourhood of any particular dyed segment, showing that $|\boldsymbol{\omega}|$ increases with time in proportion to the length of that segment.

Miscellaneous exercises on irrotational flow

5.23. Ideal fluid moves irrotationally in a simply connected region V bounded by a closed surface S, so that $\boldsymbol{u} = \nabla\phi$, where ϕ is the velocity potential. Show that
$$\nabla^2\phi = 0,$$
and that the kinetic energy
$$T = \tfrac{1}{2}\rho \int_V \boldsymbol{u}^2 \, \mathrm{d}V$$
can therefore be written in the form
$$T = \tfrac{1}{2}\rho \int_S \phi \frac{\partial \phi}{\partial n} \, \mathrm{d}S.$$

5.24. *Uniqueness of irrotational flow.* Ideal fluid moves in a bounded simply connected region V, and the normal component of velocity $\boldsymbol{u} \cdot \boldsymbol{n}$ is given (as $f(\boldsymbol{x}, t)$, say) at each point of the boundary of V. Show that there is at most one irrotational flow in V which satisfies the boundary condition.

[This explains why such flows cannot, typically, satisfy a no-slip condition as well. The theorem may, in addition, be extended to encompass unbounded simply connected regions of irrotational flow, as in the case of a sphere moving through a fluid at rest at infinity.]

5.25. *Kelvin's minimum energy theorem.* Consider the various smooth velocity fields $\boldsymbol{u}(\boldsymbol{x}, t)$ in a simply connected region V that satisfy (i) $\nabla \cdot \boldsymbol{u} = 0$ and (ii) the condition $\boldsymbol{u} \cdot \boldsymbol{n} = f(\boldsymbol{x}, t)$ on S, the boundary of V. (We suspend, then, for the present, all consideration of whether or not the velocity fields would be dynamically possible.) Show that the (unique) irrotational flow has less kinetic energy than any of the others.

5.26. In §5.5 the problem of irrotational flow past a rigid sphere was formulated, and solved, in terms of the Stokes stream function Ψ. Re-work the problem in terms of the velocity potential ϕ, which satisfies the axisymmetric version of Laplace's equation (5.4), i.e.
$$\frac{1}{r^2}\frac{\partial}{\partial r}\left(r^2 \frac{\partial \phi}{\partial r}\right) + \frac{1}{r^2 \sin\theta}\frac{\partial}{\partial \theta}\left(\sin\theta \frac{\partial \phi}{\partial \theta}\right) = 0.$$
Check that the 'slip velocity' on the sphere is eqn (5.22), as before. Show that the pressure distribution on the sphere is symmetric, fore and aft, so that the drag on the sphere is zero.

5.27. A sphere of radius a moves in a straight line with speed $U(t)$ through inviscid incompressible fluid which is at rest at infinity. Explain why, at the instant the sphere passes the origin,

$$\frac{\partial \phi}{\partial r} = U(t)\cos \theta \qquad \text{on } r = a,$$

where r and θ are spherical polar coordinates, the polar axis ($\theta = 0$) being in the direction in which the sphere is moving. Show that at the instant in question

$$\phi = -\frac{U(t)a^3}{2r^2} \cos \theta.$$

Calculate the kinetic energy of the instantaneous fluid motion, and show, by considering the rate of working of the sphere on the fluid, that the sphere experiences a drag force

$$D = \tfrac{1}{2} M \frac{dU}{dt},$$

where M denotes the mass of liquid displaced by the sphere.

5.28. Two plane rigid boundaries $\theta = \pm \Omega t$ are rotating with equal and opposite angular velocities Ω, and there is inviscid fluid in the region between them, $0 < r < \infty$, $-\Omega t < \theta < \Omega t$. The flow is irrotational, so a velocity potential $\phi(r, \theta, t)$ exists which satisfies the 2-D version of eqn (5.4) in cylindrical polar coordinates, i.e.

$$\frac{1}{r} \frac{\partial}{\partial r}\left(r \frac{\partial \phi}{\partial r}\right) + \frac{1}{r^2} \frac{\partial^2 \phi}{\partial \theta^2} = 0.$$

Use the method of separation of variables to find the velocity potential $\phi(r, \theta, t)$, and then use eqn (4.9) to find the stream function $\psi(r, \theta, t)$. Sketch the streamlines at time t. Find the pressure p on the boundaries as a function of r and t.

Show that the whole solution breaks down when the angle between the boundaries increases to π, but that until that time the origin is a stagnation point for the flow.

[This last result is of practical significance in connexion with the 'fling' in Fig. 5.3.]

5.29. Ideal fluid occupies the gap $a < r < b$ between two infinitely long cylinders, which are fixed. The irrotational flow between them is

$$\boldsymbol{u} = \frac{\Gamma}{2\pi r} \boldsymbol{e}_\theta,$$

where Γ is a constant. 'As there is no normal velocity on either bounding surface, $r = a$ or $r = b$, we find from the last result in Exercise 5.23 that the kinetic energy is zero.' This is evidently absurd. Explain the fallacy, and show how to use the last result in Exercise 5.23 correctly to give the kinetic energy of the flow.

6 The Navier–Stokes equations

6.1 Introduction

In Book II of the *Principia* (1687) Newton writes:

> SECTION IX
> *The circular motion of fluids*
> HYPOTHESIS
>
> The resistance arising from the want of lubricity in the parts of a fluid is, other things being equal, proportional to the velocity with which the parts of the fluid are separated from one another.
>
> PROPOSITION LI. THEOREM XXXIX
>
> If a solid cylinder infinitely long, in an uniform and infinite fluid, revolves with an uniform motion about an axis given in position, and the fluid be forced round by only this impulse of the cylinder, and every part of the fluid continues uniformly in its motion: I say, that the periodic times of the parts of the fluid are as their distances from the axis of the cylinder.

This is the essence of what Newton has to say about viscous flow. The hypothesis, of course, gets the subject off to a good start, but it is contained and applied wholly within a section on the *circular* motion of fluids, and it is immediately followed by a proposition which is false; the final statement implies that u_θ is independent of r, whereas the correct conclusion, on the basis of Newton's own hypothesis, is that u_θ is inversely proportional to r (see Exercise 2.8). This error gives one small indication of how rudimentary fluid mechanics was at the time, even in the hands of a great master.

Indeed, setting viscous effects aside for a moment, it was not until about 1743, when John Bernoulli published his *Hydraulica*, that the concept of internal pressure was used with clarity and confidence in the study of moving fluids. Furthermore, in spite of all Newton's work, the full generality of the basic principles of mechanics did not emerge until 1752, when Euler advanced

> *The principle of linear momentum*: the total force on a body is equal to the rate of change of the total momentum of the body,

with the clear understanding that the term 'body' might be applied to each and every part of a continuous medium such as a fluid or elastic solid. In 1755 Euler combined this with the concept of internal pressure to obtain his equations of motion for an inviscid fluid (1.12), the achievement being all the greater because he was having to formulate the calculus of partial derivatives as he went along. It was Euler, too, who put forward in 1775

> *The principle of moment of momentum*: the total torque on a body about some fixed point is equal to the rate of change of the moment of momentum of the body about that same point.

He recognized this at the time as an equally general, but quite independent, law of mechanics (see Truesdell 1968).

The next key steps were taken in 1822, when Cauchy introduced the concept of the *stress tensor*, and combined it with Euler's laws of mechanics to construct a general theoretical framework for the motion of any continuous medium. To study, say, a Newtonian viscous fluid it became necessary only to add the appropriate *constitutive relation* describing its physical properties. Yet it was not until 1845, a full 158 years after the *Principia*, that Stokes extended Newton's original hypothesis in a wholly rational way to obtain that constitutive relation, so deriving what we now term the Navier–Stokes equations.†

6.2. The stress tensor

In this section and the next we describe Cauchy's theory. While we use freely the term 'fluid' in what follows, the formalism applies equally well to any continuous deformable medium.

---

† In recognition of the fact that Navier obtained the correct equations of motion (rather earlier than Stokes), but by making assumptions about the molecular basis of viscous effects which have not stood the test of time.

The Navier–Stokes equations 203

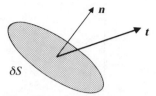

Fig. 6.1. The stress vector.

The stress vector

Let x denote the position vector of some fixed point in the fluid, and let δS be a small geometrical surface element, unit normal n, drawn through x. Consider the force exerted on this surface *by* the fluid *towards which* n is directed.

We assume that this force is

$$t \, \delta S, \qquad (6.1)$$

where the *stress vector* t, so defined, depends on the surface element in question only through its normal n. For an inviscid fluid, for example, $t = -p(x, t)n$ (see eqn (1.10)), but more generally we expect t to have components both tangential and normal to δS.

Definition of nine local quantities T_{ij}

The nine elements T_{ij} of the *stress tensor* are defined at any point, relative to rectangular Cartesian coordinates, as follows:

T_{ij} is the i-component of stress on a surface element δS which has a normal n pointing in the j-direction (6.2)

(see Fig. 6.2).

The stress on a small surface element of arbitrary orientation

Consider the stress t on a small surface element δS with unit normal n. We wish to demonstrate that the components t_i of the stress are given in terms of the components T_{ij} of the stress tensor by

$$t_i = T_{ij} n_j, \qquad (6.3)$$

204 *The Navier–Stokes equations*

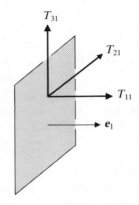

Fig. 6.2. Three components of the stress tensor T_{ij}.

where summation over $j = 1, 2, 3$ is understood by virtue of the repeated suffix.

To do this we take δS to be the large face of the tetrahedron in Fig. 6.3, and apply the principle of linear momentum to the fluid that momentarily occupies the tetrahedron. Consider the i-component of force on the fluid element. That exerted by the surrounding fluid on the main face is $t_i \, \delta S$. The i-component of stress exerted by the surrounding fluid on the face which is

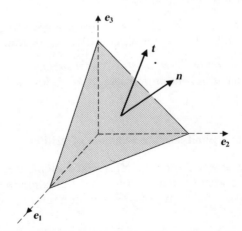

Fig. 6.3. Definition sketch for the proof of eqn (6.3).

normal to e_1 is $-T_{i1}$, because the normal n to that surface is pointing in the $-e_1$ direction, according to the conventions established above in the definition of the stress vector and stress tensor. Now, the area of the face which is normal to e_1 is, by vector algebra or elementary geometry, $n_1 \, \delta S$, where n denotes the unit outward normal to the large face. The i-component of force on the face which is normal to e_1 is therefore $-T_{i1} n_1 \, \delta S$. A similar argument holds for the remaining two faces. The i-component of the force exerted on the element by the surrounding fluid is therefore

$$(t_i - T_{ij} n_j) \, \delta S,$$

summation over j being understood.

This force, together with a body force $\rho \mathbf{g} \, \delta V$, will be equal to the mass $\rho \, \delta V$ of the element multiplied by its (finite) acceleration. Now let the linear dimension L of the tetrahedron tend to zero, while maintaining the orientation n of its large surface. As δV is proportional to L^3 and δS is proportional to L^2, it follows that $t_i = T_{ij} n_j$, as claimed above.

6.3. Cauchy's equation of motion

Having developed the notion of the stress tensor, Cauchy obtained the general equation of motion for any continuous medium:

$$\rho \frac{\mathrm{D} u_i}{\mathrm{D} t} = \frac{\partial T_{ij}}{\partial x_j} + \rho g_i. \tag{6.4}$$

To establish this we consider the ith component of force exerted, by the surrounding fluid, on some dyed blob of fluid with surface S. This is

$$\int_S t_i \, \mathrm{d}S = \int_S T_{ij} n_j \, \mathrm{d}S = \int_V \frac{\partial T_{ij}}{\partial x_j} \, \mathrm{d}V, \tag{6.5}$$

where we have used eqn (6.3) together with the divergence theorem. If we consider a *small* blob of fluid, then, $\partial T_{ij}/\partial x_j$ will be almost constant throughout it, and the surrounding fluid will exert on it a force having an i-component which is $\partial T_{ij}/\partial x_j$ multiplied by the volume of the blob δV. If there is also a body force $\mathbf{g}$ per unit mass, equating the total force on the small blob

Reynolds's transport theorem

to the rate of change of its momentum gives eqn (6.4), bearing in mind that the mass $\rho \, \delta V$ of the blob is conserved.

Reynolds's transport theorem

This theorem is about rates of change of volume integrals over finite 'dyed' blobs of fluid, and it provides, in particular, a pleasing alternative derivation of eqn (6.4). The theorem states that

$$\frac{d}{dt}\int_{V(t)} G \, dV = \int_{V(t)} \left(\frac{DG}{Dt} + G \nabla \cdot \boldsymbol{u}\right) dV, \qquad (6.6a)$$

where $G(\boldsymbol{x}, t)$ is any scalar or vector function and $V(t)$ denotes the region of space occupied by a finite, deforming blob of fluid.

A strict proof of this result may be found in Exercise 6.13. For the present we simply cast it into a different and more obvious form by writing $G = \rho F$. Then it follows that for any function $F(\boldsymbol{x}, t)$:

$$\frac{d}{dt}\int_{V(t)} F\rho \, dV = \int_{V(t)} \frac{DF}{Dt} \rho \, dV \qquad (6.6b)$$

(see Exercise 6.5). This is no surprise; the rate of change of the quantity $F\rho \, \delta V$ following a small element δV is DF/Dt multiplied by $\rho \, \delta V$, because the mass $\rho \, \delta V$ of any particular element is conserved.

Alternative derivation of Cauchy's equation

The principle of linear momentum, applied to a finite blob of dyed fluid, gives

$$\frac{d}{dt}\int_{V(t)} \rho u_i \, dV = \int_{S(t)} t_i \, dS + \int_{V(t)} \rho g_i \, dV$$

and on applying Reynolds's transport theorem (6.6b) to the left-hand side and eqn (6.5) to the right we obtain

$$\int_{V(t)} \left(\rho \frac{Du_i}{Dt} - \frac{\partial T_{ij}}{\partial x_j} - \rho g_i\right) dV = 0.$$

This being true for arbitrary $V(t)$ we deduce—provided that the

Summary

The development so far is valid for *any* continuous medium, and:

(i) the stress components t_i on a surface element with normal n may be written

$$t_i = T_{ij}n_j \tag{6.7}$$

where T_{ij} are the elements of a *stress tensor*;

(ii) the principle of linear momentum takes the form

$$\rho \frac{Du_i}{Dt} = \frac{\partial T_{ij}}{\partial x_j} + \rho g_i. \tag{6.8}$$

It is also the case, in fact, that the principle of moment of momentum (§6.1) implies

$$T_{ij} = T_{ji},$$

save in circumstances which, from a practical point of view, are most exceptional (see Exercise 6.14).

What we do not know at this stage, and what we cannot possibly know without deciding what kind of deformable medium we are working with, is how to calculate the elements T_{ij}.

6.4. A Newtonian viscous fluid: the Navier–Stokes equations

We now restrict attention to an incompressible fluid, for which

$$\nabla \cdot \boldsymbol{u} = 0,$$

and at this point it is possible to take

$$T_{ij} = -p\,\delta_{ij} + \mu\left(\frac{\partial u_j}{\partial x_i} + \frac{\partial u_i}{\partial x_j}\right) \tag{6.9}$$

as the constitutive relation *defining* an incompressible, Newtonian viscous fluid of viscosity μ. Notably, the stress tensor is symmetric, i.e. $T_{ij} = T_{ji}$. In view of this symmetry, eqn (6.9)

208 The Navier–Stokes equations

amounts to six, rather than nine equations:

$$T_{11} = -p + 2\mu \frac{\partial u_1}{\partial x_1}, \qquad T_{22} = -p + 2\mu \frac{\partial u_2}{\partial x_2},$$

$$T_{33} = -p + 2\mu \frac{\partial u_3}{\partial x_3}, \qquad T_{23} = \mu\left(\frac{\partial u_3}{\partial x_2} + \frac{\partial u_2}{\partial x_3}\right),$$

$$T_{31} = \mu\left(\frac{\partial u_1}{\partial x_3} + \frac{\partial u_3}{\partial x_1}\right), \qquad T_{12} = \mu\left(\frac{\partial u_2}{\partial x_1} + \frac{\partial u_1}{\partial x_2}\right).$$

The physical significance of the quantity p, called the pressure, is simply that $-p$ is the mean of the three normal stresses at a point, i.e.

$$p = -\tfrac{1}{3}(T_{11} + T_{22} + T_{33})$$

(see Fig. 6.2).

On substituting eqn (6.9) into Cauchy's equation of motion we obtain, in the case of constant viscosity μ,

$$\rho \frac{Du_i}{Dt} = -\frac{\partial p}{\partial x_i} + \mu \frac{\partial}{\partial x_j}\left(\frac{\partial u_j}{\partial x_i} + \frac{\partial u_i}{\partial x_j}\right) + \rho g_i$$

$$= -\frac{\partial p}{\partial x_i} + \mu \frac{\partial}{\partial x_i}\left(\frac{\partial u_j}{\partial x_j}\right) + \mu \frac{\partial^2 u_i}{\partial x_j^2} + \rho g_i.$$

But

$$\frac{\partial^2}{\partial x_j^2} = \frac{\partial^2}{\partial x_1^2} + \frac{\partial^2}{\partial x_2^2} + \frac{\partial^2}{\partial x_3^2},$$

and for an incompressible fluid

$$\partial u_j / \partial x_j = \nabla \cdot \boldsymbol{u} = 0,$$

whence the Navier–Stokes equations

$$\rho \frac{D\boldsymbol{u}}{Dt} = -\nabla p + \rho \boldsymbol{g} + \mu \nabla^2 \boldsymbol{u}, \tag{6.10}$$

$$\nabla \cdot \boldsymbol{u} = 0, \tag{6.11}$$

as claimed in eqn (2.3).

Using the vector identity (A.10) we may rewrite eqn (6.10) as

$$\rho \frac{D\boldsymbol{u}}{Dt} = -\nabla p + \rho \boldsymbol{g} - \mu \nabla \wedge (\nabla \wedge \boldsymbol{u}) \tag{6.12}$$

and this can be more convenient when working in non-Cartesian coordinate systems.

We also observe that on combining eqns (6.7) and (6.9) the stress vector may be written

$$t = -p\boldsymbol{n} + \mu[2(\boldsymbol{n} \cdot \nabla)\boldsymbol{u} + \boldsymbol{n} \wedge (\nabla \wedge \boldsymbol{u})]. \qquad (6.13)$$

We leave the proof as an exercise (Exercise 6.1).

Where does eqn (6.9) come from?

Stokes (1845) deduced eqn (6.9) from three elementary hypotheses. On writing $T_{ij} = -p\delta_{ij} + T_{ij}^D$ these amount essentially to:

(i) each T_{ij}^D should be a linear function of the velocity gradients $\partial u_1/\partial x_1$, $\partial u_1/\partial x_2$, etc.;

(ii) each T_{ij}^D should vanish if the flow involves no deformation of fluid elements;

(iii) the relationship between T_{ij}^D and the velocity gradients should be isotropic, as the physical properties of the fluid are assumed to show no preferred direction.

We do not pursue the argument in detail here (see Exercise 6.11), but try instead to indicate by example how eqn (6.9) conforms to each of the above hypotheses. With regard to (i), which is the most natural extension of Newton's original proposal, there is little to do beyond observe that in eqn (6.9) the quantities T_{ij}^D are indeed linear functions of the quantities $\partial u_i/\partial x_j$.

With regard to (ii), consider first a fluid element in 2-D flow, as in Fig. 6.4, where we have displayed the velocity components of the fluid particles at B and C relative to those of the particle at A. Plainly, the distance between the particles at A and B is momentarily increasing with time if $\partial u_1/\partial x_1 > 0$ and decreasing if $\partial u_1/\partial x_1 < 0$. Thus the terms $2\mu \, \partial u_1/\partial x_1$ and $2\mu \, \partial u_2/\partial x_2$ in the 2-D version of eqn (6.9) have a simple physical interpretation in terms of the stretching (or shrinking) of fluid elements, and they vanish if the fluid is moving without deformation. Similarly, we see that the fluid line element AB is momentarily rotating with angular velocity $\partial u_2/\partial x_1$, while the fluid line element AC is rotating with angular velocity $-\partial u_1/\partial x_2$. The angle between AB

210 The Navier–Stokes equations

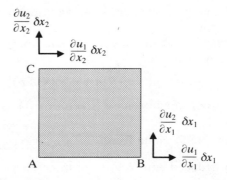

Fig. 6.4. Velocity components at two points of a fluid element, relative to those at A.

and AC is therefore momentarily decreasing with time at a rate $\partial u_2/\partial x_1 + \partial u_1/\partial x_2$. The so-called 'shear stress' term $\mu(\partial u_2/\partial x_1 + \partial u_1/\partial x_2)$ in the 2-D version of eqn (6.9) therefore also has a simple physical interpretation, and again vanishes if the fluid is moving without deformation. We say more about (ii) in the subsection which follows.

With regard to (iii) let us consider the simple example of a 2-D shear flow

$$u_1 = \beta x_2, \qquad u_2 = 0$$

over a rigid plane boundary $x_2 = 0$. In this case

$$T_{11} = -p, \qquad T_{22} = -p, \qquad T_{12} = \mu\beta.$$

The tangential stress on the boundary is

$$t_1 = T_{1j}n_j = T_{12}n_2 = T_{12} = \mu\beta. \tag{6.14}$$

Note that the terms $2\mu\, \partial u_1/\partial x_1$ and $2\mu\, \partial u_2/\partial x_2$ are zero.

But suppose that, somewhat perversely, we carry out the whole calculation of the tangential stress on the boundary not with reference to the obvious coordinate system but with reference to the coordinates x_1', x_2' shown in Fig. 6.5 instead. The velocity components relative to these coordinates are

$$u_1' = \frac{\beta}{2}(x_1' + x_2'), \qquad u_2' = -\frac{\beta}{2}(x_1' + x_2'),$$

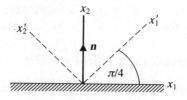

Fig. 6.5. The two coordinate systems.

and if, as is being claimed, the relationship (6.9) is isotropic, then it must take exactly the same form relative to the new axes, i.e.

$$T'_{11} = -p + 2\mu \frac{\partial u'_1}{\partial x'_1}, \qquad T'_{22} = -p + 2\mu \frac{\partial u'_2}{\partial x'_2},$$

$$T'_{12} = \mu \left(\frac{\partial u'_2}{\partial x'_1} + \frac{\partial u'_1}{\partial x'_2} \right).$$

The purpose of the present calculation is to check that this does, indeed, lead to the same expression (6.14) for the stress on the boundary. Thus

$$T'_{11} = -p + \mu\beta, \qquad T'_{22} = -p - \mu\beta, \qquad T'_{12} = 0.$$

Now

$$t'_i = T'_{ij} n'_j,$$

where n'_j are the components of the unit normal to the boundary relative to the new axes. This gives

$$t'_i = T'_{i1} n'_1 + T'_{i2} n'_2 = \frac{1}{\sqrt{2}} (T'_{i1} + T'_{i2}),$$

whence

$$t'_1 = \frac{1}{\sqrt{2}} (-p + \mu\beta), \qquad t'_2 = \frac{1}{\sqrt{2}} (-p - \mu\beta),$$

and finally

$$t_1 = \frac{1}{\sqrt{2}} (t'_1 - t'_2) = \mu\beta \qquad (6.15)$$

as before. As it happens, in this (crazy) formulation of the

212 The Navier–Stokes equations

problem the 'shear stress' T'_{12} is actually zero, and eqn (6.15) originates wholly from the terms $2\mu\, \partial u'_1/\partial x'_1$ and $2\mu\, \partial u'_2/\partial x'_2$.

The general deformation of a fluid element

We now look more deeply at (ii), and at this point it is useful to define the rate-of-strain tensor

$$e_{ij} = \tfrac{1}{2}\left(\frac{\partial u_i}{\partial x_j} + \frac{\partial u_j}{\partial x_i}\right), \tag{6.16}$$

in which case the constitutive relation for an incompressible Newtonian viscous fluid is

$$T_{ij} = -p\delta_{ij} + 2\mu e_{ij}. \tag{6.17}$$

In the foregoing discussion we have provided some evidence that e_{ij} vanishes if there is no deformation of fluid elements. We now explore this notion further.

Let the fluid velocity at some fixed point be $\boldsymbol{u}_\text{P}$. By Taylor's theorem the velocity at a point Q a small distance $\boldsymbol{s}$ from P is, to first order in $\boldsymbol{s}$,

$$\boldsymbol{u}_\text{Q} = \boldsymbol{u}_\text{P} + (\boldsymbol{s} \cdot \nabla)\boldsymbol{u}, \tag{6.18}$$

the derivatives in this expression being evaluated at P. We are interested in how $\boldsymbol{u}_\text{Q}$ depends, locally, on $\boldsymbol{s}$, and the key to this lies in rewriting eqn (6.18) as

$$\boldsymbol{u}_\text{Q} = \boldsymbol{u}_\text{P} + \tfrac{1}{2}(\nabla \wedge \mathbf{u}) \wedge \boldsymbol{s} + \tfrac{1}{2}\nabla_s(e_{ij}s_i s_j), \tag{6.19}$$

where $\nabla \wedge \boldsymbol{u}$ and e_{ij} are evaluated at P (Exercise 6.7). Here ∇_s denotes the operator $\boldsymbol{e}_k\, \partial/\partial s_k$, i.e. the ∇ operator with respect to the variable $\boldsymbol{s}$.

Now, the term $\tfrac{1}{2}(\nabla \wedge \boldsymbol{u}) \wedge \boldsymbol{s}$ is of the form '$\boldsymbol{\Omega} \wedge \boldsymbol{x}$' and

Fig. 6.6. Definition sketch for eqn (6.18).

represents a local rigid-body rotation with angular velocity $\frac{1}{2}(\nabla \wedge \boldsymbol{u})$. Thus the vorticity $\nabla \wedge \boldsymbol{u}$ (or, more precisely, one half of it) acts as a measure of the extent to which a fluid element is spinning, just as we observed in §1.4 in a strictly 2-D context.

To see that the term $\frac{1}{2}\nabla_s(e_{ij}s_is_j)$ represents a pure straining motion, i.e. one involving stretching/squashing in mutually perpendicular directions but no overall rotation, note first that it denotes a vector field which is everywhere normal to surfaces of constant $e_{ij}s_is_j$. To picture these surfaces consider first a simple 2-D example in which

$$\boldsymbol{u} = (\alpha x_1, -\alpha x_2, 0). \qquad (6.20)$$

In this case

$$e_{ij} = \begin{pmatrix} \alpha & 0 & 0 \\ 0 & -\alpha & 0 \\ 0 & 0 & 0 \end{pmatrix}$$

(which is untypical, in that e_{ij} is the same, no matter which x_P we choose), and

$$e_{ij}s_is_j = \begin{pmatrix} s_1 & s_2 & s_3 \end{pmatrix} \begin{pmatrix} \alpha & 0 & 0 \\ 0 & -\alpha & 0 \\ 0 & 0 & 0 \end{pmatrix} \begin{pmatrix} s_1 \\ s_2 \\ s_3 \end{pmatrix}$$
$$= \alpha(s_1^2 - s_2^2).$$

Thus the cross-sections of surfaces of constant $e_{ij}s_is_j$ are, in this case, as in Fig. 6.7. More generally, we note that as e_{ij} is

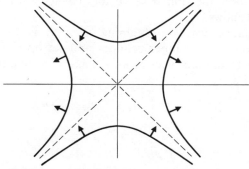

Fig. 6.7. Surfaces of constant $e_{ij}s_is_j$ in a pure straining motion.

symmetric principal axes can always be found with respect to which it is diagonal, and with respect to those axes the quantity $e_{ij}s_is_j$ is

$$e'_{11}s'^2_1 + e'_{22}s'^2_2 + e'_{33}s'^2_3.$$

Together with the incompressibility condition ($e'_{11} + e'_{22} + e'_{33} = 0$) this implies that surfaces of constant $e_{ij}s_is_j$ are hyperboloids, and the associated motions are accordingly simple 3-D equivalents of the kind shown in Fig. 6.7.

Thus eqn (6.19) does indeed decompose the flow in the neighbourhood of any point P into a pure translation (first term), a rotary flow involving no deformation (second term) and a flow involving deformation but no rotation (third term).

Finding the components of the stress vector t in cylindrical or spherical polar coordinates

If we are solving a flow problem in cylindrical or spherical polar coordinates we need a quick and effective way of calculating the stress vector t.

Consider, for example, the flow

$$\boldsymbol{u} = u_\theta(r)\boldsymbol{e}_\theta$$

between two rotating cylinders, as in eqn (2.31). One way of obtaining the stress t at any point on the inner cylinder is to use the expression (6.13), as in Exercise 6.4. This method is quite effective, although the calculation of $(\boldsymbol{n} \cdot \nabla)\boldsymbol{u}$ requires careful attention to how the unit base vectors change with position, as Exercise 6.9 shows.

An alternative way of obtaining t_θ, say, is to pick some particular point of the inner cylinder and set up Cartesian axes coincident with the unit vectors $\boldsymbol{e}_r$, $\boldsymbol{e}_\theta$, and $\boldsymbol{e}_z$ at that point. Then we want t_2, i.e.

$$t_2 = T_{2j}n_j = T_{21}n_1 + T_{22}n_2 + T_{23}n_3,$$

for which

$$t_\theta = T_{\theta r}n_r + T_{\theta\theta}n_\theta + T_{\theta z}n_z \tag{6.21}$$

is no more than an alternative notation. In the present instance

$n = e_r$, and on using eqn (6.17):

$$t_\theta = T_{\theta r} = T_{r\theta} = 2\mu e_{r\theta}.$$

To find $e_{r\theta}$ we may turn to eqn (A.36), and as $u_r = 0$ on the cylinder we obtain

$$t_\theta = \mu r \frac{\mathrm{d}}{\mathrm{d}r}\left(\frac{u_\theta}{r}\right),$$

as in Exercise 6.4.

This is effective, but it requires some understanding of where expressions such as eqn (A.36) come from. To this end, note that

$$2e_{r\theta} = 2e_{12} = \frac{\partial u_2}{\partial x_1} + \frac{\partial u_1}{\partial x_2} = (\boldsymbol{e}_1 \cdot \nabla)(\boldsymbol{u} \cdot \boldsymbol{e}_2) + (\boldsymbol{e}_2 \cdot \nabla)(\boldsymbol{u} \cdot \boldsymbol{e}_1)$$

$$= [(\boldsymbol{e}_1 \cdot \nabla)\boldsymbol{u}] \cdot \boldsymbol{e}_2 + [(\boldsymbol{e}_2 \cdot \nabla)\boldsymbol{u}] \cdot \boldsymbol{e}_1, \quad (6.22)$$

the final step following because (in marked contrast to $\boldsymbol{e}_r, \boldsymbol{e}_\theta, \boldsymbol{e}_z$) the unit vectors $\boldsymbol{e}_1, \boldsymbol{e}_2$, and $\boldsymbol{e}_3$ are all constant. Thus

$$2e_{r\theta} = [(\boldsymbol{e}_r \cdot \nabla)\boldsymbol{u}] \cdot \boldsymbol{e}_\theta + [(\boldsymbol{e}_\theta \cdot \nabla)\boldsymbol{u}] \cdot \boldsymbol{e}_r$$

$$= \left[\frac{\partial}{\partial r}(u_r \boldsymbol{e}_r + u_\theta \boldsymbol{e}_\theta + u_z \boldsymbol{e}_z)\right] \cdot \boldsymbol{e}_\theta$$

$$+ \left[\frac{1}{r}\frac{\partial}{\partial \theta}(u_r \boldsymbol{e}_r + u_\theta \boldsymbol{e}_\theta + u_z \boldsymbol{e}_z)\right] \cdot \boldsymbol{e}_r. \quad (6.23)$$

Now, the unit vectors $\boldsymbol{e}_r$ and $\boldsymbol{e}_\theta$ change with θ according to eqn (A.29), so

$$2e_{r\theta} = \frac{\partial u_\theta}{\partial r} + \frac{1}{r}\left[\frac{\partial u_r}{\partial \theta}\boldsymbol{e}_r + u_r \boldsymbol{e}_\theta + \frac{\partial u_\theta}{\partial \theta}\boldsymbol{e}_\theta - u_\theta \boldsymbol{e}_r + \frac{\partial u_z}{\partial \theta}\boldsymbol{e}_z\right] \cdot \boldsymbol{e}_r$$

$$= \frac{\partial u_\theta}{\partial r} + \frac{1}{r}\frac{\partial u_r}{\partial \theta} - \frac{u_\theta}{r}$$

$$= r\frac{\partial}{\partial r}\left(\frac{u_\theta}{r}\right) + \frac{1}{r}\frac{\partial u_r}{\partial \theta},$$

which is the last of the expressions (A.36).

6.5. Viscous dissipation of energy

Consider the kinetic energy

$$T = \tfrac{1}{2} \int_V \rho u_i^2 \, dV \qquad (6.24)$$

of a dyed blob of fluid V with surface S. The rate of change of T is

$$\frac{dT}{dt} = \int_V u_i \frac{Du_i}{Dt} \rho \, dV,$$

(see eqn (6.6b)), and by virtue of Cauchy's equation (6.4) we may rewrite this as

$$\frac{dT}{dt} = \int_V \rho u_i g_i \, dV + \int_V u_i \frac{\partial T_{ij}}{\partial x_j} \, dV. \qquad (6.25)$$

Now,

$$\int_V u_i \frac{\partial T_{ij}}{\partial x_j} \, dV = \int_V \frac{\partial}{\partial x_j}(u_i T_{ij}) \, dV - \int_V T_{ij} \frac{\partial u_i}{\partial x_j} \, dV,$$

and

$$\int_V \frac{\partial}{\partial x_j}(u_i T_{ij}) \, dV = \int_S u_i T_{ij} n_j \, dS = \int_S u_i t_i \, dS,$$

where we have used eqn (6.3) and the divergence theorem (A.13). Furthermore,

$$T_{ij} \frac{\partial u_i}{\partial x_j} = \tfrac{1}{2}\left(T_{ij} \frac{\partial u_i}{\partial x_j} + T_{ji} \frac{\partial u_j}{\partial x_i}\right) = \tfrac{1}{2} T_{ij}\left(\frac{\partial u_i}{\partial x_j} + \frac{\partial u_j}{\partial x_i}\right)$$

as (i) summation over $i = 1, 2, 3$ and $j = 1, 2, 3$ is understood, and (ii) T_{ij} is symmetric. Using the relation (6.9) for an incompressible Newtonian viscous fluid, together with $\nabla \cdot \mathbf{u} = 0$, we see that

$$T_{ij} \frac{\partial u_i}{\partial x_j} = \tfrac{1}{2} \mu \left(\frac{\partial u_i}{\partial x_j} + \frac{\partial u_j}{\partial x_i}\right)^2 = 2\mu e_{ij}^2$$

(see eqn (6.16)). Thus

$$\frac{dT}{dt} = \int_V \rho \mathbf{u} \cdot \mathbf{g} \, dV + \int_S \mathbf{t} \cdot \mathbf{u} \, dS - 2\mu \int_V e_{ij}^2 \, dV. \qquad (6.26)$$

The first term on the right-hand side represents the rate of decrease of potential energy of the 'dyed' fluid, while the second term represents the rate at which the surrounding fluid is doing work on the dyed fluid via the surface stresses t. Not all this goes into increasing the kinetic energy of the dyed fluid; viscous stresses within the blob are evidently dissipating energy at a rate

$$2\mu e_{ij}^2 \tag{6.27}$$

per unit volume which, written out in full, is

$$2\mu(e_{11}^2 + e_{22}^2 + e_{33}^2 + 2e_{23}^2 + 2e_{31}^2 + 2e_{12}^2).$$

This viscous dissipation of energy is zero only if $e_{ij} = 0$ for all i and j, i.e. if there is no deformation of fluid elements.

Exercises

6.1. We may deduce from eqns (6.7) and (6.9) that

$$t_i = -pn_i + \mu n_j \left(\frac{\partial u_i}{\partial x_j} + \frac{\partial u_j}{\partial x_i} \right).$$

Show that this identical to

$$t = -pn + \mu[2(n \cdot \nabla)u + n \wedge (\nabla \wedge u)],$$

by expanding this expression using the suffix notation and the summation convention.

6.2. Use eqn (6.13) and various vector identities to show that the net force exerted on a finite blob of fluid by the surrounding fluid is

$$\int_S t \, dS = \int_V (-\nabla p + \mu \nabla^2 u) \, dV,$$

where S is the surface of the blob and V the region occupied by the blob. Deduce that if the blob is small the net force on it, excluding gravity, is $-\nabla p + \mu \nabla^2 u$ per unit volume, in agreement with eqn (6.10).

6.3. Verify that in the case of a simple shear flow

$$u = [u(y), 0, 0]$$

eqn (6.13) reduces, when $n = (0, 1, 0)$, to

$$t = \left[\mu \frac{du}{dy}, -p, 0 \right].$$

6.4. Show that in the case of a purely rotary flow

$$u = u_\theta(r) e_\theta$$

218 *The Navier–Stokes equations*

eqn (6.13) reduces, when $n = e_r$, to

$$t = -pe_r + \mu r \frac{d}{dr}\left(\frac{u_\theta}{r}\right)e_\theta,$$

and note that the second term vanishes in the case of uniform rotation, $u_\theta \propto r$, for there is then no deformation of fluid elements.

Use this result to calculate the torque exerted on the inner cylinder by the flow (2.31) and (2.32).

6.5. Use Reynolds's transport theorem (6.6a) to provide an alternative derivation of the conservation of mass equation

$$\frac{D\rho}{Dt} + \rho \nabla \cdot u = 0$$

(cf. Exercise 1.1). Then use this equation to deduce eqn (6.6b) from eqn (6.6a).

6.6. Show that the terms $\mu(\partial u_j/\partial x_i + \partial u_i/\partial x_j)$ of the stress tensor (6.9) are zero for the uniformly rotating flow $u = \Omega \wedge x$, Ω being a constant vector.

6.7. Expand eqn (6.19) using the suffix notation and summation convention:

$$u_Q = u_P + \tfrac{1}{2}\left[\left(e_i \wedge \frac{\partial u}{\partial x_i}\right) \wedge s + e_k \frac{\partial}{\partial s_k}(e_{ij}s_i s_j)\right]$$

etc., to show that eqn (6.19) is equivalent to eqn (6.18).

6.8. Separate the shear flow $u = (\beta x_2, 0, 0)$ of Fig. 1.4 into its local (i) translation, (ii) rotation, and (iii) pure straining parts, using eqn (6.19). Find the directions of the principal axes of e_{ij}, and verify that this decomposition of the flow can be represented schematically as in Fig. 6.8.

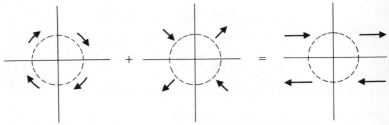

Fig. 6.8. The decomposition of a uniform shear flow.

6.9. Consider a 2-D viscous flow

$$\boldsymbol{u} = u_r(r, \theta)\boldsymbol{e}_r,$$

as might occur in a converging or diverging channel (see, e.g. Exercise 7.6). Use both methods described at the end of §6.4 to show that the stress exerted by the fluid in $\theta > 0$ on that in $\theta < 0$ is

$$\boldsymbol{t} = \frac{\mu}{r}\frac{\partial u_r}{\partial \theta}\boldsymbol{e}_r + \left(-p + \frac{2\mu u_r}{r}\right)\boldsymbol{e}_\theta.$$

[Note that the normal component of stress is not due to the pressure p alone.]

6.10. Verify by direct calculation the expression for $e_{\theta\phi}$ in the spherical polar formulae (A.44).

6.11. If T^D_{ij} is a linear function of e_{11}, e_{12}, etc., then we may write

$$T^D_{ij} = c_{ijkl}e_{kl}.$$

It is shown in books on tensor analysis (e.g. Bourne and Kendall 1977, §8.3) that the most general fourth-order *isotropic* tensor is of the form

$$c_{ijkl} = A\delta_{ij}\delta_{kl} + B\delta_{ik}\delta_{jl} + C\delta_{il}\delta_{jk},$$

where A, B, and C are scalars. Use this to show that

$$T^D_{ij} = \lambda e_{kk}\delta_{ij} + 2\mu e_{ij},$$

where λ and μ are scalars.

Show that if p is defined, as in §6.4, so that

$$p = -\tfrac{1}{3}T_{ii},$$

then

$$T_{ij} = -(p + \tfrac{2}{3}\mu\nabla\cdot\boldsymbol{u})\delta_{ij} + \mu\left(\frac{\partial u_i}{\partial x_j} + \frac{\partial u_j}{\partial x_i}\right),$$

which reduces to eqn (6.9) when the fluid is incompressible.

[With a compressible fluid some care is needed in distinguishing between the mechanical pressure, defined above, and the thermodynamic pressure (see Batchelor 1967, p. 154).]

6.12. Observe that if a flow $\boldsymbol{u}$ is irrotational, the viscous term is zero in the equation of motion (6.12).

Consider now the flow

$$\boldsymbol{u} = \frac{\Omega a^2}{r}\boldsymbol{e}_\theta, \qquad r \geq a,$$

driven by a rotating cylinder at $r = a$, as in Exercise 2.8. 'The flow is irrotational in $r \geq a$; therefore the viscous term is zero; therefore the viscous forces are zero; and so the torque on the cylinder is zero.' But it is not. What is wrong with the argument?

6.13. Let $x = x(X, t)$ denote some fluid motion, as in Exercises 1.7, 5.18, and 5.21, and let J denote the determinant

$$J = \begin{vmatrix} \dfrac{\partial x_1}{\partial X_1} & \dfrac{\partial x_1}{\partial X_2} & \dfrac{\partial x_1}{\partial X_3} \\ \dfrac{\partial x_2}{\partial X_1} & \dfrac{\partial x_2}{\partial X_2} & \dfrac{\partial x_2}{\partial X_3} \\ \dfrac{\partial x_3}{\partial X_1} & \dfrac{\partial x_3}{\partial X_2} & \dfrac{\partial x_3}{\partial X_3} \end{vmatrix}.$$

Establish *Euler's identity*

$$DJ/Dt = J \nabla \cdot \boldsymbol{u},$$

and use this to give a proof of Reynolds's transport theorem (6.6a).

6.14. If we apply the principle of moment of momentum (§6.1) to a finite 'dyed' blob of some continuous medium occupying a region $V(t)$ we obtain†

$$\frac{\mathrm{d}}{\mathrm{d}t} \int_{V(t)} \boldsymbol{x} \wedge \rho \boldsymbol{u} \, \mathrm{d}V = \int_{S(t)} \boldsymbol{x} \wedge \boldsymbol{t} \, \mathrm{d}S + \int_{V(t)} \boldsymbol{x} \wedge \rho \boldsymbol{g} \, \mathrm{d}V.$$

Use Reynolds's transport theorem, together with eqns (6.7) and (6.8), to write this in the form

$$\int_{V(t)} x_k \boldsymbol{e}_k \wedge \frac{\partial T_{ij}}{\partial x_j} \boldsymbol{e}_i \, \mathrm{d}V = \int_{S(t)} x_k \boldsymbol{e}_k \wedge T_{ij} n_j \boldsymbol{e}_i \, \mathrm{d}S$$

where summation over 1, 2, 3 is implied for i, j, and k. Re-cast this equation into the form

$$\boldsymbol{e}_k \wedge \boldsymbol{e}_i \int_{V(t)} T_{ik} \, \mathrm{d}V = 0,$$

and hence deduce that, *subject to the proviso in the footnote,*

$$T_{ij} = T_{ji},$$

i.e. the stress tensor must be symmetric, *whatever the nature of the deformable medium in question.* (This famous requirement, to which eqn (6.9) conforms, is due to Cauchy.)

---

† There is a proviso here, namely that the net torque on the blob is due simply to the moment of the stresses $\boldsymbol{t}$ on its surface and the moment of the body force $\boldsymbol{g}$ per unit mass. This is very generally the case, but there are exotic exceptions, as when the medium consists of a suspension of ferromagnetic particles, each being subject to the torque of an applied magnetic field (see Chap. 8 of Rosensweig 1985).

7 Very viscous flow

7.1. Introduction

The character of a steady viscous flow depends strongly on the relative magnitude of the terms $(\boldsymbol{u} \cdot \nabla)\boldsymbol{u}$ and $\nu \nabla^2 \boldsymbol{u}$ in the equation of motion

$$(\boldsymbol{u} \cdot \nabla)\boldsymbol{u} = -\frac{1}{\rho} \nabla p + \nu \nabla^2 \boldsymbol{u} + \boldsymbol{g}. \quad (7.1)$$

We are here concerned with the 'very viscous' case in which the $(\boldsymbol{u} \cdot \nabla)\boldsymbol{u}$ term is negligible. There are two rather different ways in which this can happen.

First, the Reynolds number may be very small, i.e.

$$R = \frac{UL}{\nu} \ll 1. \quad (7.2)$$

On the basis of the estimates (2.5) we then expect the *slow flow equations*

$$0 = -\nabla p + \mu \nabla^2 \boldsymbol{u},$$
$$\nabla \cdot \boldsymbol{u} = 0 \quad (7.3)$$

to provide a good description of the flow, in the absence of body forces.

We discuss the uniqueness and *reversibility* of solutions to these equations in §7.4, and some implications for the propulsion of biological micro-organisms follow in §7.5. In §7.3 we explore the so-called corner eddies that can occur at low Reynolds number, as in the superbly symmetric example of Fig. 7.1(a). First, however, we investigate in §7.2 the classical problem of slow flow past a sphere, and it is worth taking a moment to consider the kind of practical circumstances in which slow flow theory might apply in that case.

Suppose, for instance, that we tow a sphere of diameter $D = 1$ cm through stationary fluid at the quite modest speed

222 Very viscous flow

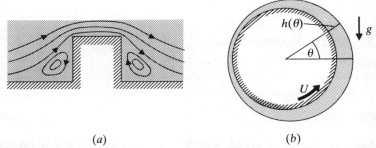

(a)   (b)

Fig. 7.1. Two very viscous flows: (a) flow at low Reynolds number past a square block on a plate; (b) a thin film of syrup on the outside of a rotating cylinder.

$U = 2$ cm s^{-1}. Then according to Table 2.1 the Reynolds number UD/ν will be about 200 for water, 2 for olive oil, 0.1 for glycerine, and 0.002 for golden syrup. Now, if we move the sphere at a speed of only 0.2 cm s^{-1}, all these values will be reduced by a factor of 10, but they are still not spectacularly small. Our point, then, is that while Reynolds numbers of order 10^8 or 10^9 are not at all uncommon in nature, to get a genuinely small Reynolds number takes a bit more effort.

A second, quite different, way in which the $(\boldsymbol{u} \cdot \nabla)\boldsymbol{u}$ term may be negligible in eqn (7.1) involves motion in a *thin film* of liquid, and in this case the 'conventional' Reynolds number need not be small. The key idea is, instead, that the velocity gradients across the film are so strong, on account of its small thickness, that viscous forces predominate. Thus if L denotes the length of the film, and h a typical thickness, the term $(\boldsymbol{u} \cdot \nabla)\boldsymbol{u}$ turns out to be negligible if

$$\frac{h^2}{L^2} \ll \frac{\nu}{UL}, \qquad (7.4)$$

as we show in §7.6. The resulting *thin-film equations* are even simpler than eqn (7.3), and provide the opportunity of tackling some problems which would otherwise be unapproachable by elementary analysis. One example will be well known to patrons of Dutch pancake houses: if you dip a wooden spoon into syrup, withdraw it, and hold it horizontal, you can prevent the syrup from draining off the handle by rotating the spoon (Fig. 7.1(*b*)). Moffatt (1977) used thin-film theory to show that there is a steady flow solution if

$$U > 2.014 \frac{g\bar{h}^2}{\nu}, \tag{7.5}$$

where $\bar{h}$ is the mean thickness of the film. If the peripheral speed U of the handle is below this critical value there is no steady solution, and the liquid slowly drains off.

In the second half of this chapter we look at a number of thin-film flows of this kind, one of the most notable being that in a *Hele-Shaw cell* (§7.7). In this quite elementary apparatus it is possible to simulate many 2-D irrotational flow patterns that would, on account of boundary layer separation, be wholly unobservable at high Reynolds number.

7.2. Low Reynolds number flow past a sphere

We now seek a solution to the slow flow equations (7.3) for uniform flow past a sphere, and using appropriate spherical polar coordinates we therefore want an axisymmetric flow

$$\boldsymbol{u} = [u_r(r, \theta), u_\theta(r, \theta), 0].$$

We may automatically satisfy $\nabla \cdot \boldsymbol{u} = 0$ by introducing a Stokes stream function $\Psi(r, \theta)$ such that

$$u_r = \frac{1}{r^2 \sin \theta} \frac{\partial \Psi}{\partial \theta}, \qquad u_\theta = -\frac{1}{r \sin \theta} \frac{\partial \Psi}{\partial r} \tag{7.6}$$

(cf. §5.5). Then

$$\nabla \wedge \boldsymbol{u} = \left[0, 0, -\frac{1}{r \sin \theta} E^2 \Psi\right],$$

where E^2 denotes the differential operator

$$E^2 = \frac{\partial^2}{\partial r^2} + \frac{\sin \theta}{r^2} \frac{\partial}{\partial \theta} \left(\frac{1}{\sin \theta} \frac{\partial}{\partial \theta}\right).$$

Writing eqn (7.3) in the form

$$\nabla p = -\mu \nabla \wedge (\nabla \wedge \boldsymbol{u}),$$

224 Very viscous flow

(see eqn (6.12)) we obtain

$$\frac{\partial p}{\partial r} = \frac{\mu}{r^2 \sin\theta} \frac{\partial}{\partial \theta} E^2\Psi,$$

$$\frac{1}{r}\frac{\partial p}{\partial \theta} = \frac{-\mu}{r \sin\theta} \frac{\partial}{\partial r} E^2\Psi,$$

and eliminating the pressure by cross-differentiation we find that $E^2(E^2\Psi) = 0$, i.e.

$$\left[\frac{\partial^2}{\partial r^2} + \frac{\sin\theta}{r^2}\frac{\partial}{\partial \theta}\left(\frac{1}{\sin\theta}\frac{\partial}{\partial \theta}\right)\right]^2 \Psi = 0. \tag{7.7}$$

The boundary conditions are

$$\frac{\partial \Psi}{\partial r} = \frac{1}{r}\frac{\partial \Psi}{\partial \theta} = 0 \qquad \text{on } r = a,$$

together with the condition that as $r \to \infty$ the flow becomes uniform with speed U:

$$u_r \sim U\cos\theta \quad \text{and} \quad u_\theta \sim -U\sin\theta \quad \text{as } r \to \infty.$$

This infinity condition may be written

$$\Psi \sim \tfrac{1}{2} U r^2 \sin^2\theta \qquad \text{as } r \to \infty,$$

which suggests trying a solution to eqn (7.7) of the form

$$\Psi = f(r)\sin^2\theta.$$

This turns out to be possible provided that

$$\left(\frac{d^2}{dr^2} - \frac{2}{r^2}\right)^2 f = 0.$$

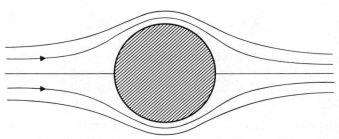

Fig. 7.2. Low Reynolds number flow past a sphere.

This equation is homogeneous in r and has solutions of the form r^α provided that

$$[(\alpha - 2)(\alpha - 3) - 2][\alpha(\alpha - 1) - 2] = 0,$$

so that

$$f(r) = \frac{A}{r} + Br + Cr^2 + Dr^4,$$

where A, B, C, and D are arbitrary constants. The condition of uniform flow at infinity implies that $C = \tfrac{1}{2}U$ and $D = 0$. The constants A and B are then determined by applying the boundary conditions on $r = a$, which reduce to $f(a) = f'(a) = 0$. We thus find that

$$\Psi = \tfrac{1}{4}U\left(2r^2 + \frac{a^3}{r} - 3ar\right)\sin^2\theta. \qquad (7.8)$$

The streamlines are symmetric fore and aft of the sphere (Fig. 7.2).

A quantity of major interest is the drag D on the sphere. By computing $E^2\Psi = \tfrac{3}{2}Uar^{-1}\sin^2\theta$ and then integrating the equations above for the pressure p we obtain

$$p = p_\infty - \tfrac{3}{2}\frac{\mu U a}{r^2}\cos\theta,$$

where p_∞ denotes the pressure as $r \to \infty$. The stress components on the sphere are

$$t_r = T_{rr} = -p + 2\mu\frac{\partial u_r}{\partial r},$$

$$t_\theta = T_{r\theta} = \mu r\frac{\partial}{\partial r}\left(\frac{u_\theta}{r}\right) + \frac{\mu}{r}\frac{\partial u_r}{\partial \theta},$$

$$t_\phi = T_{r\phi} = 0,$$

(see eqn (A.44) and §6.4). Having found Ψ, we may calculate u_r and u_θ, and hence

$$t_r = -p_\infty + \tfrac{3}{2}\frac{\mu U}{a}\cos\theta, \qquad t_\theta = -\tfrac{3}{2}\frac{\mu U}{a}\sin\theta.$$

226 Very viscous flow

By symmetry we expect the net force on the sphere to be in the direction of the uniform stream, and the appropriate component of the stress vector is

$$t = t_r \cos\theta - t_\theta \sin\theta = -p_\infty \cos\theta + \tfrac{3}{2}\frac{\mu U}{a}.$$

The drag on the sphere is therefore

$$D = \int_0^{2\pi}\int_0^\pi ta^2 \sin\theta \, d\theta \, d\phi = 6\pi\mu Ua. \tag{7.9}$$

Laboratory experiments confirm the approximate validity of this formula at low Reynolds number $R = Ua/\nu$. One such experiment involves dropping a steel ball into a pot of glycerine; the ball accelerates downwards until it reaches a terminal velocity U_T such that the viscous drag exactly balances the (buoyancy-reduced) effect of gravity:

$$6\pi\mu U_T a = \tfrac{4}{3}\pi a^3(\rho_{\text{sphere}} - \rho_{\text{fluid}})g.$$

Further considerations

The above theory, due to Stokes (1851), is not without its problems. Stokes himself knew that a similar analysis for 2-D flow past a circular cylinder does not work (Exercise 7.4). Later, in 1889, Whitehead attempted to improve on Stokes's theory for flow past a sphere by taking account of the $(\boldsymbol{u} \cdot \nabla)\boldsymbol{u}$ term as a small correction, but his 'correction' to the flow inevitably became unbounded as $r \to \infty$.

In 1910 Oseen identified the source of the difficulty. The basis for the neglect of the $(\boldsymbol{u} \cdot \nabla)\boldsymbol{u}$ term at low Reynolds number lies in eqn (2.6), where the ratio of $|(\boldsymbol{u} \cdot \nabla)\boldsymbol{u}|$ to $|\nu \nabla^2 \boldsymbol{u}|$ is estimated to be of order UL/ν. But L here denotes the characteristic length scale of the flow, i.e. a typical distance over which $\boldsymbol{u}$ changes by an amount of order U. Now, in the immediate vicinity of the sphere, L will be of order a, so if the Reynolds number based on the radius of the sphere $R = Ua/\nu$ is small, then the term $(\boldsymbol{u} \cdot \nabla)\boldsymbol{u}$ will certainly be negligible in that vicinity. The trouble is that the further we go from the sphere, the larger L becomes, for the flow becomes more and more uniform. Inevitably, then, sufficiently far from the sphere the neglect of

Very viscous flow

the $(\mathbf{u} \cdot \nabla)\mathbf{u}$ term becomes unjustified, and the basis for using eqn (7.3) as an approximation breaks down. Compare this with what often happens in flow problems at *high* Reynolds number; the viscous terms are small throughout most of the flow but inevitably become important in boundary layers, where velocity gradients are untypically high. Here the viscous terms are assumed to be *large*, but inevitably cease to dominate in regions of the flow where velocity gradients are untypically low.

Oseen provided an ingenious (partial) resolution of the difficulty, but it was not until 1957 that Proudman and Pearson thoroughly clarified the whole issue by using the method of *matched asymptotic expansions,* which subsequently proved to be one of the most effective techniques in theoretical fluid mechanics. We provide here only the briefest sketch of this work, hoping simply to convey some idea of what the 'matching' entails. For this purpose it is helpful to work with dimensionless variables

$$r' = r/a, \qquad \mathbf{u}' = \mathbf{u}/U, \qquad \Psi' = \Psi/Ua^2, \qquad (7.10)$$

based on the sphere radius a and the speed at infinity U. Then, dropping primes in what follows, substitution of eqn (7.6) into the full Navier–Stokes equations gives, on eliminating the pressure p:

$$E^2(E^2\Psi) = \frac{R}{r^2 \sin\theta}\left(\frac{\partial\Psi}{\partial\theta}\frac{\partial}{\partial r} - \frac{\partial\Psi}{\partial r}\frac{\partial}{\partial\theta} + 2\cot\theta\frac{\partial\Psi}{\partial r} - \frac{2}{r}\frac{\partial\Psi}{\partial\theta}\right)E^2\Psi, \qquad (7.11)$$

where

$$R = Ua/\nu. \qquad (7.12)$$

It is emphasized that eqn (7.11) as it stands is exact; if we neglect the terms on the right-hand side, on the grounds that R is small, we recover eqn (7.7).

Proudman and Pearson obtained the solution to eqn (7.11) in two parts. Near the sphere they found

$$\Psi \doteq \tfrac{1}{4}(r-1)^2 \sin^2\theta\left[(1+\tfrac{3}{8}R)\left(2+\frac{1}{r}\right) - \tfrac{3}{8}R\left(2+\frac{1}{r}+\frac{1}{r^2}\right)\cos\theta\right]. \qquad (7.13)$$

More precisely, this is the sum of the first two terms ($O(1)$ and

$O(R)$) in an asymptotic expansion for Ψ that is valid in the limit $R \to 0$ with r fixed. If we just take the very first term we have

$$\Psi \doteq \tfrac{1}{4}(r-1)^2\left(2 + \frac{1}{r}\right)\sin^2\theta = \tfrac{1}{4}\left(2r^2 + \frac{1}{r} - 3r\right)\sin^2\theta,$$

i.e. Stokes's solution (7.8). The more accurate representation in eqn (7.13) permits a more accurate calculation of the drag on the sphere:

$$D \doteq 6\pi\mu Ua(1 + \tfrac{3}{8}R).$$

Far away from the sphere, Proudman and Pearson found that

$$\Psi \doteq \tfrac{1}{2}\frac{r_*^2}{R^2}\sin^2\theta - \tfrac{3}{2}\frac{1}{R}(1 + \cos\theta)[1 - e^{-\frac{1}{2}r_*(1-\cos\theta)}], \quad (7.14)$$

where

$$r_* = Rr, \tag{7.15}$$

r of course denoting the dimensionless distance from the origin (i.e. r' in eqn (7.10)). More precisely, eqn (7.14) is the sum of the first two terms ($O(R^{-2})$ and $O(R^{-1})$) in an asymptotic expansion for Ψ that is valid in the limit $R \to 0$ *with r_* fixed*. Thus by 'far away' from the sphere we mean at a distance of order R^{-1} or greater as $R \to 0$.

The precise sense in which the two solutions (7.13) and (7.14) 'match' is as follows. Suppose we take eqn (7.13), rewrite it in terms of the scaled variable r_*, and then expand the result for small R, keeping r_* fixed. We obtain

$$\Psi \doteq \frac{r_*^2}{4R^2}\sin^2\theta\left[2 + \tfrac{3}{4}R(1 - \cos\theta) - \frac{3R}{r_*}\right],$$

on keeping just the first two terms ($O(R^{-2})$ and $O(R^{-1})$). By the same token—and the symmetry here is to be noted—we take eqn (7.14), rewrite it in terms of the original (but dimensionless) variable r, and then expand the result for small R, keeping r fixed. The result is

$$\Psi \doteq \frac{r^2}{4}\sin^2\theta\left[2 + \tfrac{3}{4}R(1 - \cos\theta) - \frac{3}{r}\right],$$

again on keeping just the first two terms. In view of eqn (7.15) the two expressions just obtained are identical. This gives some

7.3. Corner eddies

We have already seen in Fig. 7.1(a) an example of 2-D slow flow past a symmetric obstacle in which eddies occur symmetrically fore and aft of the body. Another example, of a uniform shear flow over a ridge in the form of a circular arc, is shown in Fig. 7.3. Why do these low Reynolds number eddies occur?

The answer appears to lie in the *corners*; if the internal angle is not too small, i.e. marginally less than 180°, as in Fig. 7.3(a), then a simple flow in and out of each corner is possible, but if the corner angle falls below 146.3°, as in Fig. 7.3(b) (where it is 90°), then a simple flow of that kind is not possible, and corner eddies occur instead. Indeed, as we probe deeper and deeper into each corner we find, in theory, not just one eddy but an infinite sequence of nested, alternately rotating eddies. The scale in Fig. 7.3(b) is too small to show more than the first of each sequence; we sketch in Fig. 7.4 an example where two eddies may be seen. The flow is driven by the rotating cylinder on the right; the Reynolds number based on the peripheral speed of the cylinder and the length of the wedge is 0.17. Theoretically, each eddy is 1000 times weaker than the next; even with a 90-minute exposure time the experiment in Fig. 7.4 (by Taneda 1979) failed to detect the third eddy.

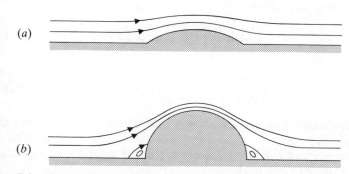

Fig. 7.3. Simple shear flow over circular bumps at low Reynolds number (after Higdon 1985).

230 *Very viscous flow*

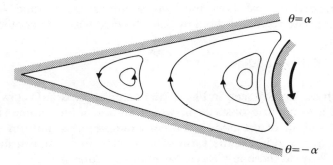

Fig. 7.4. Corner eddies (after Taneda 1979).

Some quite elementary theoretical considerations go a long way in this problem (Moffatt 1964). First we employ a stream function representation

$$u = \partial\psi/\partial y, \qquad v = -\partial\psi/\partial x,$$

so that

$$\nabla \wedge \boldsymbol{u} = (0, 0, -\nabla^2\psi).$$

On taking the curl of the slow flow equation (7.3) we then obtain the biharmonic equation

$$\nabla^2(\nabla^2\psi) = 0. \tag{7.16}$$

To tackle flows such as that in Fig. 7.4 it is convenient to use cylindrical polar coordinates, in which case

$$u_r = \frac{1}{r}\frac{\partial\psi}{\partial\theta}, \qquad u_\theta = -\frac{\partial\psi}{\partial r}, \tag{7.17}$$

and

$$\left(\frac{\partial^2}{\partial r^2} + \frac{1}{r}\frac{\partial}{\partial r} + \frac{1}{r^2}\frac{\partial^2}{\partial \theta^2}\right)^2 \psi = 0. \tag{7.18}$$

Now, the homogeneous way in which r occurs in the differential operator suggests a class of elementary solutions of the form $r^\lambda f(\theta)$, as occurs in the separable solutions of Laplace's equation, and this leads to

$$\psi = r^\lambda[A\cos\lambda\theta + B\sin\lambda\theta + C\cos(\lambda-2)\theta + D\sin(\lambda-2)\theta].$$

For flows such as that in Fig. 7.4 we want u_r to change sign when θ changes sign. As $u_r = r^{-1} \partial \psi / \partial \theta$ we therefore choose $B = D = 0$ in the above expression and concentrate on

$$\psi = r^\lambda [A \cos \lambda \theta + C \cos(\lambda - 2)\theta].$$

This is a stream function satisfying eqn (7.18) for any value of λ, but the boundary conditions $u_r = u_\theta = 0$ on $\theta = \pm \alpha$ demand that

$$A \cos \lambda \alpha + C \cos(\lambda - 2)\alpha = 0,$$

$$A\lambda \sin \lambda \alpha + C(\lambda - 2)\sin(\lambda - 2)\alpha = 0,$$

and these imply that

$$\lambda \tan \lambda \alpha = (\lambda - 2)\tan(\lambda - 2)\alpha.$$

With a little manipulation this may instead be written in the form

$$\frac{\sin x}{x} = -\frac{\sin 2\alpha}{2\alpha}, \qquad (7.19)$$

where x denotes $2(\lambda - 1)\alpha$. Given α, this particular form allows us to easily extract information about the roots λ.

The main issue is whether or not there are real roots λ. For, consider u_θ as a function of r on the centre line $\theta = 0$. It varies with r essentially as $r^{\lambda - 1}$. If λ is real and greater than unity then u_θ is zero at $r = 0$ and of one sign for $r > 0$, so the flow is of the simple form shown in Fig. 7.3(a), in and out of the corner. But if $\lambda = p + iq$ the solutions $r^\lambda f(\theta)$ will be complex, and as eqn (7.18) is linear their real and imaginary parts will individually satisfy the equation. If we look at the real part, then, we find on the centreline $\theta = 0$ that

$$u_\theta = \mathscr{R}[cr^{\lambda - 1}] = \mathscr{R}[cr^{p-1+iq}] = \mathscr{R}[cr^{p-1}e^{iq \log r}],$$

where c is some complex constant. Thus u_θ will be of the form

$$\mathscr{A} r^{p-1} \cos(q \log r + \varepsilon),$$

where $\mathscr{A}$ and ε are real. Clearly u_θ now changes sign with r; indeed, it changes sign with increasing rapidity as $r \to 0$, because $\log r \to -\infty$ as $r \to 0$. This behaviour is clearly indicative of the infinite sequence of eddies described above.

Returning to eqn (7.19), then, we plot $(\sin x)/x$ against x (Fig. 7.5). For given α we find any real roots $x = 2(\lambda - 1)\alpha$ as follows.

232 Very viscous flow

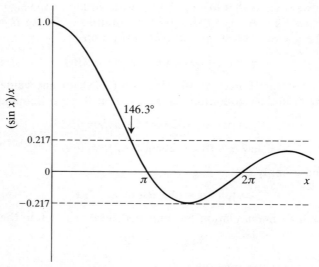

Fig. 7.5. Graph for determining the critical angle below which corner eddies occur.

Fix 2α such that $0 < 2\alpha < 2\pi$, and use the graph to read off the corresponding value of $\sin 2\alpha/2\alpha$; call it $\mathscr{B}$, say. Then use the graph again to find the value(s) of x for which $(\sin x)/x$ is $-\mathscr{B}$. This can always be done for the larger values of 2α in the range—and certainly when $2\alpha > \pi$—but there comes a point when this can no longer be done, and at that point $(\sin 2\alpha)/2\alpha$ is minus the value of $(\sin x)/x$ at the first (and deepest) minimum in Fig. 7.5, that value being -0.2172. The angle 2α in question thus turns out to be

$$2\alpha_C \doteqdot 146.3°; \tag{7.20}$$

for corner angles less than this λ is necessarily complex, and corner eddies occur.

The foregoing analysis is, of course, an entirely local one; we have paid scant attention to the mechanism (such as the roller in Fig. 7.4) that actually drives the flow, the hope being that, sufficiently far into the corner, this will not matter too much. Eddies certainly arise in all sorts of 2-D slow flows with sharp corners of angle less than 146.3° (Hasimoto and Sano 1980).

7.4. Uniqueness and reversibility of slow flows

Let there be viscous fluid in some region V which is bounded by a closed surface S. Let $\boldsymbol{u}$ be given as $\boldsymbol{u} = \boldsymbol{u}_B(\boldsymbol{x})$, say, on S. Then there is at most one solution of the slow flow equations (7.3) which satisfies that boundary condition.

To prove this, suppose there is another flow, $\boldsymbol{u}^*$, which also satisfies the slow flow equations (with corresponding pressure field p^*) and has $\boldsymbol{u}^* = \boldsymbol{u}_B(\boldsymbol{x})$ on S. Consider the 'difference flow' $\boldsymbol{v} = \boldsymbol{u}^* - \boldsymbol{u}$ and corresponding 'difference pressure' $P = p^* - p$. By hypothesis, $\boldsymbol{v}$ is not identically zero in V.

As the slow flow equations are linear we obtain, on subtraction,

$$0 = -\nabla P + \mu \nabla^2 \boldsymbol{v}, \qquad \nabla \cdot \boldsymbol{v} = 0,$$

with $\boldsymbol{v} = 0$ on S. In component form these equations become

$$0 = -\frac{\partial P}{\partial x_i} + \mu \frac{\partial^2 v_i}{\partial x_j^2}, \qquad \frac{\partial v_i}{\partial x_i} = 0,$$

where we are using the suffix notation and the summation convention (see, e.g., Bourne and Kendall 1977). Multiplying the first of these equations by v_i (which is equivalent to taking the dot product with $\boldsymbol{v}$) we obtain

$$0 = -\frac{\partial}{\partial x_i}(P v_i) + \mu v_i \frac{\partial^2 v_i}{\partial x_j^2},$$

because $\partial v_i / \partial x_i = 0$. Integrating over V and using the divergence theorem (A.13) gives

$$0 = -\int_S P v_i n_i \, dS + \mu \int_V v_i \frac{\partial^2 v_i}{\partial x_j^2} \, dV.$$

The first term vanishes, as $\boldsymbol{v} = 0$ on S. Thus

$$\mu \int_V \left[\frac{\partial}{\partial x_j} \left(v_i \frac{\partial v_i}{\partial x_j} \right) - \left(\frac{\partial v_i}{\partial x_j} \right)^2 \right] dV = 0.$$

Using the divergence theorem again:

$$\mu \int_S v_i \frac{\partial v_i}{\partial x_j} n_j \, dS - \mu \int_V \left(\frac{\partial v_i}{\partial x_j} \right)^2 dV = 0.$$

The first term again vanishes, as $\boldsymbol{v} = 0$ on S. Thus

$$\int_V (\partial v_i / \partial x_j)^2 \, dV = 0.$$

234 *Very viscous flow*

The integrand here consists of the sum of nine terms, because summation over both $i = 1, 2, 3$ and $j = 1, 2, 3$ is understood. Each one of these terms will be positive unless it is zero. To avoid violating the equation, then, $\partial v_i/\partial x_j$ must be zero for all i and j, so v is a constant. But v is zero on S, so v is identically zero in V. This contradicts the original hypothesis that u and u^* are different, and therefore that hypothesis is false. This completes the proof.

Reversibility

Let us take u_B to be some particular function $f_1(x)$ on S. Let the unique velocity field satisfying eqn (7.3) and the boundary condition be $u_1(x)$, and let $p_1(x)$ denote the corresponding pressure field, which is determined to within an inconsequential additive constant. Suppose we then change the boundary condition to $u_B = -f_1(x)$ instead. It is obvious by inspection of the slow flow equations (7.3) that $-u_1(x)$ constitutes *a* solution to this 'reversed' problem—the associated pressure field being $c - p_1(x)$, where c is a constant— but by invoking the uniqueness theorem we see it to be the *only* solution. Thus, *inasmuch as the slow flow equations hold, 'reversed' boundary conditions lead to reversed flow*.

This, then, is the explanation for the unusual behaviour in the concentric cylinder experiment of Fig. 2.6, though it has to be said that with more general boundary geometries it is typically the case that only *some* particles of a very viscous fluid return almost to their original position in this way (see the excellent photographs in Chaiken *et al.* 1986 and in Ottino 1989*a*). The reason that other particles do not is that their paths are extremely sensitive to tiny disturbances, and it is of course never possible in practice to *exactly* reverse the boundary conditions.

7.5. Swimming at low Reynolds number

One of the more exotic experiments in fluid dynamics involves a mechanical fish† (Fig. 7.6(*a*)). The fish consists of a cylindrical

---

† This experiment, and the one in Fig. 2.6, can be seen in the film *Low Reynolds Number Flows* by G. I. Taylor, one of an excellent series produced in the U.S.A. in the 1960s by the National Committee for Fluid Mechanics Films. (See Drazin and Reid 1981, p. 515 or Tritton 1988, p. 498 for further details.)

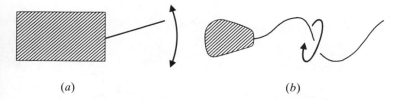

Fig. 7.6. (a) A mechanical fish. (b) A swimming spermatozoan.

body with a plane tail which flaps to and fro, powered by a battery. It swims happily in water but makes no progress whatsoever in corn syrup, the difference being that the Reynolds number is large in the first case but small in the second, so that the fish becomes a victim of the reversibility noted in §7.4. Loosely speaking, whatever is achieved by one flap of the tail is immediately undone by the 'return' flap.

This difficulty disappears if the plane tail is replaced by a rotating helical coil, as in Fig. 7.6(b), and the fish then swims in the syrup. Spermatozoa use this mechanism, sending helical waves down their tails. More generally, the trick in swimming at low Reynolds number is to do something which is not time-reversible (Childress 1981, pp. 16–21). The flapping of the tail in Fig. 7.6(a) is time-reversible, because if we film it, and run the film backwards, we see the same flapping as before, save for a half-cycle phase difference.

The swimming of a thin flexible sheet

A simple model for the ciliary propulsion of certain biological micro-organisms involves a thin extensible sheet which flexes itself in such a way that

$$x_s = x, \qquad y_s = a \sin(kx - \omega t), \tag{7.21}$$

where (x_s, y_s) denote the coordinates of any particle of the sheet (Fig. 7.7). A wave therefore travels down the sheet with speed $c = \omega/k$ while, in this particular example, the particles of the sheet move in the y-direction only, with velocity $dy_s/dt = -\omega a \cos(kx - \omega t)$. Such a flexing motion is not time-reversible ('running the film backwards' would result in the wave travelling in the opposite direction), and in the case when a/λ is small, $\lambda = 2\pi/k$ being the wavelength, we shall demonstrate that the

Fig. 7.7. The mean flow generated, at low Reynolds number, by a flexible sheet.

oscillatory flexing of the sheet induces not only an oscillatory flow, but also a steady flow component

$$U = 2\pi^2 (a/\lambda)^2 c \qquad (7.22)$$

in the x-direction. Viewed from a different frame, then, the sheet swims to the left, at speed U, through fluid which is, on average, at rest.

We first introduce a stream function ψ such that

$$u = \partial \psi / \partial y, \qquad v = -\partial \psi / \partial x, \qquad (7.23)$$

and need to solve the slow flow equation

$$\left(\frac{\partial^2}{\partial x^2} + \frac{\partial^2}{\partial y^2} \right)^2 \psi = 0 \qquad (7.24)$$

(see eqn (7.16)) subject to the condition $\boldsymbol{u} = \boldsymbol{u}_s$ on the sheet:

$$\left. \begin{array}{l} \partial \psi / \partial y = 0 \\ \partial \psi / \partial x = \omega a \cos(kx - \omega t) \end{array} \right\} \quad \text{on } y = a \sin(kx - \omega t), \quad (7.25)$$

together with suitable conditions as $y \to \infty$. Now, t appears only in the boundary conditions as a parameter. For convenience we solve the problem at $t = 0$; the flow at any other time can be obtained simply by replacing kx in our solution by $kx - \omega t$.

It is convenient to introduce non-dimensional variables

$$x' = kx, \qquad y' = ky, \qquad \psi' = k\psi/\omega a, \qquad (7.26)$$

and if we make these substitutions in eqns (7.24) and (7.25), *and then drop primes to simplify the notation,* we have

$$\left(\frac{\partial^2}{\partial x^2}+\frac{\partial^2}{\partial y^2}\right)^2 \psi = 0, \tag{7.27}$$

with

$$\left.\begin{array}{l}\partial\psi/\partial y = 0 \\ \partial\psi/\partial x = \cos x\end{array}\right\} \quad \text{on } y = \varepsilon \sin x \tag{7.28}$$

as our non-dimensional formulation of the problem, where

$$\varepsilon = ka. \tag{7.29}$$

We now make the assumption that ε is small, and expand $\partial\psi/\partial y$ and $\partial\psi/\partial x$ in eqn (7.28) in a Taylor series about $y = 0$:

$$\begin{aligned}\left.\frac{\partial\psi}{\partial y}\right|_{y=0} + \varepsilon \sin x \left.\frac{\partial^2\psi}{\partial y^2}\right|_{y=0} + \ldots &= 0, \\ \left.\frac{\partial\psi}{\partial x}\right|_{y=0} + \varepsilon \sin x \left.\frac{\partial^2\psi}{\partial y \partial x}\right|_{y=0} + \ldots &= \cos x.\end{aligned} \tag{7.30}$$

Next we seek a solution in powers of ε:

$$\psi = \psi_1 + \varepsilon\psi_2 + \varepsilon^2\psi_3 \ldots, \tag{7.31}$$

where the ψ_n are independent of ε. By substituting eqn (7.31) into eqn (7.27) and the boundary conditions (7.30) and equating coefficients of successive powers of ε to zero, we obtain a succession of problems for the ψ_n, each depending on the solutions to the earlier ones. Thus the problem for ψ_1 is

$$\left(\frac{\partial^2}{\partial x^2}+\frac{\partial^2}{\partial y^2}\right)^2 \psi_1 = 0,$$
$$\partial\psi_1/\partial y = 0, \quad \partial\psi_1/\partial x = \cos x \quad \text{on } y = 0, \tag{7.32}$$

the problem for ψ_2 is

$$\left(\frac{\partial^2}{\partial x^2}+\frac{\partial^2}{\partial y^2}\right)^2 \psi_2 = 0,$$
$$\frac{\partial\psi_2}{\partial y} + \frac{\partial^2\psi_1}{\partial y^2}\sin x = 0, \quad \frac{\partial\psi_2}{\partial x} + \frac{\partial^2\psi_1}{\partial y \partial x}\sin x = 0 \quad \text{on } y = 0, \tag{7.33}$$

and so on. As far as the first problem is concerned, solutions of

238 *Very viscous flow*

the biharmonic equation with the correct x-dependence are

$$\psi_1 = [(A + By)e^{-y} + (C + Dy)e^{y}]\sin x,$$

but we must have $C = D = 0$ in order that the velocity be bounded as $y \to \infty$, and the boundary conditions then give

$$\psi_1 = (1 + y)e^{-y} \sin x. \tag{7.34}$$

Turning to the problem for ψ_2, the boundary conditions (7.33) become

$$\partial \psi_2 / \partial y = \sin^2 x, \qquad \partial \psi_2 / \partial x = 0 \quad \text{on } y = 0. \tag{7.35}$$

We rewrite $\sin^2 x$ as $\frac{1}{2}(1 - \cos 2x)$, which forces not only a contribution $(E + Fy)e^{-2y} \cos 2x$ but also a contribution *independent of x*. The most general solution of the biharmonic equation which is a function of y alone is $Ay^3 + By^2 + Cy + D$, and in order that the velocity be bounded as $y \to \infty$ we must have $A = B = 0$. The additive constant D is of no significance and may be set equal to zero, and on adjusting E, F, and C to fit the boundary conditions (7.35) we obtain

$$\psi_2 = \tfrac{1}{2} y - \tfrac{1}{2} y e^{-2y} \cos 2x. \tag{7.36}$$

Combining eqns (7.31), (7.34), and (7.36):

$$\partial \psi / \partial y = -y e^{-y} \sin x + \varepsilon [\tfrac{1}{2} + (y - \tfrac{1}{2}) e^{-2y} \cos 2x] + \ldots, \tag{7.37}$$

but we need to remember that all variables here should really have primes (which were dropped), and on turning back to eqn (7.26) we find that the actual, dimensional, horizontal flow velocity is therefore

$$u = \partial \psi / \partial y = -\varepsilon \omega y e^{-ky} \sin(kx - \omega t) + \varepsilon^2 c [\tfrac{1}{2} + (ky - \tfrac{1}{2}) e^{-2ky} \cos 2(kx - \omega t)] + \ldots. \tag{7.38}$$

The steady term, $\tfrac{1}{2} \varepsilon^2 c$, is precisely eqn (7.22).

7.6. Flow in a thin film

Let viscous fluid be in steady flow between two rigid boundaries $z = 0$ and $z = h(x, y)$. Let U be a typical horizontal flow speed and let L be a typical horizontal length scale of the flow. Suppose, in addition, that

$$h \ll L. \tag{7.39}$$

Very viscous flow 239

Now, the no-slip condition must be satisfied at $z = 0$ and $z = h$, so u will change by an amount of order U over a z-distance of order h. Thus $\partial u/\partial z$ will be of order U/h, and likewise $\partial^2 u/\partial z^2$ will be of order U/h^2. The horizontal gradients of u, on the other hand, will be much weaker; $\partial u/\partial x$ will be of order U/L and $\partial^2 u/\partial x^2$ will be of order U/L^2. In view of eqn (7.39), then, the viscous term in the equation of motion (7.1) may be well approximated as follows:

$$\nu \nabla^2 u \doteq \nu \frac{\partial^2 u}{\partial z^2}.$$

We now ask in what circumstances this term greatly exceeds the term $(u \cdot \nabla)u$ in eqn (7.1). Order of magnitude estimates of the components of the two terms are as follows:

$$(u \cdot \nabla)u \sim \frac{U^2}{L}\left(1, 1, \frac{h}{L}\right),$$

$$\nu \frac{\partial^2 u}{\partial z^2} \sim \frac{\nu U}{h^2}\left(1, 1, \frac{h}{L}\right),$$

the z-components being smaller than the others because the incompressibility condition

$$\frac{\partial u}{\partial x} + \frac{\partial v}{\partial y} + \frac{\partial w}{\partial z} = 0$$

implies that $\partial w/\partial z$ is of order U/L and hence that w is of order Uh/L. These estimates show that the term $(u \cdot \nabla)u$ may be neglected if

$$\frac{UL}{\nu}\left(\frac{h}{L}\right)^2 \ll 1. \tag{7.40}$$

This, together with eqn (7.39), forms the basis of *thin film theory*, which will occupy the remainder of this chapter. We note, in particular, how the conventional Reynolds number UL/ν need not be small. Indeed, UL/ν is often quite large in practice; the condition (7.40) can still be met, so that viscous forces predominate, provided that h/L is small enough.

The reduction of the Navier–Stokes equations under eqns (7.39) and (7.40) is dramatic; with the term $(u \cdot \nabla)u$ absent and

240 Very viscous flow

the term $\nu \nabla^2 \mathbf{u}$ greatly simplified, the equations become, in the absence of body forces:

$$\frac{\partial p}{\partial x} = \mu \frac{\partial^2 u}{\partial z^2}, \qquad \frac{\partial p}{\partial y} = \mu \frac{\partial^2 v}{\partial z^2}, \qquad \frac{\partial p}{\partial z} = \mu \frac{\partial^2 w}{\partial z^2},$$

$$\frac{\partial u}{\partial x} + \frac{\partial v}{\partial y} + \frac{\partial w}{\partial z} = 0. \tag{7.41}$$

Furthermore, because w is smaller than the horizontal flow speed by a factor of order h/L, it follows from these equations that $\partial p/\partial z$ is much smaller than the horizontal pressure gradients. Thus p is, to a first approximation, a function of x and y alone. This means that the first two equations may be trivially integrated with respect to z (a most unusual circumstance) to give

$$u = \frac{1}{2\mu} \frac{\partial p}{\partial x} z^2 + Az + B,$$

$$v = \frac{1}{2\mu} \frac{\partial p}{\partial y} z^2 + Cz + D, \tag{7.42}$$

where $\partial p/\partial x$, $\partial p/\partial y$, A, B, C, and D are all functions of x and y only.

A final point worth noting concerns the stress tensor

$$T_{ij} = -p\delta_{ij} + \mu\left(\frac{\partial u_i}{\partial x_j} + \frac{\partial u_j}{\partial x_i}\right). \tag{7.43}$$

We infer from eqn (7.41) that

$$p = O(\mu U L/h^2), \tag{7.44}$$

and note that the largest of the second group of terms in eqn (7.43) is of order $\mu U/h$. Thus in a thin-film flow ($h \ll L$)

$$T_{ij} \doteq -p\delta_{ij}, \tag{7.45}$$

and tangential stresses at a rigid boundary are small compared with normal stresses.

In the next few sections the simplifications of thin-film theory allow us to tackle some problems that would otherwise be quite formidable. While some of the studies are quite recent, we begin

Very viscous flow 241

with one of the oldest and most remarkable examples of thin-film flow, first investigated experimentally by Hele-Shaw in 1898.

7.7. Flow in a Hele-Shaw cell

Suppose that the upper and lower boundaries are both flat and parallel, so that h is a constant. Imagine fluid being driven in the gap between them, by horizontal pressure gradients, past cylindrical objects having the z-axis as a generator, as in Fig. 7.8.

Applying the no-slip condition at both $z = 0$ and $z = h$ we find from eqn (7.42) that

$$u = -\frac{1}{2\mu} \frac{\partial p}{\partial x} z(h - z),$$
$$v = -\frac{1}{2\mu} \frac{\partial p}{\partial y} z(h - z). \qquad (7.46)$$

The fact that p is a function of x and y only is most important here, for while the flow speed depends on z (being greatest mid-way between the planes) the ratio v/u does not. This means that the direction of the flow is independent of z, so the streamline pattern is independent of z. Furthermore—and most remarkably—eliminating p from eqn (7.46) gives

$$\frac{\partial v}{\partial x} - \frac{\partial u}{\partial y} = 0. \qquad (7.47)$$

Thus at any given z the flow past a cylinder of some cross-section will correspond to the 2-D *irrotational* flow past that cylinder (see Chapter 4). There is, however, one important distinction: the

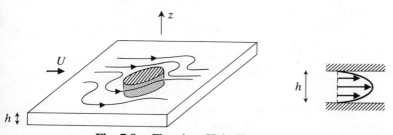

Fig. 7.8. Flow in a Hele-Shaw cell.

242 Very viscous flow

circulation Γ round *any* closed curve C lying in a horizontal plane, whether enclosing the cylinder or not, must be zero. This is because

$$\Gamma = \int_C u \, dx + v \, dy = -\frac{1}{2\mu} z(h-z) \int_C \frac{\partial p}{\partial x} dx + \frac{\partial p}{\partial y} dy$$

$$= -\frac{1}{2\mu} z(h-z)[p]_c, \tag{7.48}$$

and p is a single-valued function of position.

So, if we place a flat plate at an angle of attack α to the oncoming stream (i.e. $y = -x \tan \alpha$, $0 < x < L \cos \alpha$), then on looking down the z-axis the streamline pattern will appear exactly as in Fig. 4.6(a) and not as in Fig. 4.6(b). The fluid smoothly negotiates both sharp ends, and photographs of flows such as this really need to be seen to be believed. Some of the best, by D. H. Peregrine, are on pp. 8–10 of Van Dyke (1982), but Hele-Shaw's original photographs (1898) are well worth seeing, particularly as in three cases he puts his thin-film photographs side by side with those of the corresponding separated flow at high Reynolds number (see Fig. 7.9).

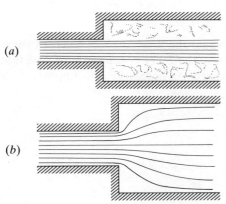

Fig. 7.9. Flow into a rectangular opening: (a) at high Reynolds number; (b) in a Hele-Shaw cell (as in Figs 13 and 14 of Hele-Shaw 1898).

7.8. An adhesive problem

It is a matter of common experience that it takes a large force F to pull a disc of radius a away from a rigid plane, if the two are separated by a thin film of viscous liquid (Fig. 7.10).

In view of the changing thickness $h(t)$ we anticipate an unsteady flow

$$u = u_r(r, z, t)e_r + u_z(r, z, t)e_z,$$

though we assume that the terms $(u \cdot \nabla)u$ and $\partial u/\partial t$ are both negligible in the equation of motion (2.3), so that the unsteadiness enters the problem only through the changing boundary conditions. (We shall verify this a posteriori.) We infer from eqn (7.41) that in the thin-film approximation

$$\frac{\partial p}{\partial r} = \mu \frac{\partial^2 u_r}{\partial z^2},$$

p being a function of r and t only. Integrating twice with respect to z, and applying the no-slip condition $u_r = 0$ at $z = 0$ and at $z = h(t)$, we obtain

$$u_r = \frac{1}{2\mu} \frac{\partial p}{\partial r} z(z - h).$$

The incompressibility condition $\nabla \cdot u = 0$ here takes the form

$$\frac{1}{r} \frac{\partial}{\partial r}(ru_r) + \frac{\partial u_z}{\partial z} = 0.$$

Substituting for u_r, integrating with respect to z, and applying

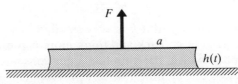

Fig. 7.10. Pulling a circular disc away from a rigid plane.

244 Very viscous flow

the boundary condition $u_z = 0$ on $z = 0$ gives

$$u_z = -\frac{1}{2\mu r}\frac{\partial}{\partial r}\left(r\frac{\partial p}{\partial r}\right)\left(\frac{z^3}{3} - \frac{hz^2}{2}\right).$$

The boundary condition $u_z = dh/dt$ at $z = h(t)$ then implies

$$\frac{\partial}{\partial r}\left(r\frac{\partial p}{\partial r}\right) = \frac{12\mu r}{h^3}\frac{dh}{dt}. \tag{7.49}$$

Integrating,

$$\frac{\partial p}{\partial r} = \frac{6\mu r}{h^3}\frac{dh}{dt} + \frac{C(t)}{r},$$

but we must choose $C(t) = 0$ to prevent a singularity at $r = 0$. A further integration gives

$$p = \frac{3\mu}{h^3}\frac{dh}{dt}r^2 + D(t).$$

Now, in view of eqn (7.45), we must have p equal to p_0, atmospheric pressure, at $r = a$, so

$$p - p_0 = \frac{3\mu}{h^3}\frac{dh}{dt}(r^2 - a^2).$$

Furthermore, the upward force exerted by the fluid on the disc is essentially

$$\int_0^{2\pi}\int_0^a (p - p_0)r\,dr\,d\theta = -\frac{3\pi}{2}\frac{\mu a^4}{h^3}\frac{dh}{dt}.$$

This is negative, of course, if $dh/dt > 0$; it then represents a suction force which makes the disc adhere to the plane. This force

$$\frac{3\pi}{2}\frac{\mu a^4}{h^3}\frac{dh}{dt} \tag{7.50}$$

is clearly very large indeed if h is very small.

Finally, we need to go back and think more carefully about the conditions under which the thin-film equations are valid. The given parameters at any time t in this problem are essentially a, h, dh/dt, and ν. The vertical velocity is of order dh/dt, and by virtue of $\nabla \cdot \mathbf{u} = 0$ the horizontal velocity is of order $ah^{-1}\,dh/dt$. Thus the conditions (7.39) and (7.40) are

$$h \ll a, \qquad h\frac{dh}{dt} \ll v. \tag{7.51}$$

We leave it as a short exercise to verify that the term $\partial \mathbf{u}/\partial t$ in eqn (2.3) is negligible in these same circumstances, as claimed above.

7.9. Thin-film flow down a slope

Consider the 2-D problem in which a layer of viscous fluid spreads down a slope, under gravity (Fig. 7.11). In the thin-film approximation

$$\begin{aligned}0 &= -\frac{1}{\rho}\frac{\partial p}{\partial x} + v\frac{\partial^2 u}{\partial z^2} + g\sin\alpha, \\ 0 &= -\frac{1}{\rho}\frac{\partial p}{\partial z} \qquad\quad - g\cos\alpha,\end{aligned} \tag{7.52}$$

and on integrating the second of these,

$$p = -\rho g z \cos\alpha + f(x, t).$$

On the free surface $z = h(x, t)$ the condition that the normal stress be equal to the atmospheric pressure p_0 reduces essentially to $p = p_0$, by virtue of eqn (7.45), so

$$p = \rho g [h(x, t) - z]\cos\alpha + p_0.$$

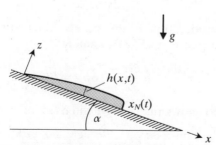

Fig. 7.11. Thin-film flow down a slope.

246 Very viscous flow

The condition that the tangential stress be zero at the free surface reduces, in the thin-film approximation, to

$$\mu \frac{\partial u}{\partial z} = 0 \qquad \text{on } z = h(x, t). \tag{7.53}$$

The equation of motion becomes

$$\nu \frac{\partial^2 u}{\partial z^2} = -g \sin \alpha + g \frac{\partial h}{\partial x} \cos \alpha. \tag{7.54}$$

Now, $\partial h/\partial x$ is small, by virtue of the thin-film approximation, so unless α is very small (or zero—see Exercise 7.13) the last term may be neglected, and

$$\nu \frac{\partial^2 u}{\partial z^2} = -g \sin \alpha.$$

This is easily integrated, and on applying eqn (7.53) together with the no-slip condition on $z = 0$ we find

$$u = \frac{g \sin \alpha}{\nu} (hz - \tfrac{1}{2} z^2). \tag{7.55}$$

The incompressibility condition now gives

$$\frac{\partial w}{\partial z} = -\frac{\partial u}{\partial x} = -\frac{g \sin \alpha}{\nu} \frac{\partial h}{\partial x} z,$$

and on integration and application of the boundary condition $w = 0$ at $z = 0$ we find

$$w = -\frac{g \sin \alpha}{2\nu} \frac{\partial h}{\partial x} z^2.$$

The final consideration is the purely kinematic condition at the free surface (see eqn (3.18)), namely

$$w = \frac{\partial h}{\partial t} + u \frac{\partial h}{\partial x} \qquad \text{on } z = h(x, t).$$

Now, eqn (7.55) shows that $u = gh^2 \sin \alpha / 2\nu$ on $z = h(x, t)$, so

$$-\frac{g \sin \alpha}{2\nu} \frac{\partial h}{\partial x} h^2 = \frac{\partial h}{\partial t} + \frac{gh^2 \sin \alpha}{2\nu} \frac{\partial h}{\partial x}.$$

The evolution equation for $h(x, t)$ is therefore

$$\frac{\partial h}{\partial t} + \frac{g \sin \alpha}{\nu} h^2 \frac{\partial h}{\partial x} = 0. \tag{7.56}$$

The solution of this equation is

$$h = f\left(x - \frac{g \sin \alpha}{\nu} h^2 t\right),$$

where f is an arbitrary function of a single variable, so any particular value of h propagates down the slope with speed $gh^2 \sin \alpha / \nu$. Larger values of h therefore travel faster (cf. finite-amplitude shallow-water wave theory in §3.9, especially Fig. 3.16).

Consider now the evolution of a finite 2-D blob of liquid, so that at any time t it occupies the region $0 < x < x_N(t)$, where $x_N(t)$ denotes the position of the 'nose' of the blob (Fig. 7.11). As larger values of h travel faster, the back of the blob will acquire a gentler slope as time goes on, while the front will steepen. Now, in practice, surface tension effects are important at the nose and tend to counteract such steepening. In fact, Huppert (1986) finds that nose effects can be largely ignored in determining the spreading of the blob as a whole. As time goes on, the main part of the blob approaches the following simple similarity solution of eqn (7.56):

$$h = \left(\frac{\nu}{g \sin \alpha}\right)^{\frac{1}{2}} \frac{x^{\frac{1}{2}}}{t^{\frac{1}{2}}}, \tag{7.57}$$

more or less regardless of the initial conditions (see Exercise 7.10). On coupling this with the condition that the volume of the blob as a whole must be conserved,

$$\int_0^{x_N(t)} h(x, t) \, dx = A, \tag{7.58}$$

we obtain

$$x_N(t) = \left(\frac{9A^2 g \sin \alpha}{4\nu}\right)^{\frac{1}{3}} t^{\frac{1}{3}} \tag{7.59}$$

as the expression for the eventual rate at which the blob spreads

248 *Very viscous flow*

down the slope, A denoting its cross-sectional area. Despite the neglect of effects in the vicinity of the nose, this expression agrees well with experiment (see Huppert 1986; Fig. 20).

7.10. Lubrication theory

When a solid body is in sliding contact with another, the frictional resistance is usually comparable in magnitude to the normal force between the two bodies. If, on the other hand, there is a thin film of fluid in between, the frictional resistance may be very small. The basis for this *lubrication theory* is that, as we have already observed, typical pressures in thin-film flow are of order $\mu UL/h^2$ (see eqn (7.44)), while tangential stresses are of order $\mu U/h$, and therefore smaller by a factor of order h/L.

Slider bearing

Consider the 2-D system in Fig. 7.12, where a rigid lower boundary $z = 0$ moves with velocity U past a stationary block of length L, the space between them being occupied by viscous fluid, the pressure being p_0 at both ends of the bearing.

The first stage of the familiar 'thin-film' approach of previous sections leads to

$$u(x, z) = \left[\frac{U}{h} - \frac{z}{2\mu}\frac{dp}{dx}\right](h - z), \tag{7.60}$$

p being a function of x only. Turning to the incompressibility condition, we may express it by asserting that the volume flux Q across all cross-sections of the film must be the same, i.e.

$$Q = \int_0^{h(x)} u \, dz = \tfrac{1}{2}Uh - \frac{h^3}{12\mu}\frac{dp}{dx} \tag{7.61}$$

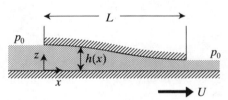

Fig. 7.12. The slider bearing.

must be independent of x. Rewriting this as an expression for dp/dx we may then integrate to obtain

$$\frac{p-p_0}{6\mu} = U \int_0^x \frac{1}{h^2(s)}\,ds - 2Q \int_0^x \frac{1}{h^3(s)}\,ds,$$

bearing in mind that $p = p_0$ at $x = 0$. But p is also equal to p_0 at $x = L$, and thus

$$Q = U \int_0^L \frac{1}{h^2(s)}\,ds \Big/ 2 \int_0^L \frac{1}{h^3(s)}\,ds. \qquad (7.62)$$

In the special case of a plane slider bearing, with $h(x)$ varying linearly between the values h_1 at $x = 0$ and h_2 at $x = L$, it turns out that $Q = Uh_1 h_2/(h_1 + h_2)$, and hence that

$$\frac{p-p_0}{6\mu UL} = \frac{(h_1 - h)(h_2 - h)}{(h_2^2 - h_1^2)h^2}. \qquad (7.63)$$

As $h(x)$ lies between h_1 and h_2 it is clear that p will be greater than p_0 throughout the film, so there will be a net upward force on the block to support a load, if $h_2 < h_1$, i.e. if the width of the lubricating layer decreases in the direction of flow, as in Fig. 7.12.

Flow between eccentric rotating cylinders

A related problem involves flow in the narrow gap between a fixed outer cylinder $r = a(1 + \varepsilon)$ and a slightly smaller, off-set, inner cylinder of radius a which rotates with peripheral velocity U (see Fig. 7.13). This is a simple model for an axle rotating in its housing.

Some elementary geometry shows that the width of the gap between the two (circular) cylinders is approximately

$$h(\theta) = a\varepsilon(1 - \lambda \cos \theta). \qquad (7.64)$$

The small parameter ε acts as a measure of the smallness of the gap, while the parameter λ may be taken between 0 and 1 and acts as a measure of the eccentricity of the two cylinders. With $\lambda = \frac{1}{2}$, for example, the gap is substantially smaller at $\theta = 0$ than at $\theta = \pi$, as is the case in Fig. 7.13. With $\lambda = 1$ the cylinders touch at $\theta = 0$; with $\lambda = 0$ they are coaxial.

250 *Very viscous flow*

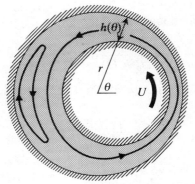

Fig. 7.13. Flow between eccentric rotating cylinders, with λ in excess of the critical value (7.70).

As the gap is small, curvature effects may essentially be neglected, and the analysis above for a plane slider bearing can be used simply by replacing x by $a\theta$. Thus eqn (7.60) converts directly into

$$u_\theta = \left[\frac{U}{h} - \frac{z}{2\mu a}\frac{\mathrm{d}p}{\mathrm{d}\theta}\right](h-z), \tag{7.65}$$

where z denotes distance across the gap, measured radially outwards from the inner cylinder. Likewise, eqn (7.61) converts to

$$Q = \tfrac{1}{2}Uh - \frac{h^3}{12\mu a}\frac{\mathrm{d}p}{\mathrm{d}\theta}. \tag{7.66}$$

As Q is constant this may be integrated to find p, and the condition that p be the same at $\theta = 0$ as at $\theta = 2\pi$ gives an expression equivalent to eqn (7.62), the integrals being between 0 and 2π. Knowing $h(\theta) = a\varepsilon(1 - \lambda\cos\theta)$, the two integrals may be evaluated (most easily by contour integration and the residue calculus), and the result is an explicit expression for Q:

$$Q = Ua\varepsilon\left(\frac{1-\lambda^2}{2+\lambda^2}\right). \tag{7.67}$$

A quantity of practical interest is the net force on, say, the inner cylinder. Now $h(\theta)$ is an even function of θ, so $\mathrm{d}p/\mathrm{d}\theta$ is also, by virtue of eqn (7.66), and p itself is therefore an odd function of θ. There is therefore no 'horizontal' force on the inner cylinder in Fig. 7.13; the 'upward' force is in fact

$$F = \frac{12\pi\mu U\lambda}{\varepsilon^2(1-\lambda^2)^{\frac{1}{2}}(2+\lambda^2)}. \tag{7.68}$$

The factor $(1-\lambda^2)^{\frac{1}{2}}$ in the denominator implies that the eccentricity parameter λ can, in principle, adjust itself so as to permit any external load on the inner cylinder, however large.

Returning to the flow itself, it is easy to show from eqns (7.66) and (7.67) that $dp/d\theta$ is positive at $\theta = \pi$, so that there is an adverse pressure gradient in some neighbourhood of that angle, and consequently the possibility of *reversed flow* near the stationary outer cylinder. That this does, indeed, occur can be seen by using eqns (7.65), (7.66), and (7.67) to calculate the velocity gradient $\partial u_\theta / \partial z$ on the outer cylinder:

$$\left.\frac{\partial u_\theta}{\partial z}\right|_{z=h} = \frac{2Ua\varepsilon\lambda}{h^2}\left[\frac{4\lambda^2-1}{\lambda(2+\lambda^2)} - \cos\theta\right]. \tag{7.69}$$

It is then a simple matter to show that if

$$\lambda > \tfrac{1}{2}(\sqrt{13}-3) \doteq 0.30 \tag{7.70}$$

there is a range of θ for which $\partial u_\theta/\partial z$ is positive at $z=h$, corresponding to reversed flow in the neighbourhood of the outer cylinder (see Fig. 7.13).

In practical lubrication theory this particular feature is overshadowed by other complications, but it is of some relevance to the arguments at the beginning of §8.6, and it has been clearly observed in experiments with very viscous fluids between offset rotating cylinders (see Chaiken *et al.* 1986, Fig. 3; also Aref 1986, Fig. 5 and Ottino 1989b, Fig. 7.4).

Exercises

7.1. Viscous fluid is contained between two rigid boundaries, $z=0$ and $z=h$. The lower plane is at rest, the upper plane rotates about a vertical axis with constant angular velocity Ω. The Reynolds number $R = \Omega h^2/\nu$ is small, so that the slow flow equations (7.3) provide a good approximation to the resulting flow. Use §2.4 to write these equations in

252 Very viscous flow

cylindrical polar coordinates, and show that they admit a purely rotary flow solution $\boldsymbol{u} = u_\theta(r, z)\boldsymbol{e}_\theta$ provided that

$$\left(\frac{\partial^2}{\partial r^2} + \frac{1}{r}\frac{\partial}{\partial r} - \frac{1}{r^2} + \frac{\partial^2}{\partial z^2}\right)u_\theta = 0. \tag{7.71}$$

Write down the boundary conditions which u_θ must satisfy at $z = 0$ and $z = h$. Hence seek a solution of the form $u_\theta = rf(z)$. Show that the θ-component of stress, t_θ, on the upper plane is $-\mu\Omega r/h$.

Suppose instead that both upper and lower boundaries are horizontal discs of radius a. If end effects are neglected, show that the external torque on the upper disc needed to sustain the flow is

$$\tau = \tfrac{1}{2}\pi\mu\Omega\frac{a^4}{h}.$$

7.2. A rigid sphere of radius a is immersed in an infinite expanse of viscous fluid. The sphere rotates with constant angular velocity Ω. The Reynolds number $R = \Omega a^2/\nu$ is small, so that the slow flow equations

$$\nabla p = -\mu\nabla \wedge (\nabla \wedge \boldsymbol{u}), \qquad \nabla \cdot \boldsymbol{u} = 0$$

apply (see eqns (7.3) and (6.12)). Using spherical polar coordinates (r, θ, ϕ) with $\theta = 0$ as the rotation axis, show that a purely rotary flow $\boldsymbol{u} = u_\phi(r, \theta)\boldsymbol{e}_\phi$ is possible provided that

$$\frac{\partial^2}{\partial r^2}(ru_\phi) + \frac{1}{r}\frac{\partial}{\partial \theta}\left[\frac{1}{\sin\theta}\frac{\partial}{\partial \theta}(u_\phi\sin\theta)\right] = 0. \tag{7.72}$$

(This is, of course, just eqn (7.71) written in terms of different coordinates; u_ϕ here means the same thing as u_θ in Exercise 7.1.)

Write down the boundary conditions which u_ϕ must satisfy at $r = a$ and as $r \to \infty$, and hence seek an appropriate solution to eqn (7.72), thus finding

$$u_\phi = \frac{\Omega a^3}{r^2}\sin\theta.$$

Show that the ϕ-component of stress on $r = a$ is $t_\phi = -3\mu\Omega\sin\theta$, and deduce that the torque needed to maintain the rotation of the sphere is

$$\tau = 8\pi\mu\Omega a^3.$$

[In practice, in both the above situations there will be a small *secondary circulation*, of order R, in addition to the rotary flow. In the case of Exercise 7.1 we have already seen that the full Navier–Stokes equations do not admit a purely rotary flow solution (Exercise 2.11).]

7.3. Consider uniform slow flow past a spherical bubble of radius a by modifying the analysis of §7.2 accordingly, i.e. by replacing the no-slip condition on $r = a$ by the condition of no tangential stress ($t_\theta = 0$) on $r = a$. Show, in particular, that

$$\Psi = \tfrac{1}{2}U(r^2 - ar)\sin^2\theta$$

and that the normal component of stress on $r = a$ is $t_r = 3\mu U a^{-1} \cos\theta$. Hence show that the drag on the bubble is

$$D = 4\pi\mu U a$$

in the direction of the free stream (cf. eqn (7.9)).

[A similar but rather more involved problem is the uniform slow flow past a spherical drop of different fluid, of viscosity $\bar{\mu}$, say. This involves solving the slow flow equations separately outside and inside $r = a$, with $u_r = 0$ at $r = a$, and tangential stresses continuous at $r = a$. The drag on the drop is

$$D = 4\pi\mu U a\left(\frac{\mu + \tfrac{3}{2}\bar{\mu}}{\mu + \bar{\mu}}\right);$$

the limit $\bar{\mu}/\mu \to 0$ gives the 'bubble' result, while the limit $\bar{\mu}/\mu \to \infty$ gives a drag identical to that for a rigid sphere (see eqn (7.9)).]

7.4. Consider uniform slow flow past a circular cylinder, and show that the problem reduces to

$$\left(\frac{\partial^2}{\partial r^2} + \frac{1}{r}\frac{\partial}{\partial r} + \frac{1}{r^2}\frac{\partial^2}{\partial \theta^2}\right)^2 \psi = 0,$$

with $\partial\psi/\partial r = \partial\psi/\partial\theta = 0$ on $r = a$ and

$$\psi \sim Ur\sin\theta \qquad \text{as } r \to \infty.$$

Show that seeking a solution of the form $\psi = f(r)\sin\theta$ leads to

$$\psi = \left[Ar^3 + Br\log r + Cr + \frac{D}{r}\right]\sin\theta \qquad (7.73)$$

and thus fails, in that for no choice of the arbitrary constants can all the boundary conditions be satisfied.

[There is, as stated in §7.2, no solution of *any* form to the problem as posed, but this takes rather more proving. Proudman and Pearson (1957) show that the equivalent dimensionless expression to eqn (7.13) in the neighbourhood of the cylinder $r = 1$ is

$$\psi \doteq \left[\left(r\log r - \tfrac{1}{2}r + \frac{1}{2r}\right)\sin\theta\right]\Big/\left[\log\!\left(\frac{8}{R}\right) - \gamma + \tfrac{1}{2}\right],$$

254 *Very viscous flow*

where γ is Euler's constant:
$$\gamma = \lim_{n\to\infty}\left[1 + \tfrac{1}{2} + \tfrac{1}{3}\ldots + \frac{1}{n} - \log n\right] \doteq 0.58.$$

Thus ψ is of the form (7.73) at moderate distances from the cylinder, but after applying the boundary conditions on the cylinder the remaining constants in eqn (7.73) cannot be obtained by appealing directly to the boundary condition at infinity; they have to be obtained by matching the partial solution so obtained to one valid far from the cylinder, as indicated in the text.]

7.5. Two infinite plates, $\theta = \pm\Omega t$, are hinged together at $r = 0$ and are moving apart with angular velocities $\pm\Omega$ as in Fig. 7.14. Between them the space $-\Omega t < \theta < \Omega t$, $0 < r < \infty$ is filled with viscous fluid. Write down the boundary conditions satisfied by u_r and u_θ at $\theta = \pm\Omega t$, and hence write down boundary conditions for the stream function $\psi(r, \theta, t)$.

Assuming that the slow flow equations apply, show that
$$\left(\frac{\partial^2}{\partial r^2} + \frac{1}{r}\frac{\partial}{\partial r} + \frac{1}{r^2}\frac{\partial^2}{\partial \theta^2}\right)^2 \psi = 0,$$

and solve this equation, subject to the boundary conditions, to obtain
$$\psi = -\tfrac{1}{2}\Omega r^2\left(\frac{\sin 2\theta - 2\theta \cos 2\Omega t}{\sin 2\Omega t - 2\Omega t \cos 2\Omega t}\right). \tag{7.74}$$

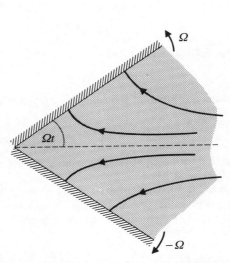

Fig. 7.14. Slow flow in an opening wedge.

Show that when $2\Omega t = \pi/2$ the instantaneous streamlines are rectangular hyperbolae.

Use the above expression for ψ to give rough order of magnitude estimates for the terms $\partial \boldsymbol{u}/\partial t$, $(\boldsymbol{u}\cdot\nabla)\boldsymbol{u}$ and $\nu\nabla^2\boldsymbol{u}$ in the full equations of motion. Deduce that the slow flow equations only provide a good approximation if

$$\Omega r^2/\nu \ll 1,$$

so that no matter how large the viscosity those equations must become invalid if r is sufficiently large.

[Note that eqn (7.74) blows up when the angle between the walls satisfies $\tan 2\Omega t = 2\Omega t$, i.e. when $2\Omega t \doteq 257.45°$. Moffatt and Duffy (1980) demonstrate that for greater angles than this the simple flow solution (7.74) will, in practice, be irrelevant, and the real flow will depend, *inter alia*, on the manner in which the plates are hinged in the neighbourhood of $r = 0$. They do this by investigating an ingenious composite problem which embraces both this exercise and Exercise 7.6.]

7.6. Consider 2-D flow into a convergent channel, $-\alpha < \theta < \alpha$, the fluid being extracted through a narrow gap between the walls at $r = 0$. Seek a solution to the slow flow equations and the boundary conditions in which the streamlines are purely radial. Show that

$$u_r = \frac{C}{r}(\cos 2\theta - \cos 2\alpha),$$

where C is a constant, and show that the slow flow equations are approximately valid if $C/\nu \ll 1$.

Deduce that for $2\alpha < 180°$ the flow is radially inward everywhere (if $C < 0$), but that for $2\alpha > 180°$ the flow is radially inward in some places and radially outward in others.

7.7. Consider the 2-D problem of flow in a corner, as in §7.3, but suppose that two counter-rotating rollers generate the flow far from the corner, so that it is appropriate instead to focus on solutions of eqn (7.18) in which u_r is an even function of θ, thus:

$$\psi = r^\lambda[B \sin \lambda\theta + D \sin(\lambda - 2)\theta].$$

Show that the boundary conditions imply, in place of eqn (7.19):

$$\frac{\sin x}{x} = \frac{\sin 2\alpha}{2\alpha},$$

where x denotes $2(\lambda - 1)\alpha$. Use Fig. 7.5 to deduce that corner eddies occur if $2\alpha < 159°$.

7.8. Suppose that the extensible sheet in §7.5 is engaged instead in a worm-like squirming motion, so that in place of eqn (7.21):

$$x_s = x_0 + a \sin(kx_0 - \omega t), \qquad y_s = 0,$$

where x_0 is the mean position of any particular particle of the sheet. Show that this induces a steady flow component

$$U = -2\pi^2 (a/\lambda)^2 c$$

in the x-direction, i.e. in the opposite direction to eqn (7.22).

[Examples exist among micro-organisms of both type of propulsion (Childress 1981, p. 67), in opposite senses for a given direction of the body wave, just as slow-flow theory predicts.]

7.9. Viscous fluid occupies the 2-D region $0 < r < a$, $0 < \theta < \alpha(t)$ between two flat plates, $\theta = 0$ and $\theta = \alpha(t)$, the plates being hinged along the line $r = 0$. If α is very small, use thin-film theory to show that

$$\frac{d}{dr}\left(r^3 \frac{dp}{dr}\right) = \frac{12\mu r}{\alpha^3} \frac{d\alpha}{dt},$$

and deduce that the fluid exerts a torque

$$\tfrac{3}{2} \frac{\mu a^2}{\alpha^3} \frac{d\alpha}{dt}$$

on the upper plate, in a sense such as to oppose opening of the gap.

7.10. Verify that eqn (7.57) is a solution of eqn (7.56):

$$\frac{\partial h}{\partial t} + \frac{g \sin \alpha}{\nu} h^2 \frac{\partial h}{\partial x} = 0.$$

Disregarding effects at the nose, obtain an exact expression for $h(x, t)$ when, at $t = 0$, $h = \beta x$ for $0 < x < L$, where β is a positive constant. Then show that $h(x, t)$ takes the form (7.57) as $t \to \infty$.

7.11. A plate is drawn out of a bath and held vertically to allow the thin film of liquid to drain off. If the film is initially of uniform thickness h_0 show that

$$h(x, t) = \left(\frac{\nu}{g}\right)^{\frac{1}{2}} \frac{x^{\frac{1}{2}}}{t^{\frac{1}{2}}} \qquad \text{for } 0 < x < \frac{gh_0^2}{\nu} t,$$

$$= h_0 \qquad \text{for } x > \frac{gh_0^2}{\nu} t,$$

where x denotes distance downward from the upper edge of the plate.

7.12. Consider the axisymmetric version of the problem in §7.9, i.e. thin-film flow down the outside of a cone, its vertex upward. Let x denote distance down the surface from the vertex, let z denote the coordinate normal to the surface of the cone, and let u and w denote the corresponding velocity components. Then, according to thin-film theory, the only modification to the 2-D analysis of §7.9 is in respect of conservation of volume, i.e. the incompressibility condition. Explain, in particular, why $\partial u/\partial x + \partial w/\partial z = 0$ is replaced by

$$\frac{1}{x}\frac{\partial}{\partial x}(xu) + \frac{\partial w}{\partial z} = 0$$

in the thin-film approximation.

Show that in place of eqn (7.56) we have

$$\frac{\partial h}{\partial t} + \frac{g \sin \alpha}{3vx}\frac{\partial}{\partial x}(xh^3) = 0,$$

where α is the angle made with the horizontal by a line of steepest descent down the cone. By following the approach of §7.9 deduce that the nose of a fixed amount of fluid eventually progresses down the cone at a rate proportional to $t^{\frac{1}{5}}$.

[I am grateful to Professor H. E. Huppert for pointing out a serious error in the original version of this problem.]

7.13. Modify the analysis of §7.9 to deal with the 2-D spreading of a thin layer of viscous fluid on a *horizontal* plate, and show that the height $h(x, t)$ of the free surface evolves according to

$$\frac{\partial h}{\partial t} = \frac{g}{3v}\frac{\partial}{\partial x}\left(h^3 \frac{\partial h}{\partial x}\right). \tag{7.75}$$

Show, too, that the equivalent axisymmetric problem, for the spreading of a thin circular drop on a horizontal plate, gives rise to the equation

$$\frac{\partial h}{\partial t} = \frac{g}{3v}\frac{1}{r}\frac{\partial}{\partial r}\left(rh^3 \frac{\partial h}{\partial r}\right). \tag{7.76}$$

[Solving either of these subject to initial conditions is more difficult than solving eqn (7.56), but some remarkable 'waiting-time' solutions have recently been found, in which a viscous drop spends some time adjusting its shape before increasing its area of contact with the plate (Tayler 1986, pp. 193–197).]

7.14. Consider the large-time axisymmetric spreading of a thin viscous drop on a horizontal surface by looking for a similarity solution to eqn

258 Very viscous flow

(7.76) of the form

$$h(r, t) = f(t)F(\eta), \qquad \text{where } \eta = r/k(t).$$

Let $r_N(t)$ denote the radius of the drop at time t, and let $\eta_N = r_N(t)/k(t)$. In contrast to flow down a slope, effects at the nose are most important here, so we insist that h is zero at $r = r_N(t)$, i.e.

$$F(\eta_N) = 0.$$

Consequently, η_N is a constant. Use this fact, together with the fact that the volume V of the drop must remain constant, to show that $f(t)$ is proportional to the inverse square of $k(t)$. By substitution in eqn (7.76) deduce that $k(t) = ct^{\frac{1}{8}}$, where c is a constant which is at our disposal.

Show that on choosing

$$k(t) = (gV^3/3\nu)^{\frac{1}{8}} t^{\frac{1}{8}}$$

the differential equation for $F(\eta)$ is

$$(\eta F^3 F')' + \tfrac{1}{8}(\eta^2 F' + 2\eta F) = 0.$$

Integrate this, subject to suitable conditions, to find $F(\eta)$ and hence determine

$$\eta_N = \left(\frac{2^{10}}{3^4 \pi^3}\right)^{\frac{1}{8}} \doteq 0.894,$$

so that

$$r_N(t) = 0.894 (gV^3/3\nu)^{\frac{1}{8}} t^{\frac{1}{8}}.$$

[The agreement with experiment is good (Huppert 1982; see in particular his Fig. 6). In a later paper, Huppert (1986) relates these results to the spreading of volcanic lava, and discusses several other interesting applications of fluid dynamics to geological problems.]

7.15. A viscous layer of small thickness $h(\theta)$ is on the outside of a circular cylinder which rotates with peripheral speed U about a horizontal axis (Fig. 7.1(b)). We wish to find the minimum value of U for which such a steady solution exists.

Explain why, according to thin-film theory,

$$\nu \frac{\partial^2 u}{\partial z^2} = g \cos \theta,$$

with

$$u = U \quad \text{on } z = 0; \qquad \partial u/\partial z = 0 \quad \text{on } z = h(\theta),$$

where z denotes distance normal to the cylinder, and $u(\theta, z)$ is the velocity of the fluid in the θ-direction. Solve the equation, and calculate the volume flux

$$Q = \int_0^{h(\theta)} u \, dz$$

across any section θ = constant. Explain why Q must be a constant, and show that

$$\frac{1}{H^2} - \frac{1}{H^3} = \frac{gQ^2}{3\nu U^3} \cos\theta,$$

where $H(\theta) = Uh(\theta)/Q$. Deduce that there is no satisfactory solution for $H(\theta)$ unless

$$gQ^2/3\nu U^3 \leq \tfrac{4}{27},$$

and show that when this inequality is satisfied, $h(\theta)$ is as in Fig. 7.1(b), i.e. largest when $\theta = 0$ and smallest when $\theta = \pi$.

In the limiting case, $H(\theta)$ satisfies

$$\frac{1}{H^2} - \frac{1}{H^3} = \tfrac{4}{27} \cos\theta,$$

and it is then found numerically that

$$\int_0^{2\pi} H(\theta) \, d\theta = 6.641.$$

Use this to show that a steady solution is possible only if

$$U \geq 2.014(g\bar{h}^2/\nu),$$

as claimed in eqn (7.5).

7.16. Work through the lubrication theory of §7.10, establishing eqns (7.63) and (7.70), and the earlier results on which they depend.

7.17. The Hele-Shaw flow (7.46) is, at any constant z, identical to the irrotational 2-D flow (without circulation) past the obstacle in question, and such a flow involves a certain slip velocity on the obstacle itself. Yet there can be no such slip, as the fluid is viscous, and this implies that eqn (7.46) cannot properly represent the flow in some region adjacent to the obstacle.

Give an order of magnitude estimate of the thickness of this (highly unusual) 'boundary layer' adjacent to the obstacle.

8 Boundary layers

8.1. Prandtl's paper

In August 1904 the Third International Congress of Mathematicians took place at Heidelberg, and Prandtl presented a paper 'On the motion of fluids of very small viscosity'. He was no great figure at the time, and was given just ten minutes in the programme. The published version of his paper (1905) has $7\frac{1}{2}$ pages of text and two pages of photographs.† At the heart of it Prandtl addresses the flow of a fluid of small viscosity past a solid body. He affirms that there is no slip on the boundary, or wall, of the body itself, but:

... if the viscosity is very small, and the path of the fluid along the wall not too long, then the velocity will assume its usual value very close to the wall. In the thin transition layer the sharp changes of velocity produce notable effects, despite the small coefficient of viscosity.

These problems are best tackled by making an approximation in the governing differential equation. If μ is taken to be of second order in smallness, then the thickness of the transition layer becomes of the first order in smallness, and so too do the normal components of velocity. The pressure difference across the layer may be neglected, as may be any bending of the streamlines. The pressure distribution of the free fluid will be impressed on the transition layer.

For the two-dimensional problem, with which I have hitherto been solely concerned, we get, at any particular position (x—tangential coordinate, y—normal coordinate, u and v the corresponding velocity components), the differential equation

$$\rho\left(u\frac{\partial u}{\partial x} + v\frac{\partial u}{\partial y}\right) = -\frac{dp}{dx} + \mu\frac{\partial^2 u}{\partial y^2},$$

to which must be added

$$\frac{\partial u}{\partial x} + \frac{\partial v}{\partial y} = 0.$$

---

† I would like to thank Mr J. F. Acheson for his help in translating this paper.

Let us assume that, as is usual, dp/dx is given as a function of x, and, further, that the velocity u is given as a function of y at some initial value of x. Then we can determine numerically, from each u, the associated $\partial u/\partial x$, and with one of the known algorithms we can then proceed, step by step, in the x-direction. A difficulty exists, however, in various singularities which appear on the fixed boundary. The simplest case of the flow situations under discussion is that of water streaming along a thin flat plate. Here a reduction in variables is possible; we can write $u = f(y/x^{\frac{1}{2}})$. By numerical integration of the resulting differential equation we obtain an expression for the drag

$$D = 1.1 \times b\sqrt{\mu\rho l u_0^3}$$

(b breadth, l length of the plate, u_0 velocity of the undisturbed water relative to the plate). The velocity profile is shown in [Fig. 8.1].

For practical purposes the most important result of these investigations is that in certain cases, and at a point wholly determined by the external conditions, *the flow separates from the wall* [Fig. 8.2]. A fluid layer which is set into rotation by friction at the wall thus pushes itself out into the free fluid where, in causing a complete transformation of the motion, it plays the same role as a Helmholtz surface of discontinuity. A change in the coefficient of viscosity μ produces a change in the thickness of the vortex layer (this thickness being proportional to $\sqrt{(\mu l/\rho u)}$), but everything else remains unchanged, so that one may, if one so wishes, take the limit $\mu \to 0$ and still obtain the same flow picture.

Separation can only occur if there is an increase in pressure along the wall in the direction of the stream . . .

The amount of insight packed into this part of Prandtl's paper is staggering, and much of the present chapter will be spent filling in the details, particularly with regard to the derivation of the

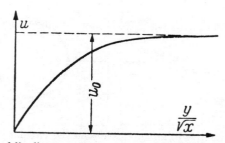

Fig. 8.1. Prandtl's diagram of the velocity profile in the boundary layer on a flat plate.

262 *Boundary layers*

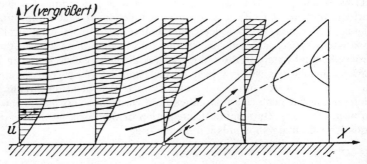

Fig. 8.2. The sketch of boundary-layer separation in Prandtl's 1905 paper.

boundary layer equations (§8.2) and their solution in the case of flow past a flat plate (§8.3).

After the passage quoted above, Prandtl emphasizes how the flow of a fluid of small viscosity must be dealt with in two interacting parts, namely an inviscid flow obeying Helmholtz's vortex theorems and thin boundary layers in which viscous effects are important. The motion in the boundary layers is regulated by the pressure gradient in the mainstream flow but, on the other hand, the character of the mainstream flow is, in turn, markedly influenced by any separation that may occur.

Prandtl goes on to discuss some particular cases, including the impulsively started motion of a circular cylinder (Fig. 8.3). He finally reports some experiments undertaken in a hand-operated

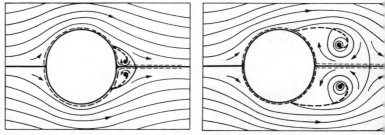

Fig. 8.3. Flow relative to an impulsively moved circular cylinder at two different times (from Prandtl 1905). Dashed lines indicate layers of strong vorticity.

Boundary layers 263

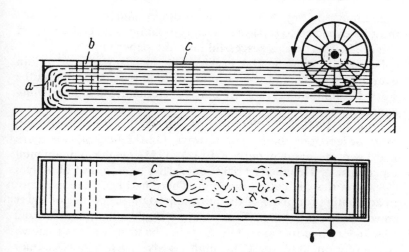

Fig. 8.4. Prandtl's hand-operated flow tank.

water tank (Fig. 8.4). These include flow past a wall, flow past a circular arc at zero incidence, and flow past a circular cylinder. In the last case he demonstrates that even a very small amount of suction into a slit on one side of the cylinder is enough to prevent separation of the boundary layer on that side (Fig. 8.5). He notes, too, a most interesting consequence of this, because 'the speed must decrease in the broadening aperture through which the water flows, and therefore the pressure must rise'. A substantial adverse pressure gradient will therefore be impressed on the boundary layer on the corresponding side wall of the tank

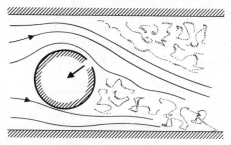

Fig. 8.5. Sketch of the final photograph in Prandtl's paper.

(uppermost in Fig. 8.5), and accordingly *that* boundary layer must be expected to separate. Such separation is indeed observed (Fig. 8.5), and on this successful note the paper ends.†

For all its fundamental insights, the paper was scarcely an overnight success, and several years were to pass before Prandtl's work became widely known outside Germany, let alone fully appreciated.

Prior to 1900, ideal flow theory and viscous flow theory had more or less gone their separate ways. On the inviscid side there had been the great papers of Euler (1755) on the fundamental equations, of Helmholtz (1858) on vortex motion and of Kelvin (1869) on the circulation theorem. There had been success, too, in accounting for many of the most important properties of water waves and sound waves. There was no doubt, then, that inviscid flow theory had its value. On the other hand, it was well known that uniform flow past a 'bluff' body—such as a circular cylinder—bore little resemblance *at the rear of the body* to the predictions of inviscid flow theory.

Viscous flow theory effectively began with the pioneering paper of Stokes (1845), who not only laid down the equations of motion but obtained many of the elementary exact solutions that are to be found in Chapter 2. He followed this with another important paper (1851) on what we would now call low Reynolds number flow (§7.2), and when Hele-Shaw performed his remarkable experiments with irrotational streamline patterns (1898; see §7.7) it was Stokes again who produced the relevant viscous theory. The other pioneering work of the time was by Reynolds, notably his beautiful experiments in 1883 on transition in flow down a pipe (§9.1).

Yet a major problem remained: that of accounting for the motion of a fluid of small viscosity past a solid body. Prandtl was not alone, of course, in addressing the matter. As early as 1872 Froude had conducted experiments on the drag on a thin flat plate towed through still water, and had attributed that drag to

---

† Prandtl's paper is not exclusively concerned with boundary layers; its subject is the motion of a fluid with very small viscosity. He points out, quite early in the paper, a wholly different way in which small viscosity can be significant, namely through its cumulative effect, over a long time interval, in a region of closed streamlines (see §5.10).

the layers of fluid in intense shear near the plate. He had found, too, that the drag varied not in proportion to the length l of the plate, but at a slower rate. Lanchester later proposed, independently of Prandtl, that the drag should be proportional to $\mu^{1/2}u_0^{3/2}$. He also discussed separation, and affirmed correctly that on a rotating cylinder in a uniform stream separation would be delayed on one side and hastened on the other. He published all this and much more, in his *Aerodynamics* of 1907, although just how much earlier the work was done is not entirely clear.

If, nigh on a hundred years later, the concept of a boundary layer and its separation seem to have been a long time in the making, it is worth recalling that there were at least two factors which clouded the issue at the time. First, there was substantial uncertainty about whether the correct boundary condition was one of no slip. It is one thing to find Stokes unsure about the matter on pp. 96–99 of his 1845 paper, but it is quite another to find this uncertainty continuing right up to the turn of the century, with some investigators convinced of the no-slip condition only in the case of slow flow (see Goldstein 1969, and pp. 676–680 of Goldstein 1938). Second, it was known that when ideal flow theory predicts a negative value of the absolute pressure at any point in a liquid, the formation of bubbles of vapour, known as *cavitation,* may be expected. Thus when irrotational flow past sharp corners (e.g. Fig. 4.6(*a*)) bore little resemblance to the actual flow of real liquids such as water, there seemed to be a ready explanation: ideal flow theory implies an infinite speed at the corner, and by Bernoulli's theorem this means an infinitely negative pressure. The onset of cavitation will prevent such a singularity occurring but, in so doing, will give rise to a different and 'separated' flow (see, e.g., Batchelor 1967, pp. 497–506). In an essay on pp. 1–5 of Rosenhead (1963), Lighthill emphasizes how this obscured the possibility that there might be a quite different (viscous) explanation for flow separation, one which would obtain, indeed, for liquids or gases and whether the rigid boundary was sharp-cornered or smooth. It was this quite different explanation, along with so much else, that Prandtl was eventually to squeeze into just a few pages in 1904.

8.2. The steady 2-D boundary layer equations

We now derive the equations for a steady 2-D boundary layer adjacent to a rigid wall $y = 0$:

$$u \frac{\partial u}{\partial x} + v \frac{\partial u}{\partial y} = -\frac{1}{\rho}\frac{dp}{dx} + \nu \frac{\partial^2 u}{\partial y^2}, \qquad (8.1)$$

$$\frac{\partial u}{\partial x} + \frac{\partial v}{\partial y} = 0, \qquad (8.2)$$

$p(x)$ being a function of x alone. The boundary conditions at the wall are

$$u = v = 0 \quad \text{at } y = 0, \qquad (8.3)$$

if the wall is at rest. The boundary layer flow must also match with the inviscid mainstream in some appropriate manner. This is a matter of some subtlety, and we postpone it for the time being.

There are two key ideas involved in boundary layer theory. The first is that u and v vary much more rapidly with y, the coordinate normal to the boundary, than they do with x, the coordinate parallel to the boundary. Let U_0 denote some typical value of u, and let u change by an amount of order U_0 over an x-distance of order L, say. If δ denotes a typical value of the thickness of the boundary layer, our basic approximation is

$$\left| \frac{\partial u}{\partial y} \right| \gg \left| \frac{\partial u}{\partial x} \right|,$$

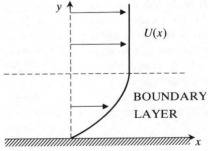

Fig. 8.6. The boundary layer.

and this amounts, by making an order of magnitude estimate of each term, to $U_0/\delta \gg U_0/L$, i.e.

$$\delta \ll L. \tag{8.4}$$

Now, the exact 2-D equations are

$$u\frac{\partial u}{\partial x} + v\frac{\partial u}{\partial y} = -\frac{1}{\rho}\frac{\partial p}{\partial x} + \nu\left(\frac{\partial^2 u}{\partial x^2} + \frac{\partial^2 u}{\partial y^2}\right),$$

$$u\frac{\partial v}{\partial x} + v\frac{\partial v}{\partial y} = -\frac{1}{\rho}\frac{\partial p}{\partial y} + \nu\left(\frac{\partial^2 v}{\partial x^2} + \frac{\partial^2 v}{\partial y^2}\right), \tag{8.5}$$

$$\frac{\partial u}{\partial x} + \frac{\partial v}{\partial y} = 0.$$

It follows at once from the last of these that $\partial v/\partial y$ is of order U_0/L, and as v is zero at $y=0$ it follows that v is of order $U_0\delta/L$ in the boundary layer. Thus v is much smaller than u. On viewing the first two equations as expressions for $\partial p/\partial x$ and $\partial p/\partial y$ respectively it then follows that

$$\left|\frac{\partial p}{\partial y}\right| \ll \left|\frac{\partial p}{\partial x}\right|,$$

which means that in the boundary layer p is, to a first approximation, a function of x alone. This justifies the use of dp/dx, rather than $\partial p/\partial x$, in eqn (8.1), and bears out Prandtl's remark that 'the pressure distribution of the free fluid will be impressed on the transition layer'. But the most dramatic simplification of eqn (8.5) arises on account of the following estimates:

$$\frac{\partial^2 u}{\partial x^2} = O\left(\frac{U_0}{L^2}\right), \qquad \frac{\partial^2 u}{\partial y^2} = O\left(\frac{U_0}{\delta^2}\right). \tag{8.6}$$

In view of eqn (8.4) the term $\partial^2 u/\partial x^2$ is negligible compared with the term $\partial^2 u/\partial y^2$, and with this major simplification of eqn (8.5) we complete our derivation of eqn (8.1).

The other key idea involved in boundary layer theory is that the rapid variation of u with y should be just sufficient to prevent the viscous term from being negligible, notwithstanding the small coefficient of viscosity ν. We may at once use this consideration

268 *Boundary layers*

to obtain an order of magnitude estimate of the boundary layer thickness. Both non-linear terms on the left-hand side of eqn (8.1) are of order U_0^2/L, and in order that the viscous term be of comparable magnitude we require that

$$\frac{U_0^2}{L} \sim \frac{\nu U_0}{\delta^2},$$

i.e.

$$\frac{\delta}{L} = O(R^{-\frac{1}{2}}). \tag{8.7}$$

The basic hypothesis (8.4) is evidently correct if the Reynolds number $R = U_0 L/\nu$ is large; the whole procedure is then self-consistent, and may indeed be put on a more formal basis (Exercise 8.1).

Equation (8.1) is also valid in the case of a curved boundary, provided that x denotes distance along the boundary and y distance normal to it. This may be demonstrated by writing the full Navier–Stokes equations in a suitable system of curvilinear coordinates (x, y); the argument is much as before, save that $\partial p/\partial y$ is then comparable in magnitude with $\partial p/\partial x$, for a substantial pressure gradient in the y-direction is required to balance the centrifugal effect of the flow round the curved surface (Rosenhead 1963, pp. 201–203; Goldstein 1938, pp. 119–120). It is still the case that within the boundary layer p is essentially a function of x alone, for although the two pressure gradients are comparable, actual changes in p across the boundary layer are still much smaller, by a factor $O(\delta/L)$, than changes in p along the boundary, simply because the boundary layer is so thin.

To actually determine the pressure distribution $p(x)$, suppose that $U(x)$ denotes the slip velocity that would arise, at $y = 0$, if the fluid were (mistakenly) treated as being entirely inviscid. The velocity at the 'edge' of the boundary layer in Fig. 8.6 will be almost equal to $U(x)$, and by Bernoulli's theorem $p + \frac{1}{2}\rho U^2$ will be constant along a streamline at the edge of the boundary layer. It follows that

$$-\frac{1}{\rho}\frac{\mathrm{d}p}{\mathrm{d}x} = U\frac{\mathrm{d}U}{\mathrm{d}x}; \tag{8.8}$$

Boundary layers

thus if $U(x)$ increases with x the pressure $p(x)$ decreases, and vice versa.

We must finally address the matter of how to ensure a 'match' between the flow velocity in the boundary layer and that in the inviscid mainstream. In the sections which follow we shall, to this end, impose on the boundary layer flow the condition

$$u \to U(x) \qquad \text{as } y/\delta \to \infty, \tag{8.9}$$

δ denoting a typical measure of the boundary layer thickness, proportional to $v^{\frac{1}{2}}$. Note that the limiting process here is $y/v^{\frac{1}{2}} \to \infty$, not $y \to \infty$, which would correspond to rocketing out of the laboratory and into the heavens. This important distinction may become clearer with the following elementary example.

An elementary differential equation with a 'boundary layer'

Consider the following problem for a function $u(y)$:

$$\varepsilon u'' + u' = 1; \qquad u(0) = 0, \quad u(1) = 2, \tag{8.10}$$

where ε denotes a small positive constant. The exact solution is easily shown to be

$$u = y + \frac{1 - e^{-y/\varepsilon}}{1 - e^{-1/\varepsilon}}. \tag{8.11}$$

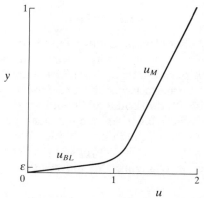

Fig. 8.7. The solution to eqn (8.10) for small ε.

Now, $e^{-1/\varepsilon}$ is extremely small, and so is $e^{-y/\varepsilon}$, for $0 < y < 1$, unless y is of order ε. The solution may therefore be approximated, in two parts, by a 'mainstream'

$$u_M = y + 1,$$

and a 'boundary layer' adjacent to $y = 0$ with thickness of order ε:

$$u_{BL} = 1 - e^{-y/\varepsilon}.$$

These two expressions represent particular limits of the full, exact solution (8.11), the first being obtained by letting $\varepsilon \to 0$ at fixed y, and the second being obtained by letting $\varepsilon \to 0$ *with y/ε fixed*. Notably,

$$\lim_{y/\varepsilon \to \infty} u_{BL} = \lim_{y \to 0} u_M,$$

and this is the equivalent statement to eqn (8.9) in this elementary example.

It is instructive to take the analogy further by returning to eqn (8.10) and proceeding on an approximate basis from the outset, exploiting the fact that ε is small. If we neglect the term $\varepsilon u''$ entirely, on this basis, we obtain

$$u_0' = 1, \quad \text{i.e. } u_0 = y + c,$$

and on making this satisfy the condition $u_0(1) = 2$ we obtain an 'outer' solution,

$$u_0(y) = y + 1.$$

This procedure thus far is comparable with treating a high Reynolds number flow as being entirely inviscid; the small parameter ε multiplies the highest derivative in the equation, and by ignoring that term we lower the order of the system and are unable to satisfy all the boundary conditions. Here an 'inner' solution, or boundary layer, is needed near $y = 0$, in order to satisfy the boundary condition there. We may recognize variations of u in this boundary layer to be much more rapid than those elsewhere by changing the independent variable in eqn (8.10) to

$$Y = y/\varepsilon.$$

With this scaling the previously negligible second derivative regains its importance:

$$\varepsilon \frac{1}{\varepsilon^2}\frac{d^2 u}{dY^2} + \frac{1}{\varepsilon}\frac{du}{dY} = 1,$$

so that to a first approximation the inner solution u_i satisfies

$$\frac{d^2 u_i}{dY^2} + \frac{du_i}{dY} = 0.$$

This is the equivalent of the boundary layer equation (8.1), in our simple example (and cf. Exercise 8.1). On making the inner solution satisfy the boundary condition $u(0) = 0$ we obtain

$$u_i = A(1 - e^{-Y}),$$

and the matching condition

$$\lim_{Y \to \infty} u_i = \lim_{y \to 0} u_0$$

determines that $A = 1$. Thus

$$u = \begin{cases} y + 1 & \text{as } \varepsilon \to 0 \text{ for fixed } y, \\ 1 - e^{-y/\varepsilon} & \text{as } \varepsilon \to 0 \text{ for fixed } y/\varepsilon, \end{cases}$$

in keeping with our deductions from the exact solution (8.11).

8.3. The boundary layer on a flat plate

On inviscid theory a uniform stream approaching a flat plate at zero angle of incidence is unaffected by the presence of the plate, so $U(x)$ is a constant. The boundary layer equations then reduce to

$$u\frac{\partial u}{\partial x} + v\frac{\partial u}{\partial y} = \nu\frac{\partial^2 u}{\partial y^2}, \qquad (8.12)$$

$$\frac{\partial u}{\partial x} + \frac{\partial v}{\partial y} = 0. \qquad (8.13)$$

We seek a similarity solution in which u is some function of the single variable

$$\eta = y/g(x). \qquad (8.14)$$

272 Boundary layers

This implies that the velocity profile at any distance x from the leading edge will be just a 'stretched out' version of the velocity profile at any other distance x, as in Fig. 2.14; this is a natural assumption if, as we shall suppose, the plate is semi-infinite, from $x = 0$ to $x = \infty$. We here take the similarity method of §2.3 a little further by not attempting to guess the function $g(x)$ in advance; we show instead how it can be left to emerge in a rational way as the calculation proceeds.

We first satisfy eqn (8.13) by introducing a stream function $\psi(x, y)$ such that

$$u = \partial \psi / \partial y, \qquad v = -\partial \psi / \partial x. \tag{8.15}$$

If we write u in the form $u = Uh(\eta)$ we may integrate to obtain

$$\psi = Ug(x) \int_0^\eta h(s) \, \mathrm{d}s + k(x).$$

But we want the plate itself to be a streamline, so that $\psi = 0$, say, at $\eta = 0$; so $k(x) = 0$. It is then more convenient to write ψ in the form

$$\psi = Ug(x)f(\eta), \qquad \text{with } f(0) = 0, \tag{8.16}$$

whence

$$u = Uf'(\eta) \tag{8.17}$$

and

$$\begin{aligned}
v = -\frac{\partial \psi}{\partial x} &= -U\left(g'f + gf' \frac{\partial \eta}{\partial x}\right) \\
&= -U\left(g'f - \frac{y}{g} f' g'\right) \\
&= U(\eta f' - f)g'.
\end{aligned} \tag{8.18}$$

Here, of course, f' denotes $f'(\eta)$, but g' denotes $g'(x)$. On substituting for u and v in eqn (8.12) we obtain

$$-U^2 f' f'' \frac{y}{g^2} g' + U^2 (\eta f' - f) g' \frac{f''}{g} = \nu U \frac{f'''}{g^2},$$

which simplifies to

$$f''' + \frac{Ugg'}{\nu} ff'' = 0.$$

Our aim is, of course, to obtain an ordinary differential equation for f as a function of η. We must therefore choose gg'—which would otherwise be a function of x—to be a constant. Clearly the choice of ν/U for this constant is convenient in that it rids the equation of all parameters of the problem, and integrating $gg' = \nu/U$ gives

$$\tfrac{1}{2}g^2 = \frac{\nu x}{U} + d,$$

where d is an arbitrary constant. Now, if g vanishes for some value of x, certain flow quantities such as

$$\partial u/\partial y = Uf''/g$$

become singular. We clearly expect some such behaviour at the leading edge, if only because on $y=0$ the velocity suddenly changes from U in $x<0$ to zero in $x>0$. We therefore choose $d=0$ to ensure that any such behaviour occurs at the leading edge. Thus $g(x) = (2\nu x/U)^{\frac{1}{2}}$ and, to sum up, we have found that

$$\psi = (2\nu Ux)^{\frac{1}{2}}f(\eta), \qquad \text{where } \eta = \frac{y}{(2\nu x/U)^{\frac{1}{2}}}, \qquad (8.19)$$

and

$$f''' + ff'' = 0. \qquad (8.20)$$

This equation must be supplemented by the boundary conditions

$$f(0) = f'(0) = 0, \qquad f'(\infty) = 1. \qquad (8.21)$$

The first of these stems from eqn (8.18), the second from eqn (8.17), and the third from the fact that u must tend to U, the mainstream value, as we leave the boundary layer (cf. eqn (8.9)).

The boundary value problem (8.20), (8.21) has to be solved numerically, and the results are shown in Fig. 8.8. The ratio u/U is 0.97 at $\eta = 3$ and 0.999936 at $\eta = 5$. According to eqn (8.19), therefore, the boundary layer thickness δ is such that

$$\delta = O\left(\frac{\nu x}{U}\right)^{\frac{1}{2}}, \qquad (8.22)$$

as indicated in Fig. 2.14. As the boundary layer thickens the horizontal stress on the plate

$$t_x = \mu\left(\frac{\partial u}{\partial y} + \frac{\partial v}{\partial x}\right)_{y=0} = \mu\left.\frac{\partial u}{\partial y}\right|_{y=0} = \mu U\left(\frac{U}{2\nu x}\right)^{\frac{1}{2}}f''(0) \qquad (8.23)$$

274 *Boundary layers*

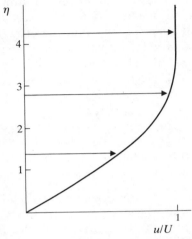

Fig. 8.8. The velocity profile in the boundary layer on a flat plate.

decreases with x. (Here we have used eqns (6.7) and (6.9), with, of course, $\boldsymbol{n} = (0, 1, 0)$.)

Application of the theory to a finite flat plate of length L

It is natural to hope that the above similarity solution will hold reasonably well for a finite plate of length L, even if behaviour of a different kind must be expected near and beyond the trailing edge. Taking into account both the top and the bottom of the plate, we obtain for the drag

$$D = 2 \int_0^L t_x \, dx = 2\sqrt{2} f''(0) \rho U^2 L R^{-\frac{1}{2}}, \qquad (8.24)$$

where $R = UL/\nu$. Thus D is proportional to $L^{\frac{1}{2}}$, rather than to L, because the velocity gradients at the plate decrease with x, corresponding to the thickening of the boundary layer. The drag is proportional to $\nu^{\frac{1}{2}}$, and vanishes as $\nu \to 0$. The numerical value of $f''(0)$ is 0.4696.

The agreement between boundary layer theory and experiment is very good, both in respect of the expression (8.24) for the drag and in respect of the details of the velocity profile. This agreement does break down, however, if the Reynolds number is very high, for the boundary layer then becomes unstable and turbulent flow ensues. The critical value of R at which this happens can be anywhere between about 10^5 and 3×10^6, depending on the level of turbulence in the oncoming stream.

8.4. High Reynolds number flow in a converging channel

We now consider high Reynolds number flow between two plane walls, $y = 0$ and $y = x \tan \alpha$, there being a narrow slit at the origin through which fluid is extracted (in the case of inflow) or injected (in the case of outflow). The low Reynolds number limit is discussed in Exercise 7.6; here we proceed on the assumption that the flow divides into an essentially inviscid mainstream with thin viscous boundary layers on the walls, as in Fig. 8.9.

It is natural to look first for a mainstream flow which is purely radial, i.e. $\boldsymbol{u} = u_r(r, \theta)\boldsymbol{e}_r$; we leave it as an exercise to show that the inviscid equations of motion then demand that

$$\boldsymbol{u} = -\frac{Q}{r}\boldsymbol{e}_r,$$

where Q is a constant, positive in the case of inflow.

We now analyse the boundary layer on $y = 0$. For this purpose, then,

$$U(x) = -Q/x, \tag{8.25}$$

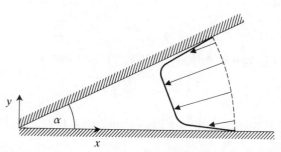

Fig. 8.9. Flow in a converging channel at high Reynolds number.

Boundary layers

and eqns (8.1) and (8.8) give

$$u\frac{\partial u}{\partial x} + v\frac{\partial u}{\partial y} = -\frac{Q^2}{x^3} + v\frac{\partial^2 u}{\partial y^2}. \quad (8.26)$$

Following the approach of §8.3 we try a similarity solution in which

$$u = -\frac{|Q|}{x}f'(\eta), \quad \text{where } \eta = \frac{y}{g(x)}.$$

The choice is, of course, guided by the mainstream boundary condition, and we shall require

$$f'(\infty) = \begin{cases} 1 & \text{in the case of inflow,} \\ -1 & \text{in the case of outflow.} \end{cases}$$

With the assumed form of solution

$$v\frac{\partial^2 u}{\partial y^2} = -v\frac{|Q|}{x}\frac{f'''}{g^2},$$

and we see at once that the only chance of converting eqn (8.26) into an ordinary differential equation for $f(\eta)$ is if $g(x)$ is proportional to x. The choice

$$g(x) = v^{\frac{1}{2}}x/|Q|^{\frac{1}{2}}$$

begins to clear eqn (8.26) of the parameters of the problem, for the right-hand side is then

$$-\frac{Q^2}{x^3}(1+f''').$$

Now, with

$$u = -\frac{|Q|}{x}f'(\eta) \quad \text{and} \quad \eta = \frac{|Q|^{\frac{1}{2}}}{v^{\frac{1}{2}}}\frac{y}{x}, \quad (8.27)$$

the stream function ψ is $-v^{\frac{1}{2}}|Q|^{\frac{1}{2}}f(\eta)$, and hence

$$v = -\frac{\partial \psi}{\partial x} = -\frac{|Q|^{\frac{1}{2}}v^{\frac{1}{2}}}{x}\eta f',$$

which satisfies $v = 0$ on $y = 0$. Substituting into eqn (8.26) we obtain

$$f'^2 = 1 + f'''.$$

The no-slip condition $u = 0$ on $y = 0$ is satisfied provided that $f'(0) = 0$.

We see that this problem is, in fact, just a second-order problem in $F(\eta) = f'(\eta)$:

$$F'' + 1 - F^2 = 0, \tag{8.28}$$

subject to

$$F(0) = 0 \quad \text{and} \quad F(\infty) = \begin{cases} 1 & \text{for inflow} \\ -1 & \text{for outflow} \end{cases} \tag{8.29}$$

The case of inflow: $F(\infty) = 1$

We may multiply eqn (8.28) by F' and integrate once to obtain

$$\tfrac{1}{2} F'^2 + F - \tfrac{1}{3} F^3 = \text{constant}.$$

Now $F(\infty) = 1$, and we may easily deduce that $F'(\infty) = 0$, so

$$\tfrac{1}{2} F'^2 + F - \tfrac{1}{3} F^3 = \tfrac{2}{3}.$$

This may be written

$$F'^2 = \tfrac{2}{3}(1 - F)^2 (2 + F),$$

which calls for the substitution $2 + F = G^2$, whence

$$G' = \pm \frac{1}{\sqrt{6}} (3 - G^2).$$

The further substitution $G = \sqrt{3} \tanh H$ leads to

$$H' = \pm \frac{1}{\sqrt{2}},$$

so

$$F = -2 + 3 \tanh^2 \left(\frac{\eta}{\sqrt{2}} + C \right), \tag{8.30}$$

where C is an arbitrary constant. This expression satisfies $F(\infty) = 1$, but to satisfy $F(0) = 0$ we need to choose C such that

$$\tanh C = \pm (\tfrac{2}{3})^{\frac{1}{2}}, \quad \text{i.e. } C \doteq \pm 1.14.$$

The two possible choices for C correspond to two different kinds of velocity profile in the boundary layer. The first case,

$C = 1.14$, corresponds to the kind of profile anticipated in Fig. 8.9, and this is what is typically observed in experiments (Goldstein 1938, p. 371). The second case, with $C = -1.14$, involves reversed flow close to the wall in the boundary layer. We have here, then, an example of non-uniqueness of flow at high Reynolds number (see §9.7); indeed, the non-uniqueness of high Reynolds number flow in a converging channel is far greater than a boundary layer treatment of the problem suggests (see Exercise 8.7).

As in Fig. 8.8, the ratio u/U becomes very close to 1 within an $O(1)$ distance η from the boundary, so, according to eqn (8.27), the boundary layer thickness is

$$\delta \sim \frac{\nu^{\frac{1}{2}}}{|Q|^{\frac{1}{2}}} x. \tag{8.31}$$

This decreases as x decreases, i.e. as the mainstream flow (8.25) increases. If the angle α is $O(1)$ the condition that the boundary layers be thin is

$$Q/\nu \gg 1, \tag{8.32}$$

and this is the opposite extreme to that in Exercise 7.6.

The case of outflow: $F(\infty) = -1$

It is possible to show that *there is no solution of the boundary layer equations in this case*; the whole supposition that the flow divides into an inviscid mainstream and thin viscous boundary layers is false (see Exercises 8.6 and 8.7).

8.5. Rotating flows controlled by boundary layers

In their most passive form, boundary layers simply effect a smooth transition between a given mainstream flow and no-slip at the boundary. There are occasions, however, when boundary layers exert a controlling influence on the flow as a whole, and one case amenable to elementary analysis arises when a fluid is almost, but not quite, in a state of uniform rotation.

Suppose, for instance, that viscous fluid occupies the space between two rigid boundaries $z = 0$ and $z = L$, the lower boundary rotating about the z-axis with angular velocity Ω, the upper boundary rotating about the same axis with angular velocity $\Omega(1+\varepsilon)$, where ε is small. Between $z=0$ and $z=L$ the fluid must somehow achieve the slight change in angular velocity implied by the boundary conditions. If the Reynolds number

$$R = \Omega L^2/\nu \qquad (8.33)$$

is large, we expect thin viscous layers on both boundaries and an essentially inviscid 'interior' flow in between. But what happens in the inviscid interior? The flow there is certainly not known in advance; it is, instead, largely controlled by the boundary layers.

Almost-uniform rotation: the basic equations

If a fluid is almost rotating with some uniform angular velocity $\mathbf{\Omega}$ it is convenient to write down the Navier–Stokes equations relative to a frame of reference which rotates at that angular velocity:

$$\frac{\partial \mathbf{u}}{\partial t} + (\mathbf{u}\cdot\nabla)\mathbf{u} + 2\mathbf{\Omega}\wedge\mathbf{u} + \mathbf{\Omega}\wedge(\mathbf{\Omega}\wedge\mathbf{x}) = -\frac{1}{\rho}\nabla p + \nu\nabla^2\mathbf{u},$$
$$\nabla \cdot \mathbf{u} = 0. \qquad (8.34)$$

Here $\mathbf{u}$ denotes the fluid velocity *relative to the rotating frame*, and $\partial \mathbf{u}/\partial t$ denotes the rate of change of $\mathbf{u}$ at a fixed position $\mathbf{x}$ in that frame. The familiar expression for the acceleration (first two terms) has been augmented by a Coriolis (third) term and a centrifugal (fourth) term in the normal way, as in elementary particle mechanics (e.g. Smith and Smith 1968, Chapter 8). In fact the vector identity

$$\mathbf{\Omega}\wedge(\mathbf{\Omega}\wedge\mathbf{x}) = -\nabla[\tfrac{1}{2}(\mathbf{\Omega}\wedge\mathbf{x})^2] \qquad (8.35)$$

($\mathbf{\Omega}$ being constant) enables us to clear away the centrifugal term by defining a 'reduced pressure'

$$p_R = p - \tfrac{1}{2}\rho(\mathbf{\Omega}\wedge\mathbf{x})^2. \qquad (8.36)$$

We shall not continue with the suffix, but will understand p to denote reduced pressure in all that follows.

We are interested in relative flows $\mathbf{u}$ which are weak compared to the basic rotation of the system as a whole. If we let U denote a typical value of $|\mathbf{u}|$, and let L denote a typical length scale of the flow, the dimensionless parameter $U/\Omega L$ will therefore be small (of order ε in the specific problem mentioned above). Now,

$$|(\boldsymbol{u} \cdot \nabla)\boldsymbol{u}| = O(U^2/L) \quad \text{and} \quad |2\boldsymbol{\Omega} \wedge \boldsymbol{u}| = O(\Omega U),$$

and *we now assume that $U/\Omega L$ is so small that the term $(\boldsymbol{u} \cdot \nabla)\boldsymbol{u}$ may be neglected in comparison with the Coriolis term $2\boldsymbol{\Omega} \wedge \boldsymbol{u}$.* The equations governing the small departure $\boldsymbol{u}$ from a state of uniform rotation with angular velocity $\boldsymbol{\Omega}$ are then

$$\frac{\partial \boldsymbol{u}}{\partial t} + 2\boldsymbol{\Omega} \wedge \boldsymbol{u} = -\frac{1}{\rho}\nabla p + \nu \nabla^2 \boldsymbol{u}, \tag{8.37}$$

$$\nabla \cdot \boldsymbol{u} = 0. \tag{8.38}$$

Within this framework we consider the flow at large Reynolds number $R = \Omega L^2/\nu$, and assume the main part of that flow to be essentially inviscid.

Steady, inviscid flow

Let us take Cartesian coordinates (x, y, z) fixed in the rotating frame with the z-axis parallel to the rotation axis, so that $\boldsymbol{\Omega} = (0, 0, \Omega)$. We deduce from eqns (8.37) and (8.38) that a steady, inviscid flow $\boldsymbol{u}_I = (u_I, v_I, w_I)$ satisfies

$$-2\Omega v_I = -\frac{1}{\rho}\frac{\partial p_I}{\partial x}, \qquad 2\Omega u_I = -\frac{1}{\rho}\frac{\partial p_I}{\partial y}, \quad (8.39, 8.40)$$

$$0 = -\frac{1}{\rho}\frac{\partial p_I}{\partial z}, \qquad \frac{\partial u_I}{\partial x} + \frac{\partial v_I}{\partial y} + \frac{\partial w_I}{\partial z} = 0. \quad (8.41, 8.42)$$

Clearly p_I is independent of z. It follows immediately from eqns (8.39) and (8.40) that u_I and v_I are independent of z also. Moreover, on substituting eqns (8.39) and (8.40) into (8.42) we see that $\partial w_I/\partial z = 0$. It follows that $\boldsymbol{u}_I$ *is independent of z.* This far-reaching result is known as *the Taylor–Proudman theorem.*

Ekman boundary layers

Let us now turn to the particular problem of steady flow between two differentially rotating rigid boundaries at $z = 0$ and $z = L$ (Fig. 8.10). If R is large, the flow in the 'interior' will be

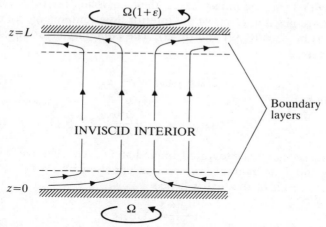

Fig. 8.10. The secondary flow between differentially rotating boundaries.

essentially inviscid, and therefore subject to the Taylor–Proudman theorem, but there will be thin viscous layers on both boundaries.

Consider the boundary layer on $z = 0$. If we assume, in the normal way, that variations of $\boldsymbol{u}$ with z are much more rapid than those with x or y, we find that eqns (8.37) and (8.38) reduce to

$$-2\Omega v = -\frac{1}{\rho}\frac{\partial p}{\partial x} + \nu \frac{\partial^2 u}{\partial z^2}, \tag{8.43}$$

$$2\Omega u = -\frac{1}{\rho}\frac{\partial p}{\partial y} + \nu \frac{\partial^2 v}{\partial z^2}, \tag{8.44}$$

$$0 = -\frac{1}{\rho}\frac{\partial p}{\partial z} + \nu \frac{\partial^2 w}{\partial z^2}, \tag{8.45}$$

$$\frac{\partial u}{\partial x} + \frac{\partial v}{\partial y} + \frac{\partial w}{\partial z} = 0. \tag{8.46}$$

From eqn (8.46) we deduce that w is much smaller than the velocity component parallel to the boundary, and the usual argument of §8.2 then leads to the conclusion that p is essentially a function of x and y only. Thus $\partial p/\partial x$ and $\partial p/\partial y$ take on throughout the boundary layer their inviscid 'interior' values, which are given in terms of the interior flow components $u_I(x, y)$, $v_I(x, y)$ by eqns (8.39) and (8.40). The boundary layer equations then become

$$-2\Omega(v - v_I) = \nu \frac{\partial^2 u}{\partial z^2}, \qquad (8.47)$$

$$2\Omega(u - u_I) = \nu \frac{\partial^2 v}{\partial z^2}, \qquad (8.48)$$

and these can be integrated immediately. An effective way of doing this is to multiply the second equation by i $(=\sqrt{-1})$ and add the result to the first, whence

$$\nu \frac{\partial^2 f}{\partial z^2} = 2\Omega i f,$$

where

$$f = u - u_I(x, y) + i[v - v_I(x, y)].$$

The general solution is

$$f = A e^{-(1+i)z_*} + B e^{(1+i)z_*},$$

where

$$z_* = (\Omega/\nu)^{\frac{1}{2}} z,$$

and A and B are arbitrary functions of x and y. To match with the interior flow we require $f \to 0$ as $z_* \to \infty$, so $B = 0$. As the rigid boundary $z = 0$ is at rest in the rotating frame, we require $u = v = 0$ there, so

$$f = -(u_I + iv_I) e^{-(1+i)z_*},$$

which implies

$$u = u_I - e^{-z_*}(u_I \cos z_* + v_I \sin z_*), \qquad (8.49)$$

$$v = v_I - e^{-z_*}(v_I \cos z_* - u_I \sin z_*). \qquad (8.50)$$

At the 'edge' of this *Ekman boundary layer*, where the flow matches with that in the interior, there is a small, but highly significant, z-component of velocity. To see this, note that

$$\left(\frac{\Omega}{\nu}\right)^{\frac{1}{2}} \frac{\partial w}{\partial z_*} = \frac{\partial w}{\partial z} = -\left(\frac{\partial u}{\partial x} + \frac{\partial v}{\partial y}\right)$$

$$= \left(\frac{\partial v_I}{\partial x} - \frac{\partial u_I}{\partial y}\right) e^{-z_*} \sin z_* - \left(\frac{\partial u_I}{\partial x} + \frac{\partial v_I}{\partial y}\right)(1 - e^{-z_*} \cos z_*).$$

Now, the final term vanishes, by virtue of eqns (8.39) and (8.40), so on integrating with respect to z_* from $z_* = 0$ to $z_* = \infty$ we find the value of w at the edge of the Ekman layer to be

$$w_E(x, y) = \tfrac{1}{2}\left(\frac{\nu}{\Omega}\right)^{\frac{1}{2}}\left(\frac{\partial v_I}{\partial x} - \frac{\partial u_I}{\partial y}\right). \quad (8.51)$$

This expression may be written

$$w_E(x, y) = \tfrac{1}{2}(\nu/\Omega)^{\frac{1}{2}}\omega_I,$$

where ω_I is the z-component of the vorticity of the interior flow.

If the boundary is rotating with angular velocity Ω_B relative to the rotating frame, the above expression generalizes to

$$w_E(x, y) = (\nu/\Omega)^{\frac{1}{2}}(\tfrac{1}{2}\omega_I - \Omega_B). \quad (8.52)$$

We leave this as a simple exercise. Similarly, if Ω_T denotes the angular velocity of a rigid upper boundary $z = L$ relative to the rotating frame, then there is a small z-component of velocity up into the boundary layer on $z = L$ of

$$w_E(x, y) = (\nu/\Omega)^{\frac{1}{2}}(\Omega_T - \tfrac{1}{2}\omega_I). \quad (8.53)$$

Determination of the 'interior' flow

We are now in a position to use what we know of the Ekman layers in Fig. 8.10 to determine the flow in the inviscid interior of the fluid. The argument is beautifully simple: the components u_I, v_I, and w_I are all independent of z, by the Taylor–Proudman theorem, so $\omega_I = \partial v_I/\partial x - \partial u_I/\partial y$ and w_I are independent of z. The expressions (8.52) and (8.53), valid at the top of the lower boundary layer and the bottom of the upper boundary layer respectively, must therefore match. So

$$\tfrac{1}{2}\omega_I - \Omega_B = \Omega_T - \tfrac{1}{2}\omega_I, \quad (8.54)$$

i.e.

$$\omega_I = \Omega_B + \Omega_T.$$

In the case of Fig. 8.10, with $\Omega_B = 0$ and $\Omega_T = \Omega\varepsilon$, this gives

$$\omega_I = \frac{\partial v_I}{\partial x} - \frac{\partial u_I}{\partial y} = \Omega\varepsilon.$$

At this point it is convenient to switch to cylindrical polar coordinates, and on assuming that the velocity field is axisymmetric we find

$$\frac{1}{r}\frac{d}{dr}(ru_{\theta I}) = \Omega\varepsilon.$$

The solution of this which is finite at $r = 0$ is

$$u_{\theta I} = \tfrac{1}{2}\Omega\varepsilon r,$$

so the fluid in the interior rotates at an angular velocity which is the mean of those of the two boundaries. *This behaviour is a direct result of the influence of the top and bottom boundary layers.*

The solution in the interior is completed by returning to eqn (8.51) to obtain

$$u_{zI} = \tfrac{1}{2}(\nu\Omega)^{\frac{1}{2}}\varepsilon,$$

and then turning to the incompressibility condition

$$\frac{1}{r}\frac{\partial}{\partial r}(ru_{rI}) + \frac{\partial u_{zI}}{\partial z} = 0 \tag{8.55}$$

in the interior, which gives $u_{rI} = 0$. The secondary flow in the interior is therefore purely in the z-direction (Fig. 8.10).

Unsteady flow: 'spin-down'

Consider now a related but unsteady problem in which two boundaries, at $z = \pm\tfrac{1}{2}L$, and the fluid between them, initially rotate with angular velocity $\Omega(1 + \varepsilon)$. Suppose that at $t = 0$ the angular velocity of the boundaries is reduced to Ω. We wish to find the manner in which the fluid spins down to its new state of uniform rotation with angular velocity Ω. The time scale on which this happens is of particular interest.

Now, Ekman layers form quickly on both boundaries, within a time of order Ω^{-1}. In between them there is an essentially inviscid interior flow in which the main rotary component, represented by u_I and v_I, is independent of z, while the small z-component w_I is not. In fact, early in the spin-down process the interior is essentially still rotating with angular velocity $\Omega(1 + \varepsilon)$, and according to eqns (8.52) and (8.53) w_I is therefore positive at the top of the lower layer and negative at the bottom of the

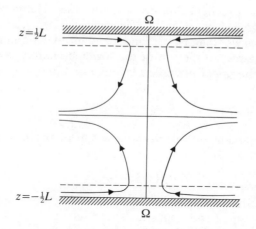

Fig. 8.11. The secondary flow during spin-down between two plane boundaries.

upper layer. This means that a tall, thin column of 'dyed' fluid in the interior is turned into a shorter, fatter one by the secondary flow as time proceeds (Fig. 8.11). The column has to conserve its angular momentum, as the interior flow is inviscid, so its angular velocity decreases with time. This is the essence of the spin-down process.

To quantify the above ideas, first eliminate the pressure between

$$\frac{\partial u_I}{\partial t} - 2\Omega v_I = -\frac{1}{\rho}\frac{\partial p_I}{\partial x}, \qquad (8.56)$$

$$\frac{\partial v_I}{\partial t} + 2\Omega u_I = -\frac{1}{\rho}\frac{\partial p_I}{\partial y}, \qquad (8.57)$$

and then use $\nabla \cdot \mathbf{u}_I = 0$ to obtain the vorticity equation

$$\frac{\partial}{\partial t}\left(\frac{\partial v_I}{\partial x} - \frac{\partial u_I}{\partial y}\right) = 2\Omega \frac{\partial w_I}{\partial z}. \qquad (8.58)$$

In the present circumstances $\partial w_I/\partial z$ is negative, vortex lines (which are essentially in the z-direction) are being compressed, and the vorticity in the interior is decreasing with time, as we have already argued. This behaviour in the inviscid interior is, of

286 Boundary layers

course, an excellent illustration of the Helmholtz vortex theorems (§5.3).

Now, because u_I and v_I are independent of z we may integrate eqn (8.58) between the top of the lower boundary layer and the bottom of the upper boundary layer to obtain essentially

$$L\frac{\partial \omega_I}{\partial t} = 2\Omega[w_I].$$

The z-component of velocity at the top of the lower boundary layer is

$$(\nu/\Omega)^{\frac{1}{2}}\tfrac{1}{2}\omega_I, \qquad (8.59)$$

by eqn (8.51), as the boundary is at rest in the rotating frame. There is an equal and opposite velocity component at the bottom of the upper boundary layer, so

$$L\frac{\partial \omega_I}{\partial t} = -2(\Omega\nu)^{\frac{1}{2}}\omega_I. \qquad (8.60)$$

Integrating with respect to t we obtain

$$\omega_I = A e^{-t/T},$$

where the 'spin-down time' T is given by

$$T = \frac{L}{2(\Omega\nu)^{\frac{1}{2}}}. \qquad (8.61)$$

Here $A(x, y)$ is an arbitrary function of x and y, but on applying the initial condition $\omega_I = 2\Omega\varepsilon$ at $t = 0$ (corresponding to rotation of the interior fluid with angular velocity $\Omega\varepsilon$, relative to the rotating frame) we find

$$\omega_I = 2\Omega\varepsilon e^{-t/T}.$$

As in the subsection above we now switch to cylindrical polar coordinates and assume axisymmetry. Thus

$$\frac{1}{r}\frac{\partial}{\partial r}(ru_{\theta I}) = 2\Omega\varepsilon e^{-t/T},$$

and the solution which is finite at $r = 0$ is

$$u_{\theta I} = \varepsilon\Omega r e^{-t/T}. \qquad (8.62)$$

This is the main result, for it displays how the excess rotation of the interior fluid decreases as time proceeds.

The solution is completed by turning to eqn (8.58) to obtain w_I and then to eqn (8.55) to obtain the radial flow u_{rI}. The results are

$$u_{rI} = \varepsilon(\Omega v)^{\frac{1}{2}}\frac{r}{L}e^{-t/T}, \qquad w_I = -2\varepsilon(\Omega v)^{\frac{1}{2}}\frac{z}{L}e^{-t/T}. \qquad (8.63)$$

The streamlines of this weak secondary flow are the solutions of

$$\frac{dz}{dr} = \frac{w_I}{u_{rI}},$$

i.e. $r^2 z = $ constant, as in Fig. 8.11.

The above analysis, and eqn (8.62) in particular, is valid for *small* decreases in rotation rate only; we cannot otherwise justify the neglect of the term $(\boldsymbol{u} \cdot \nabla)\boldsymbol{u}$ compared to the term $2\boldsymbol{\Omega} \wedge \boldsymbol{u}$ in eqn (8.34). Nevertheless, the mechanism by which a stirred cup of tea comes to rest is, essentially, that described above (see §5.3), and the time scale is, roughly, that emerging from the above theory, i.e. eqn (8.61). Typical values, e.g. $L = 4$ cm, $v = 10^{-2}$ cm^2 s^{-1}, and $\Omega = 2\pi$, lead us to expect significant changes in angular velocity within 10 seconds or so, in accord with casual observation. More serious experimental confirmation of eqn (8.62), within the limits of its validity, may be found in Fig. 1 of Greenspan and Howard (1963).

8.6. Boundary layer separation

We observed in §2.1 how a viscous boundary layer cannot, in general, subsist on a stationary rigid boundary if the pressure at the edge of the boundary layer increases substantially in the direction of flow. The reason is that the adverse pressure gradient, being the same at all levels in the boundary layer (§8.2), is usually sufficient to cause reversed flow close to the boundary, where the fluid moves sluggishly on account of the no-slip condition.† The way in which this reversed flow can lead to separation of the boundary layer as a whole is indicated in Prandtl's famous sketch (Fig. 8.2).

---

† This kind of flow reversal driven by an adverse pressure gradient also occurs in the much simpler thin-film situation of Fig. 7.13.

288 *Boundary layers*

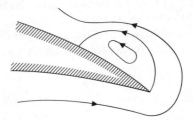

Fig. 8.12. The starting vortex.

The starting vortex of Fig. 1.1 provides a good example of the consequences of boundary layer separation. At the instant the aerofoil is jerked into motion the relative flow is irrotational and without circulation, as in Fig. 1.10(a). This flow does not persist, however, because the pressure is very low at the trailing edge, where the speed is large, and comparatively high at the stagnation point on the upper surface. As fluid passes round the trailing edge it therefore experiences an enormous adverse pressure gradient, and reversed flow deep in the boundary layer leads to separation and the generation of an anticlockwise vortex (Fig. 8.12). This vortex is then swept downstream by the main flow, leaving a negative circulation round the aerofoil (§5.1).

Even in the case of steady flow, boundary layer separation presents severe theoretical difficulties. If we try to find the separation point by using classical boundary layer theory we encounter a singularity (Exercise 8.10). This arises because the pressure gradient is effectively prescribed, via eqn (8.8), and prescribed, moreover, on the basis of the mainstream flow $U(x)$ that would obtain if there were no separation. Yet separation typically causes a major change to the inviscid mainstream flow, and hence to the pressure distribution at the edge of the boundary layer. What we really need, therefore, is an extension of classical boundary layer theory which incorporates the interaction between the mainstream pressure gradient and the separating boundary layer. Such a theory has emerged only recently, and it predicts a complicated *triple-deck* structure in the neighbourhood of the separation point (Fig. 8.13). The adverse pressure gradient drives a reversed flow in a sublayer of thickness $O(R^{-\frac{5}{8}})$. This causes a shear layer of thickness $O(R^{-\frac{1}{2}})$ to be

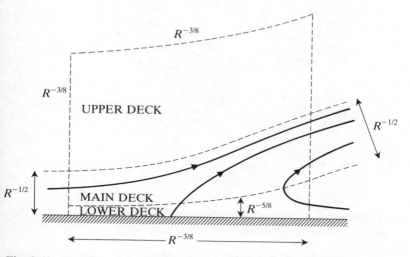

Fig. 8.13. Sketch of the triple-deck structure in the neighbourhood of a separation point on a rigid boundary.

pushed out into the mainstream, and this in turn modifies the adverse pressure gradient. This interaction between the mainstream pressure gradient and the $O(R^{-\frac{5}{8}})$ sublayer takes place on a length scale of order $R^{-\frac{3}{8}}$ around the separation point, which is large compared to the boundary layer thickness but small compared to the $O(1)$ scale of the main flow (Smith 1977; Stewartson 1981).

In the case of large-scale separation there remains the question: 'What form does the inviscid, mainstream flow take in the limit $R \to \infty$?'. For flow past a circular cylinder, for instance, the answer is certainly not the inviscid flow of Fig. 4.4(a). In fact, the only inviscid flow which can, apparently, accommodate the triple-deck structure in the neighbourhood of the separation point is one of the 'free-streamline' flows discovered by Kirchhoff in 1869 (see Smith 1979; Stewartson 1981). In this flow a streamline leaves the cylinder at an angle of approximately 55° from the forward stagnation point and divides the flow into two parts, an irrotational mainstream flow and a large stagnant wake in which the pressure p is a constant. The pressure is continuous across the dividing streamline, but the velocity is not. The free

290 *Boundary layers*

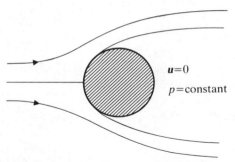

Fig. 8.14. Kirchhoff inviscid, free-streamline flow past a circular cylinder.

streamline is therefore a vortex sheet,† and it constitutes the limit, as $R \to \infty$, of the breakaway shear layer indicated in Fig. 8.13.

In practice, high Reynolds number flow past a circular cylinder is unsteady. This unsteadiness may take the form of von Kármán vortex shedding (§5.7), but at larger values of R other complications arise, one being the sudden drop in drag—known as the *drag crisis*—at $R \sim 10^5$. This happens because the boundary layer becomes turbulent ahead of the separation point, and the consequent exchange of momentum between different levels in the boundary layer helps to keep the deeper fluid going, thus postponing the onset of reversed flow. As R increases in this range the separation point shifts from the front to the rear of the cylinder, leading to a narrower wake and a decrease in the drag (see Fig. 4.12).

In a striking experiment on flow past a sphere, Prandtl showed that when R is a little too low for the drag crisis to occur naturally it can be provoked by fixing a thin wire hoop to the sphere at about 80° from the forward stagnation point, so 'tripping' the boundary layer into turbulence. Separation duly occurs on the rear portion of the sphere, and good photographs of this can be found in Goldstein (1938, p. 73), Batchelor (1967, plate 11), or van Dyke (1982, pp. 34–35). This artificial inducement of the drag crisis is exploited in cricket by seam

---

† This is what Prandtl is referring to, on p. 261, when he speaks of 'a Helmholtz surface of discontinuity'.

Boundary layers

bowlers, and in golf, where the dimples on the ball serve a similar purpose to Prandtl's trip wire.

Exercises

8.1. Rewrite the exact 2-D equations of motion in terms of the non-dimensional and scaled variables

$$x' = \frac{x}{L}, \qquad y' = \frac{y}{R^{-\frac{1}{2}}L}, \qquad u' = \frac{u}{U_0}, \qquad v' = \frac{v}{R^{-\frac{1}{2}}U_0}, \qquad p' = \frac{p}{\rho U_0^2},$$

where $R = U_0 L/\nu$. By taking the limit $R \to \infty$ *with fixed u', $\partial u'/\partial x'$,* etc., derive the boundary layer equations in their non-dimensional and scaled form:

$$u'\frac{\partial u'}{\partial x'} + v'\frac{\partial u'}{\partial y'} = -\frac{\partial p'}{\partial x'} + \frac{\partial^2 u'}{\partial y'^2},$$

$$0 = -\frac{\partial p'}{\partial y'}, \qquad \frac{\partial u'}{\partial x'} + \frac{\partial v'}{\partial y'} = 0.$$

8.2. Consider the boundary layer near the forward stagnation point on some 2-D body, such as a circular cylinder (see Fig. 8.14). The mainstream flow at the edge of the boundary layer is $U(x) = \alpha x$, where x denotes distance along the boundary measured from the stagnation point and α is a positive constant (see eqn (4.23)). Seek a similarity solution of the boundary layer equations in which

$$u(x, y) = \alpha x f'(\eta), \qquad \text{where } \eta = y/g(x),$$

and deduce that $g(x)$ must be a constant. By choosing that constant as $(\nu/\alpha)^{\frac{1}{2}}$ show that the problem can be reduced to

$$f''' + ff'' + 1 - f'^2 = 0$$

subject to

$$f(0) = f'(0) = 0, \qquad f'(\infty) = 1.$$

[Note that the boundary layer is of constant thickness, $O(\nu/\alpha)^{\frac{1}{2}}$, and that this flow is in fact an exact solution of the Navier–Stokes equations (see Exercise 2.14 and Fig. 2.13).

The corresponding problem at a *rear* stagnation point has $f'(\infty) = -1$. There is no solution to the boundary layer equations, which is in keeping with the idea that the flow given by eqn (4.35) cannot be reconciled with the no-slip condition on the rear of the cylinder by a steady boundary layer. Proving this non-existence is, however, more difficult than in Exercise 8.6 (see Rosenhead 1963, p. 251 for references).]

292 Boundary layers

8.3. Consider a general similarity solution

$$\psi = F(x)f(\eta), \qquad \eta = y/g(x)$$

to the boundary layer equations (8.1) and (8.2), where ψ denotes the stream function. Show that the mainstream condition (8.9) demands that $F(x) = cU(x)g(x)$, where c is a constant, and then, for convenience, choose c to be 1. Show that substitution in the boundary layer equations leads to

$$f'^2 - \left(1 + \frac{U}{U'}\frac{g'}{g}\right)ff'' = 1 + \frac{\nu f'''}{g^2 U'},$$

where a prime denotes differentiation with respect to the appropriate single variable in each case. Deduce that a similarity solution is only possible if

either $\quad U(x) \propto (x - x_0)^m \quad$ *or* $\quad U(x) \propto e^{\alpha x}$

where x_0, m, and α are constants.

In the case

$$U(x) = Ax^m, \qquad A > 0,$$

show that $g(x)$ is proportional to $x^{\frac{1}{2}(1-m)}$, and show that choosing

$$g(x) = \left[\frac{2\nu}{(m+1)Ax^{m-1}}\right]^{\frac{1}{2}}$$

leads to

$$f''' + ff'' + \frac{2m}{m+1}(1 - f'^2) = 0,$$

subject to

$$f(0) = f'(0) = 0, \qquad f'(\infty) = 1.$$

Verify that flow past a flat plate (§8.3) and flow near a forward stagnation point (Exercise 8.2) are special cases of these *Falkner–Skan equations*.

[It is possible to show that if $m \geq 0$, so that the flow speed $U(x)$ is increasing with x, then the problem for $f(\eta)$ has a unique solution. The velocity profile for each such m is qualitatively similar to that in Fig. 8.8.

If m is less than zero, so that the flow speed $U(x)$ is decreasing with x, there are two solutions for $f(\eta)$, provided that m is not less than -0.0904. One has a velocity profile of the 'normal' kind, with $0 < f'(\eta) < 1$, but the other has a region of reversed flow near the boundary. If m is less than -0.0904 there are no 'normal' solutions, so

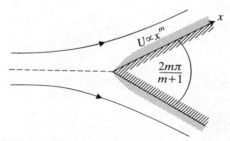

Fig. 8.15. Irrotational flow past a wedge ($0 \leq m \leq 1$).

only a very weak deceleration of the mainstream flow is needed to make reversed flow inevitable in the boundary layer. Indeed, in the more extreme case of $m = -1$ we have already seen (§8.4) that there are no solutions of the boundary layer equations at all if, as we have been assuming here, $A > 0$.

In discussing these Falkner–Skan equations we have not laid much emphasis on how the mainstream flows $U(x) \propto x^m$ might arise in practice, but one possibility, for $m \geq 0$, is indicated in Fig. 8.15.]

8.4. A two-dimensional jet emerges from a narrow slit in a wall into fluid which is at rest. If the jet is thin, so that u varies much more rapidly across the jet than along it (Fig. 8.16), the arguments of boundary layer

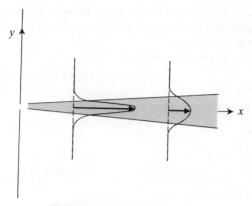

Fig. 8.16. A thin 2-D jet.

294 Boundary layers

theory apply and

$$u\frac{\partial u}{\partial x} + v\frac{\partial u}{\partial y} = \nu \frac{\partial^2 u}{\partial y^2} \tag{8.64}$$

in the jet. (The pressure gradient is zero in the jet because it is zero in the stationary fluid outside; cf. eqn (8.8).) The boundary conditions are that $u \to 0$ as we leave the jet and $\partial u/\partial y = 0$ at $y = 0$, as the motion is symmetrical about the x-axis.

By integrating eqn (8.64) across the jet, show that

$$\int_{-\infty}^{\infty} u\frac{\partial u}{\partial x}\,dy + \int_{-\infty}^{\infty} v\frac{\partial u}{\partial y}\,dy = \left[\nu\frac{\partial u}{\partial y}\right]_{-\infty}^{\infty}$$

and hence that

$$M = \rho \int_{-\infty}^{\infty} u^2\,dy = \text{constant}. \tag{8.65}$$

Seek a similarity solution for the stream function

$$\psi = F(x)f(\eta), \quad \text{where } \eta = y/g(x),$$

and show first that

$$F(x) = (3M/2\rho)^{\frac{1}{2}}[g(x)]^{\frac{1}{2}} \tag{8.66}$$

if we set

$$\int_{-\infty}^{\infty} [f'(\eta)]^2\,d\eta = \tfrac{2}{3}. \tag{8.67}$$

(The choice of $\tfrac{2}{3}$ is arbitrary, but it keeps the numerical factors relatively simple in what follows.) Then use the jet equation (8.64) to deduce that $g(x)$ must be proportional to $x^{\frac{2}{3}}$.

Show that the particular choice

$$g(x) = (3x\nu)^{\frac{2}{3}}(2\rho/3M)^{\frac{1}{3}}$$

leads to the problem

$$f''' + ff'' + f'^2 = 0$$

subject to

$$f(0) = f''(0) = 0, \quad f'(\infty) = 0.$$

Integrate twice to obtain

$$f' + \tfrac{1}{2}f^2 = \text{constant}$$

and deduce that $f = 2A \tanh(A\eta)$. Then use eqn (8.67) to determine A, and show that

$$u = \tfrac{1}{2}\left(\frac{3M^2}{4\nu\rho^2 x}\right)^{\frac{1}{3}} \operatorname{sech}^2(\tfrac{1}{2}\eta).$$

Give a rough estimate of the width of the jet at a distance x from the slit, and deduce a condition involving M/ρ, ν, and x which must be satisfied for the above boundary-layer-type treatment to be valid.

[It is possible to investigate an axisymmetric jet emerging from a small hole in a wall in a similar manner (Goldstein 1938, pp. 147–148; Rosenhead 1963, pp. 452–455; Schlichting 1979, pp. 230–234). Using cylindrical polar coordinates, the momentum flux

$$M = 2\pi\rho \int_0^\infty r u_z^2 \, dr$$

is independent of z, the distance from the hole, and the width of the jet is of order $z/(M/\rho\nu^2)^{\frac{1}{2}}$, so the boundary-layer-type treatment is valid only if $M/\rho\nu^2 \gg 1$.

This axisymmetric jet is in fact a limiting case (for large $M/\rho\nu^2$) of an exact, jet-like solution of the full Navier–Stokes equations (Rosenhead 1963, pp. 150–155; Batchelor 1967, pp. 205–211). At the opposite extreme, $M/\rho\nu^2 \ll 1$, the jet is very broad, and its Stokes stream function is, in spherical polar coordinates,

$$\Psi = \frac{M}{8\pi\mu} r \sin^2\theta,$$

which is a solution of the slow flow equations (see §7.2).

In practice, jets become unstable and turbulent at much lower Reynolds numbers than do boundary layers. Furthermore, the 'solutions' above for the jet emerging from a hole in a wall do not, of course, satisfy the no-slip condition on the wall. This matter has been put to right only quite recently, with interesting consequences for the flow outside the jet, particularly at lower Reynolds numbers (Schneider 1981, 1985; Zauner 1985, especially pp. 115 and 116).]

8.5. Consider the flow downstream of a 2-D streamlined body at high Reynolds number, so that there is a thin wake in which u varies much more rapidly with y than with downstream distance x. Suppose, too, that we are sufficiently far from the body that the flow velocity has nearly returned to its original (constant) value U, so that $u = U + u_1$, where u_1 is small (and negative). Show that in these circumstances the momentum equation approximates to

$$U \frac{\partial u_1}{\partial x} = \nu \frac{\partial^2 u_1}{\partial y^2}.$$

Boundary layers

Deduce that

$$\int_{-\infty}^{\infty} u_1 \, dy = \text{constant}.$$

Seek a similarity solution

$$u_1 = F(x)f(\eta), \qquad \eta = y/g(x),$$

so obtaining

$$u_1 = \frac{A}{x^{\frac{1}{2}}} e^{-Uy^2/4vx},$$

where A is a constant (which may be related to the drag on the body; see Batchelor 1967, pp. 348–352). Sketch the velocity profiles at two different downstream distances x.

8.6. Prove that the differential equation (8.28) has no real solution for $F(\eta)$ satisfying $F(0) = 0$ and $F(\infty) = -1$.

8.7. Consider 2-D flow in a diverging channel, $-\alpha < \theta < \alpha$, $0 < r < \infty$. To satisfy $\nabla \cdot \boldsymbol{u} = 0$ any purely radial flow must take the form

$$u_r = F(\theta)/r, \qquad u_\theta = 0.$$

Let F_0 denote $F(0)$, and define a Reynolds number

$$R = \alpha F_0 / \nu.$$

Writing $\eta = \theta/\alpha$ and $F(\theta) = F_0 f(\eta)$, show that the Navier–Stokes equations and the boundary conditions are satisfied if

$$f''' + 2\alpha R f f' + 4\alpha^2 f' = 0, \tag{8.68}$$

and

$$f(-1) = f(1) = 0, \qquad f(0) = 1. \tag{8.69}$$

Suppose now that we consider only velocity profiles which are symmetric about the centre line $\theta = 0$, so that f is an even function of η, and

$$f'(0) = 0. \tag{8.70}$$

Deduce that

$$f'^2 = (1-f)[\tfrac{2}{3}\alpha R(f^2 + f) + 4\alpha^2 f + c],$$

where

$$c = [f'(1)]^2 \geq 0$$

is an arbitrary constant.

Suppose further that of these symmetric velocity profiles we now consider only those which involve *pure outflow* and are fastest on the centreline, so that $R>0$ and $f>0$ for $-1<\eta<1$, as in Fig. 8.17(a). Show that in this case

$$\int_0^1 \frac{df}{(1-f)^{\frac{1}{2}}[\frac{2}{3}\alpha R(f^2+f)+4\alpha^2 f+c]^{\frac{1}{2}}}=1,$$

which in principle determines $c \geq 0$ in terms of α and R. Hence show that there can be no such solution if

$$\int_0^1 \frac{df}{f^{\frac{1}{2}}(1-f^2)^{\frac{1}{2}}}<(\tfrac{2}{3}\alpha R)^{\frac{1}{2}},$$

i.e. if

$$\alpha R > 10.31. \tag{8.71}$$

Show too that there can be no such solution if

$$\alpha > \pi/2, \tag{8.72}$$

no matter what the value of $R>0$. Relate these results to those of Exercises 7.6 and 8.6.

[When R is large there are many different solutions of the kind sketched in Fig. 8.17(b). The width of each 'peak' is of order $R^{-\frac{1}{2}}$, i.e. the classical boundary layer thickness, and in this way viscous effects remain important throughout the whole flow. Extensive discussions of these 'Jeffery–Hamel' flows may be found in Goldstein (1938, pp. 105–110), Rosenhead (1963, pp. 144–150), and Batchelor (1967, pp. 294–302). A more recent discussion of the whole problem, including instability and the importance of asymmetric solutions at higher Reynolds numbers, may be found in Sobey and Drazin (1986).]

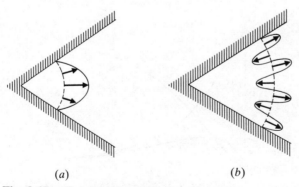

Fig. 8.17. Two possible flows in a divergent channel.

Boundary layers

8.8. Viscous fluid occupies the region above a plane rigid boundary $z = 0$ which is rotating with angular velocity Ω. Verify that there is a similarity solution to the Navier–Stokes equations of the form

$$u_r = \Omega r f(\xi), \qquad u_\theta = \Omega r g(\xi), \qquad u_z = (\nu\Omega)^{\frac{1}{2}} h(\xi),$$

where

$$\xi = z(\Omega/\nu)^{\frac{1}{2}},$$

if

$$f^2 + hf' - g^2 = f'', \qquad 2fg + hg' = g'', \qquad 2f + h' = 0,$$

the boundary conditions being

$$f = 0, \quad g = 1, \quad h = 0 \qquad \text{at } \xi = 0,$$
$$f \to 0, \quad g \to 0 \qquad \text{as } \xi \to \infty.$$

[This classical problem was first investigated by von Kármán in 1921. From the numerical solution in Fig. 8.18 we see that variations in f, g, and h take place effectively in a distance ξ of order 1, i.e. in a boundary layer of thickness $O(\nu/\Omega)^{\frac{1}{2}}$. There is a significant radially outward secondary flow in this boundary layer, and to compensate for this fluid is sucked down from above, into the boundary layer, so that as $\xi \to \infty$, h tends not to zero but to -0.88, i.e.

$$u_z \to -0.88(\nu\Omega)^{\frac{1}{2}} \qquad \text{as } z \to \infty$$

(cf. the Ekman layers in §8.5).

When there is an upper, rigid boundary at $z = L$ the Reynolds number $R = \Omega L^2/\nu$ enters the problem. In §8.5 we examined the case when R is large and the two boundaries rotate at almost the same angular velocity. The case when the upper boundary is at rest is much more difficult,

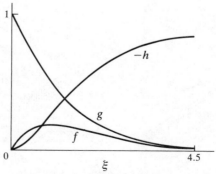

Fig. 8.18. Numerical solution of the single-disc von Kármán problem.

though the similarity equations for f, g, and h still hold, with $f = g = h = 0$ at $\xi = R^{\frac{1}{2}}$ as the upper boundary condition. This problem caused some excitement in the early 1950s when two well known figures in fluid dynamics came to different conclusions about the nature of the flow at high Reynolds number. Stewartson argued that there would be a thin boundary layer on the lower (rotating) disc, and more or less stationary fluid elsewhere, while Batchelor argued that there would be thin boundary layers on both discs, with an essentially inviscid, rotating core of fluid between them. In 1968 Mellor, Chapple, and Stokes found computer solutions of both kinds, and more beside, for Reynolds numbers R greater than about 200. This is a good example of non-uniqueness of solution to the Navier–Stokes equations (cf. §9.7), and an extensive survey of the von Kármán problem by Zandbergen and Dijkstra (1987) contains many others.]

8.9. Use eqns (8.1)–(8.3) to show that

$$\text{if } \frac{dp}{dx} > 0 \quad \text{then} \quad \left.\frac{\partial^2 u}{\partial y^2}\right|_{y=0} > 0.$$

Consider the flow in Fig. 8.2 upstream of the point at which $\partial u/\partial y$ is zero at the wall, and use the above result, together with eqn (8.8), to explain why, with the mainstream velocity $U(x)$ decreasing with x, the velocity profile in the boundary layer must have an inflection point.

8.10. Use eqns (8.1)–(8.3) to consider the flow in the immediate neighbourhood of the point $x = x_s$ at which $\partial u/\partial y$ is zero at the wall. Show, in particular, that according to those equations

$$\left.\frac{\partial u}{\partial y}\right|_{y=0} \propto (x_s - x)^{\frac{1}{2}}$$

when x is close to x_s, provided that $(\partial^4 u/\partial y^4)|_{y=0}$ is finite and non-zero at $x = x_s$.

[This square-root singularity is a consequence of treating the mainstream pressure gradient dp/dx as a prescribed quantity, instead of using triple-deck theory (Fig. 8.13) to let it respond to what is happening in the boundary layer near $x = x_s$.]

9 Instability

9.1. The Reynolds experiment

In his classic paper on the instability of flow down a pipe Reynolds (1883) writes:

The ... experiments were made on three tubes .... The diameters of these were nearly 1 inch, $\frac{1}{2}$ inch and $\frac{1}{4}$ inch. They were all about 4 feet 6 inches long, and fitted with trumpet mouthpieces, so that water might enter without disturbance. The water was drawn through the tubes out of a large glass tank [Fig. 9.1], in which the tubes were immersed, arrangements being made so that a streak or streaks of highly coloured water entered the tubes with the clear water.

The general results were as follows:

(1) When the velocities were sufficiently low, the streak of colour extended in a beautiful straight line through the tube [Fig. 9.2(a)].

(2) If the water in the tank had not quite settled to rest, at sufficiently low velocities, the streak would shift about the tube, but there was no appearance of sinuosity.

(3) As the velocity was increased by small stages, at some point in the tube, always at a considerable distance from the trumpet or intake, the colour band would all at once mix up with the surrounding water, and fill the rest of the tube with a mass of coloured water [Fig. 9.2(b)]. Any increase in the velocity caused the point of break down to approach the trumpet, but with no velocities that were tried did it reach this. On viewing the tube by the light of an electric spark, the mass of colour resolved itself into a mass of more or less distinct curls, showing eddies [Fig. 9.2(c)].

To quantify these results Reynolds used the now familiar dimensionless parameter

$$R = \frac{Ud}{\nu}, \qquad (9.1)$$

with U denoting, in this context, the mean velocity of the water down the tube and d denoting the diameter of the tube. Reynolds made it quite clear, however, that there is no single

Instability 301

Fig. 9.1. Sketch of Reynolds's dye experiment, taken from his 1883 paper.

'critical' value of R below which the flow is stable and above which it is unstable; the whole matter is more complicated. In his own words:

... the critical velocity was very sensitive to disturbance in the water before entering the tubes.... This at once suggested the idea that the condition might be one of instability for disturbances of a certain magnitude and stability for smaller disturbances.

The situation may be crudely likened to that in Fig. 9.3(c), which contrasts with the simpler examples of stable and unstable states in Figs 9.3(a,b).

By taking great care to minimize the disturbances, Reynolds was able to keep the flow stable up to values of R approaching

302 *Instability*

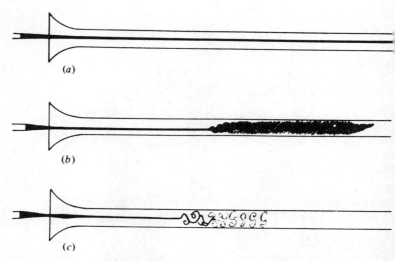

Fig. 9.2. Reynolds's drawings of the flow in his dye experiment.

13 000. Subsequently, even more refined experiments have pushed this figure up to 90 000 or more, and all the theoretical evidence to date suggests that fully developed flow down a pipe is stable to *infinitesimal* disturbances at any finite value of R, no matter how large. On the other hand, if no great care is taken to minimize disturbances, instability typically occurs when $R \sim 2000$.

Reynolds also enquired whether there is a critical value of R *below* which a previously turbulent flow reverts to a smooth or

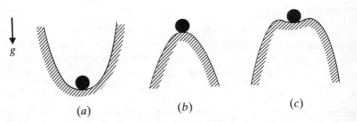

Fig. 9.3. (*a*) A stable state. (*b*) An unstable state. (*c*) A state which is stable to infinitesimal disturbances but unstable to disturbances which exceed some small threshold amplitude.

'laminar' form. His dye technique was of course useless for this purpose, and he measured instead the pressure gradient P needed to drive the flow. For turbulent flow he found $P \propto U^{1.7}$, but the dependence changed to $P \propto U$, in accord with the laminar flow theory of Exercise 2.3, when R was reduced below about 2000.

Reynolds's original apparatus still stands in the hydraulics laboratory of the Engineering Department at Manchester University. Recently, Johannesen and Lowe used it to obtain some excellent photographs of the phenomena in Fig. 9.2 (see Fig. 103 in van Dyke 1982). In accord with the above ideas, vibration from the heavy traffic on the streets of Manchester now makes the critical value of R substantially lower than the value of 13 000 obtained by Reynolds himself in the horse-and-cart days of 100 years ago.

9.2. Kelvin–Helmholtz instability

The natural first step in examining the stability of any system is to consider infinitesimal disturbances, so that all terms in the equations involving products of small perturbations may be neglected. This makes analysis simpler, although in fluid dynamics the resulting *linear stability theory* may still be difficult. In §§9.2–9.5 we focus largely on this kind of theory, keeping firmly in mind the lesson from §9.1 that there may also be instabilities that arise only if a certain threshold disturbance amplitude is exceeded.

The particular example of *Kelvin–Helmholtz instability* serves to illustrate some of the key ideas of linear theory. Let one deep layer of inviscid fluid, density ρ_2, flow with uniform speed U over another deep layer of density ρ_1 which is at rest, as in Fig. 9.4. Consider a small travelling-wave disturbance so that the interface, $y = \eta(x, t)$, has the form

$$\eta(x, t) = A e^{i(kx - \omega t)}, \qquad (9.2)$$

the real part of the right-hand side being understood. By studying the associated small-amplitude motions, neglecting all quadratically small terms, we find a *dispersion relation* for ω as a function of the wavenumber k and the parameters of the

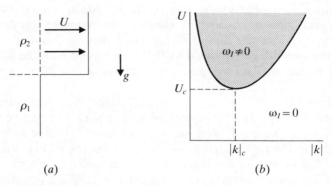

Fig. 9.4. Kelvin–Helmholtz instability: linear theory.

problem:

$$(\rho_1 + \rho_2)\omega - \rho_2 Uk$$
$$= \pm[(\rho_1 + \rho_2)|k|\{k^2 T + (\rho_1 - \rho_2)g\} - \rho_1\rho_2 U^2 k^2]^{\frac{1}{2}}, \quad (9.3)$$

where T denotes the surface tension between the two fluids. The manner in which this expression is obtained is covered in Exercise 3.6 and needs no elaboration here.

Our present concern is the use of expressions such as eqn (9.3) in drawing conclusions about the stability of a system. Suppose that $\rho_1 > \rho_2$, so that the configuration of the two fluids is bottom-heavy. Writing $\omega = \omega_R + i\omega_I$ we find two complex roots, one with $\omega_I > 0$ and the other with $\omega_I < 0$, if

$$\frac{\rho_1\rho_2 U^2}{\rho_1 + \rho_2} > (\rho_1 - \rho_2)\frac{g}{|k|} + |k|T. \quad (9.4)$$

The root with $\omega_I > 0$ is particularly significant, for $\omega_I > 0$ corresponds to exponential growth of the disturbance amplitude with time (see eqn (9.2)). If, on the other hand, eqn (9.4) is not satisfied, ω is real (see Fig. 9.4(b)).

Now, any small 2-D disturbance to the system will produce an interface displacement which may be written in the form of a Fourier integral:

$$\eta(x, t) = \int_{-\infty}^{\infty} A(k)e^{i(kx - \omega t)} dk. \quad (9.5)$$

Thus if $\omega_I > 0$ for *any* band of wavenumbers k, however small, the disturbance will not remain small as time proceeds. The system is unstable, then, if

$$\frac{\rho_1\rho_2 U^2}{\rho_1+\rho_2} > \min_k \left\{ (\rho_1-\rho_2)\frac{g}{|k|} + |k|\,T \right\}$$
$$= 2[(\rho_1-\rho_2)gT]^{\frac{1}{2}}. \qquad (9.6)$$

Gravity and surface tension therefore play a stabilizing role; the larger g or T the larger the velocity difference U between the two layers before instability occurs.

Kelvin–Helmholtz instability may also occur in a continuously stratified fluid in which the density $\rho_0(y)$ decreases with height. The buoyancy frequency N, where

$$N^2(y) = -\frac{g}{\rho_0}\frac{d\rho_0}{dy}, \qquad (9.7)$$

then acts as a measure of the stabilizing effects of the bottom-heavy density distribution (cf. §3.8), and it is possible to show that according to linear theory instability of a shear flow $U(y)$ can only occur if the *Richardson number*

$$J = \frac{N^2}{(dU/dy)^2} \qquad (9.8)$$

is less than $\frac{1}{4}$ somewhere in the flow (Exercise 9.2). The velocity gradient must therefore be sufficiently strong before instability occurs. Thorpe (1969, 1971) conducted some laboratory experiments, producing the shear flow by tilting the tank and then restoring it to the horizontal (Fig. 9.5; see also van Dyke 1982, p. 85; Tritton 1988, p. 268). The instability also occurs in the atmosphere, sometimes in the form of 'clear-air turbulence', but sometimes marked by distinctive cloud patterns (Drazin and Reid 1981, p. 21; Scorer 1972, pp. 86–99).

9.3. Thermal convection

Let viscous fluid be at rest between two horizontal rigid boundaries, $z = 0$ and $z = d$, and suppose there is a temperature difference ΔT between the two boundaries, the lower boundary being the hotter. The lower fluid will have a slightly lower

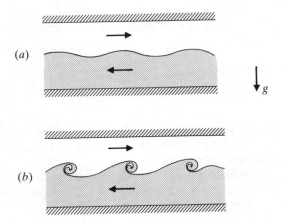

Fig. 9.5. Development of Kelvin–Helmholtz billows.

density, on account of greater thermal expansion, and the system will be slightly top-heavy. Now, if the temperature difference between the boundaries is increased by small steps the state of rest remains stable until ΔT reaches a certain critical value, whereupon an organized cellular motion sets in, with hot fluid rising in some parts of the flow, and cold fluid descending in others (Fig. 9.6).

Linear stability theory

In this section we leave our usual realm of fluids of constant density and therefore need (i) a *mass conservation equation*, (ii) a *momentum equation*, (iii) an *energy equation*, and (iv) an

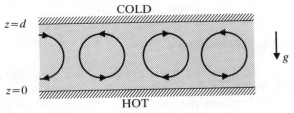

Fig. 9.6. Thermal convection.

equation of state, expressing how the density of the fluid depends on temperature and pressure. These equations are in general extremely complicated, but in dealing with the instability of a layer of viscous *liquid* heated from below many simplifying approximations may be made.

First, the density of a liquid varies slightly with temperature but only minutely with pressure, so we may take as our equation of state

$$\rho = \bar{\rho}[1 - \alpha(T - \bar{T})], \qquad (9.9)$$

where $\bar{\rho}$ is the density of the fluid at temperature $\bar{T}$, and α denotes the volume coefficient of thermal expansion. As the variation of ρ is so slight, the mass conservation equation (1.38) reduces essentially to

$$\nabla \cdot \boldsymbol{u} = 0, \qquad (9.10)$$

and if the viscosity is assumed to be constant, independent of temperature, the momentum equation takes the form

$$\rho \frac{D\boldsymbol{u}}{Dt} = -\nabla p + \rho \nu \nabla^2 \boldsymbol{u} + \rho \boldsymbol{g}. \qquad (9.11)$$

Finally, the energy equation may be taken to be

$$\frac{\partial T}{\partial t} + \boldsymbol{u} \cdot \nabla T = \kappa \nabla^2 T, \qquad (9.12)$$

κ denoting the thermal diffusivity of the fluid (i.e. the thermal conductivity divided by the product of the density and the specific heat). In the case of no motion, eqn (9.12) reduces to the classic equation of heat conduction in a solid. Equation (9.12) states, then, that the heat of a moving blob of fluid changes only as a result of the conduction of heat into that blob from the surrounding fluid; all other sources and sinks of energy (e.g. work done on the blob by stresses exerted by the surrounding fluid, dissipation of energy by viscosity) are neglected.

Now, in the undisturbed state of no motion the temperature $T_0(z)$ must satisfy eqn (9.12), so that

$$0 = \kappa \frac{d^2 T_0}{dz^2}.$$

308 *Instability*

It follows that

$$T_0(z) = T_l - \frac{z}{d}\Delta T, \tag{9.13}$$

where T_l denotes the temperature of the lower boundary. Accordingly,

$$\rho_0(z) = \bar{\rho}[1 - \alpha\{T_0(z) - \bar{T}\}], \tag{9.14}$$

and the basic hydrostatic pressure distribution may be calculated from

$$0 = -\frac{dp_0}{dz} - \rho_0(z)g. \tag{9.15}$$

In this state the fluid remains still and conducts heat upward as if it were a conducting solid.

Now disturb the system slightly, writing

$$T = T_0(z) + T_1, \qquad \rho = \rho_0(z) + \rho_1, \qquad p = p_0(z) + p_1, \qquad \boldsymbol{u} = \boldsymbol{u}_1 \tag{9.16}$$

where the variables T_1, ρ_1, p_1, and $\boldsymbol{u}_1$, all functions of x, y, z and t, are assumed small. Then linearization of eqns (9.9)–(9.12) gives

$$\rho_1 = -\alpha\bar{\rho}T_1, \qquad \nabla \cdot \boldsymbol{u}_1 = 0, \tag{9.17, 9.18}$$

$$\rho_0 \frac{\partial \boldsymbol{u}_1}{\partial t} = -\nabla p_1 + \rho_0 \nu \nabla^2 \boldsymbol{u}_1 + \rho_1 \boldsymbol{g}, \tag{9.19}$$

$$\frac{\partial T_1}{\partial t} + w_1 \frac{dT_0}{dz} = \kappa \nabla^2 T_1. \tag{9.20}$$

If we finally replace $\rho_0(z)$ in the third equation by $\bar{\rho}$, on the grounds that variations in density are small, we obtain a set of linear equations with constant coefficients.

We next obtain, by elimination, an equation for the vertical component of velocity w_1 alone. Taking the curl of eqn (9.19) gives, on using eqn (9.17):

$$\left(\frac{\partial}{\partial t} - \nu \nabla^2\right)\nabla \wedge \boldsymbol{u}_1 = -\alpha(\nabla T_1) \wedge \boldsymbol{g}.$$

Taking the curl again gives, on using eqns (A.6) and (A.10):

$$\left(\frac{\partial}{\partial t} - \nu \nabla^2\right)\nabla^2 \boldsymbol{u}_1 = \alpha[(\boldsymbol{g}\cdot\nabla)\nabla T_1 - \boldsymbol{g}\nabla^2 T_1],$$

and on taking the z-component and recognizing that $\boldsymbol{g} = (0, 0, -g)$ the result is

$$\left(\frac{\partial}{\partial t} - \nu \nabla^2\right)\nabla^2 w_1 = \alpha g\left(\frac{\partial^2}{\partial x^2} + \frac{\partial^2}{\partial y^2}\right)T_1. \qquad (9.21)$$

On using eqn (9.20) we finally obtain

$$\left(\frac{\partial}{\partial t} - \nu \nabla^2\right)\left(\frac{\partial}{\partial t} - \kappa \nabla^2\right)\nabla^2 w_1 = -\alpha g \frac{dT_0}{dz}\left(\frac{\partial^2}{\partial x^2} + \frac{\partial^2}{\partial y^2}\right)w_1. \qquad (9.22)$$

Now, derivatives with respect to x and y enter this equation only in the combination $\partial^2/\partial x^2 + \partial^2/\partial y^2$; there is no preferred horizontal direction in the problem. This permits separable solutions of the form

$$w_1 = W(z)f(x, y)e^{st} \qquad (9.23)$$

provided that $\partial^2 f/\partial x^2 + \partial^2 f/\partial y^2$ is a constant multiple of f, i.e.

$$\frac{\partial^2 f}{\partial x^2} + \frac{\partial^2 f}{\partial y^2} + a^2 f = 0, \qquad (9.24)$$

say. Then substitution of eqn (9.23) into eqn (9.22) gives

$$[\nu(D^2 - a^2) - s][\kappa(D^2 - a^2) - s](D^2 - a^2)W = \alpha g \frac{dT_0}{dz}a^2 W, \qquad (9.25)$$

where D denotes d/dz. We thus have a sixth-order ordinary differential equation for W with constant coefficients.

The boundary conditions are $u_1 = v_1 = w_1 = T_1 = 0$ at both $z = 0$ and $z = d$. The first two imply that both $\partial u_1/\partial x$ and $\partial v_1/\partial y$ are zero at $z = 0$ and $z = d$, and in view of eqn (9.18) this means that $\partial w_1/\partial z$ is also zero. As T_1 is zero there, so too is the right-hand side of eqn (9.21), and in view of eqns (9.23) and (9.24) it follows that

$$[\nu(D^2 - a^2) - s](D^2 - a^2)W = 0 \qquad \text{at } z = 0, d.$$

As $W = DW = 0$ at the boundaries this may, on expansion, be simplified, and the full set of boundary conditions is then

$$W = DW = D^4W - (2a^2 + s/\nu)D^2W = 0 \qquad \text{at } z = 0, d. \quad (9.26)$$

Together with eqn (9.25) this gives a sixth-order eigenvalue problem for s.

What happens next is straightforward enough in principle, but greatly complicated in practice by the awkward boundary conditions (9.26). To bring out the key ideas we shall apply instead the alternative conditions

$$W = D^2W = D^4W = 0 \qquad \text{at } z = 0 \text{ and } z = d,$$

which happen to arise in the artificial problem of thermal instability between two boundaries which are stress-free. These conditions loosely resemble eqn (9.26) and have the great merit that suitable eigenfunctions of eqn (9.25) satisfying them may be obtained immediately by inspection:

$$W = \sin(N\pi z/d), \qquad N = 1, 2, 3, \ldots.$$

The corresponding eigenvalues s are therefore given by

$$(s + \nu a_*^2)(s + \kappa a_*^2)a_*^2 + \alpha g \frac{dT_0}{dz} a^2 = 0,$$

with

$$a_*^2 = a^2 + \frac{N^2\pi^2}{d^2}.$$

Solving for s, we obtain

$$s = -\tfrac{1}{2}(\nu + \kappa)a_*^2 \pm \left[\tfrac{1}{4}(\nu + \kappa)^2 a_*^4 + \left\{\frac{\alpha g \, \Delta T}{d} \frac{a^2}{a_*^2} - \nu\kappa a_*^4\right\}\right]^{\frac{1}{2}},$$

where we have substituted for dT_0/dz using eqn (9.13). It is easy to show that if $\Delta T > 0$ the contents of the square root are always positive, so s is real, and the question then is whether either of the roots has $s > 0$, corresponding to exponential growth of the disturbance with time (see eqn (9.23)). The answer, evidently, is that the root with the plus sign will give $s > 0$ if

$$\frac{\alpha g \, \Delta T}{\nu\kappa d} > \frac{1}{a^2}\left(a^2 + \frac{N^2\pi^2}{d^2}\right)^3.$$

Now, a is not some parameter of the problem, but simply an unknown constant related to the horizontal length scale of the disturbance via eqn (9.24). Therefore instability occurs as soon as the left-hand side of the above inequality exceeds the minimum of the right-hand side with respect to both a and N. Clearly $N = 1$, and by differentiation we find that the minimum with respect to a^2 occurs when

$$a = a_c = \pi/\sqrt{2}d.$$

Introducing the *Rayleigh number*

$$\mathcal{R} = \frac{\alpha g \,\Delta T d^3}{\nu \kappa}, \qquad (9.27)$$

we see that $\mathcal{R} > 27\pi^4/4$ is the criterion for instability, in this somewhat artificial case of stress-free boundaries. This was in fact the case solved by Lord Rayleigh (1916a); as he surmised, it captures the essentials of the problem.

The corresponding calculation with the boundary conditions (9.26) was not carried out until rather later. The result is that instability between two rigid plane boundaries occurs when

$$\mathcal{R} > 1708.$$

There is a correspondingly different value for a_c, namely $3.1/d$.

The criterion above reveals how the various parameters of the system play a part in determining the stability of the basic state. Viscosity plays a stabilizing role; the larger the value of ν, the larger the temperature difference ΔT needed before convection sets in.

Stability to finite-amplitude disturbances

We have just answered the question: 'Is there a critical value of ΔT above which infinitesimal disturbances do not remain infinitesimal as $t \to \infty$?'. There is also, as always, the quite different question: 'Is there a critical value of ΔT *below* which the energy $\mathcal{E}$ of *any* disturbance tends to zero as $t \to \infty$?'.

In this particular (idealized) system the answer is yes, and that second critical value is the same as the first; if the Rayleigh number is less than 1708 then disturbances of arbitrary initial

312 *Instability*

magnitude die out as $t \to \infty$, whereupon the initial state of rest is restored (see, e.g., Drazin and Reid 1981, p. 464). In this respect the whole matter of stability is much simpler and more clear-cut than in the case of pipe flow (see §9.1).

Experimental results

Good agreement has been found between the measured value of the critical Rayleigh number and the theoretical prediction of 1708. But this is, just about, the extent to which linear theory accounts for the observations. According to such a theory infinitesimal disturbances grow exponentially with time when $\mathcal{R} > \mathcal{R}_c$; in practice the non-linear terms in the equations of motion quickly cease to be small and bring the exponential growth to a halt. In this manner a state of steady convection is reached, the vigour of which depends on $\mathcal{R} - \mathcal{R}_c$. We emphasize again that linear theory has nothing to say about this.

Nor does linear theory have anything to say about the pattern of convection when viewed from above, even when $\mathcal{R}$ is only very slightly above $\mathcal{R}_c$; all it yields is a critical value of a, thus fixing the general scale of the horizontal variations in the slightly supercritical case but leaving a whole multitude of possibilities for $f(x, y)$ satisfying eqn (9.24). In practice both 2-D rolls (Fig. 9.7(a)) and hexagonal cells (Fig. 9.7(b)) are quite common (van Dyke 1982, pp. 82–83), but as $\mathcal{R}$ is increased well beyond 1708 the initial state of steady convection may itself become unstable, leading to a different steady convection pattern. This in turn will typically become unstable at a still higher value of $\mathcal{R}$, leading

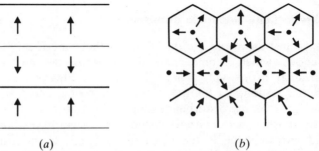

Fig. 9.7. Thermal convection viewed from above: (a) rolls; (b) hexagons.

perhaps to a time-dependent motion. The precise sequence of events as $\mathcal{R}$ increases depends critically on a variety of factors, including the value of the *Prandtl number* ν/κ, the variation of viscosity μ with temperature, and the presence of side-wall boundaries. Some idea of these developments, along with some excellent photographs, may be found in Turner (1973, pp. 207–226), Busse (1985), Craik (1985, pp. 258–272), and Tritton (1988, pp. 366–370).

Further complications arise if the upper surface of the fluid is free. It is now known, for instance, that the beautiful hexagonal cells observed by Bénard in 1900, which prompted Rayleigh to develop the above theory, were in fact driven by an altogether different mechanism involving the variation of surface tension with temperature.

To perform an elementary experiment (although many of the above complicating factors will be at work) we may follow the advice of Drazin and Reid (1981, p. 64): 'Pour corn oil in a clean frying pan (i.e. skillet), so that there is a layer of oil about 2 mm deep. Heat the bottom of the pan gently and uniformly. To visualize the instability, drop in a little powder (cocoa serves well). Sprinkling powder on the surface reveals the polygonal pattern of the steady cells. The movement of individual particles of powder may be seen, with rising near the centre of a cell and falling near the sides.'

9.4. Centrifugal instability

Let viscous fluid occupy the gap between two circular cylinders, the inner one having radius r_1 and angular velocity Ω_1, the outer one having radius r_2 and angular velocity Ω_2. The purely rotary flow

$$U_\theta(r) = Ar + \frac{B}{r}, \qquad (9.28)$$

where

$$A = \frac{\Omega_2 r_2^2 - \Omega_1 r_1^2}{r_2^2 - r_1^2}, \qquad B = \frac{(\Omega_1 - \Omega_2) r_1^2 r_2^2}{r_2^2 - r_1^2}, \qquad (9.29)$$

is an exact solution of the Navier–Stokes equations satisfying the no slip condition on the cylinders (see eqn (2.31)).

314 *Instability*

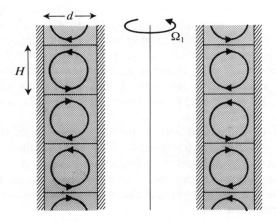

Fig. 9.8. Taylor vortices

Taylor (1923) investigated the stability of this flow to infinitesimal axisymmetric disturbances, and we present below a simplified version of his theory. If the two cylinders rotate in the same sense, so that $\Omega_1 > 0$ and $\Omega_2 > 0$, say, then instability is predicted if the angular velocity of the *inner* cylinder exceeds some critical value depending on Ω_2, r_1, r_2, and ν (see eqn (9.41) and (9.42)). The ensuing motion consists of counter rotating *Taylor vortices* superimposed on a predominantly rotary flow (Fig. 9.8). According to linear theory the magnitude of this secondary flow increases exponentially with time, but non-linear terms eventually cease to be negligible and bring this growth to a halt, so that a steady state is reached, the strength of the Taylor vortices depending on the amount by which Ω_1 exceeds the critical value (Stuart 1986).

Linear stability theory

We first write

$$\boldsymbol{u} = [u'_r, U_\theta(r) + u'_\theta, u'_z], \qquad (9.30)$$

where u'_r, u'_θ, and u'_z are small functions of r, z, and t. Likewise we write $p = p_0(r) + p'(r, z, t)$. Substituting into eqn (2.22) and neglecting quadratically small terms we obtain

$$\frac{\partial u_r'}{\partial t} - \frac{2U_\theta u_\theta'}{r} = -\frac{1}{\rho}\frac{\partial p'}{\partial r} + \nu\left(\nabla^2 u_r' - \frac{u_r'}{r^2}\right),$$

$$\frac{\partial u_\theta'}{\partial t} + u_r'\frac{dU_\theta}{dr} + \frac{u_r' U_\theta}{r} = \nu\left(\nabla^2 u_\theta' - \frac{u_\theta'}{r^2}\right),$$

$$\frac{\partial u_z'}{\partial t} = -\frac{1}{\rho}\frac{\partial p'}{\partial z} + \nu\nabla^2 u_z', \qquad (9.31)$$

$$\frac{1}{r}\frac{\partial}{\partial r}(ru_r') + \frac{\partial u_z'}{\partial z} = 0,$$

where

$$\nabla^2 = \frac{\partial^2}{\partial r^2} + \frac{1}{r}\frac{\partial}{\partial r} + \frac{\partial^2}{\partial z^2}.$$

Let us make the further simplifying assumption that there is a *narrow gap* between the two cylinders, so that $d = r_2 - r_1 \ll r_1$. Then

$$\frac{\partial^2 u_r'}{\partial r^2} = O\left(\frac{u_r'}{d^2}\right), \qquad \frac{1}{r}\frac{\partial u_r'}{\partial r} = O\left(\frac{u_r'}{r_1 d}\right), \qquad \frac{u_r'}{r^2} = O\left(\frac{u_r'}{r_1^2}\right), \quad (9.32)$$

and so on. As a result we may write

$$\left(\frac{\partial}{\partial t} - \nu\tilde{\nabla}^2\right)u_r' - \frac{2U_\theta}{r}u_\theta' = -\frac{1}{\rho}\frac{\partial p'}{\partial r},$$

$$\left(\frac{\partial}{\partial t} - \nu\tilde{\nabla}^2\right)u_\theta' + 2Au_r' = 0, \qquad (9.33)$$

$$\left(\frac{\partial}{\partial t} - \nu\tilde{\nabla}^2\right)u_z' = -\frac{1}{\rho}\frac{\partial p'}{\partial z},$$

$$\frac{\partial u_r'}{\partial r} + \frac{\partial u_z'}{\partial z} = 0,$$

where

$$\tilde{\nabla}^2 = \frac{\partial^2}{\partial r^2} + \frac{\partial^2}{\partial z^2}.$$

We may eliminate p' and u_z' between the first, third, and fourth of these to obtain

$$\left(\frac{\partial}{\partial t} - \nu \bar{\nabla}^2\right) \bar{\nabla}^2 u'_r = 2 \frac{U_\theta}{r} \frac{\partial^2 u'_\theta}{\partial z^2}. \tag{9.34}$$

Now seek normal mode solutions to eqns (9.33) and (9.34), writing

$$u'_r = \mathcal{R}[\hat{u}_r(r)\cos nz\, e^{st}],$$
$$u'_\theta = \mathcal{R}[\hat{u}_\theta(r)\cos nz\, e^{st}], \tag{9.35}$$

so that

$$[\nu(D^2 - n^2) - s]\hat{u}_\theta = 2A\hat{u}_r, \tag{9.36}$$

$$[\nu(D^2 - n^2) - s](D^2 - n^2)\hat{u}_r = 2\frac{U_\theta}{r} n^2 \hat{u}_\theta, \tag{9.37}$$

where D denotes d/dr. The boundary conditions are $u'_r = u'_\theta = u'_z = 0$ at $r = r_1, r_2$, and the first and last of these give

$$\hat{u}_r = D\hat{u}_r = 0 \qquad \text{at } r = r_1, r_2.$$

In view of these conditions, and eqn (9.37), the condition on u'_θ gives

$$D^4 \hat{u}_r - (2n^2 + s/\nu)D^2 \hat{u}_r = 0 \qquad \text{at } r = r_1, r_2.$$

Now, the coefficients of eqns (9.36) and (9.37) are constant, save for the factor U_θ/r, which is the angular velocity of the fluid at any radius r, as given by eqn (9.28). If the cylinders rotate at significantly different angular velocities this will certainly not be constant across the gap, but *suppose now that Ω_1 and Ω_2 are almost equal*. It would then seem reasonable to replace U_θ/r in eqn (9.37) by either one of them, or by their average, $\bar{\Omega}$, say. In that case, eliminating $\hat{u}_\theta$ gives

$$[\nu(D^2 - n^2) - s]^2 (D^2 - n^2)\hat{u}_r = 4A\bar{\Omega} n^2 \hat{u}_r. \tag{9.38}$$

Let us now assume that instability takes place in a non-oscillatory manner, i.e. by one of the eigenvalues s changing from a negative to a positive value. To obtain the marginal state we therefore set $s = 0$ in the above equations, and on introducing

$$x = (r - r_1)/d, \qquad a = nd, \tag{9.39}$$

the problem reduces to

$$\left(\frac{d^2}{dx^2} - a^2\right)^3 \hat{u}_r = -Ta^2 \hat{u}_r,$$

subject to

$$\hat{u}_r = \frac{d\hat{u}_r}{dx} = \frac{d^4\hat{u}_r}{dx^4} - 2a^2 \frac{d^2\hat{u}_r}{dx^2} = 0 \quad \text{at } x = 0, 1, \quad (9.40)$$

where the *Taylor number* T may be written

$$T = \frac{2(\Omega_1 r_1^2 - \Omega_2 r_2^2)\bar{\Omega} d^3}{\nu^2 r_1}, \quad (9.41)$$

on making use of the narrow gap approximation ($r_1 + r_2 \doteq 2r_1$).

For a given dimensionless axial wavenumber a there will be non-trivial solutions $\hat{u}_r$ to the problem (9.40) only for certain discrete values of T, and there will be some least eigenvalue $T_l(a)$ corresponding to each particular a. We seek the minimum of these least eigenvalues $T_l(a)$ over all values of a.

Now, remarkably, the problem (9.40) is mathematically identical to that of thermal instability (set $s = 0$ in eqns (9.25) and (9.26)), and

$$T > 1708 \quad (9.42)$$

thus emerges as the criterion for the centrifugal instability of flow in the narrow gap between two rotating cylinders (see eqn (9.28)). Likewise, the critical value of n is approximately $3.1/d$. The streamlines of the secondary flow take the form shown in Fig. 9.8; the radial flow is periodic in the z-direction with period $2\pi/n$ (see eqn (9.35)), so the height of the cells is

$$H = \pi/n = \pi d/3.1. \quad (9.43)$$

They are therefore almost square in cross-section.

While the final steps in the above analysis are only valid for the case Ω_1 almost equal to Ω_2, it turns out that eqns (9.41) and (9.42) give a remarkably good approximation to the instability criterion more generally, so long as Ω_1 and Ω_2 are of the same sign.

If we take Ω_1 and Ω_2 positive, for convenience, the criterion clearly points to the importance of Ωr^2 *decreasing* with r if instability is to occur; this decrease has to be sufficient, evidently, to overcome the stabilizing effects of viscosity.

Inviscid theory: the Rayleigh criterion

We may gain some physical insight into why a decrease of Ωr^2 is important to the instability mechanism by the following wholly inviscid argument due to von Kármán.

Inviscid rotary flow with velocity $U_\theta(r)$ need not, of course, be of the form (9.28), but in the steady state we must nonetheless have

$$-\frac{U_\theta^2}{r} = -\frac{1}{\rho}\frac{dp_0}{dr}, \qquad (9.44)$$

i.e. the centrifugal force at any radius r must be balanced by a radially inward pressure gradient. Now, if a ring of fluid at radius r_1 with circumferential velocity U_1 is displaced to r_2 ($>r_1$), where the local speed of the fluid is U_2, it will, in the absence of viscous forces, conserve its angular momentum. It will therefore acquire a new velocity U_1' such that $r_1 U_1 = r_2 U_1'$, as its mass will be conserved. But the prevailing inward pressure gradient at r_2 is just that required to hold in place a ring rotating with speed U_2. If $U_1'^2 > U_2^2$ this pressure gradient will, according to eqn (9.44), be too small to offset the centrifugal force of the displaced ring, which will move further out; if $U_1'^2 < U_2^2$ it will be more than sufficient, and the displaced ring will be forced back towards its original position. There should therefore be instability if $U_1^2 r_1^2 > U_2^2 r_2^2$ and stability if $U_1^2 r_1^2 < U_2^2 r_2^2$, and on substituting $U = \Omega r$ we deduce that a necessary and sufficient condition for stability to axisymmetric disturbances is that

$$\frac{d}{dr}(\Omega r^2)^2 \geq 0 \qquad (9.45)$$

throughout the flow. This criterion was in fact first obtained by a different (energy) argument by Lord Rayleigh (1916b), and may also be established by an elegant piece of inviscid stability analysis (Exercise 9.1).

Experimental results

Taylor (1923) carried out some experiments, increasing the angular velocity Ω_1 of the inner cylinder by small amounts until the vortices appeared. He found excellent agreement between that measured value of Ω_1 and the critical value predicted by his linear stability theory. This was something of a breakthrough, for, to quote from the introduction to his paper:

...A great many attempts have been made to discover some mathematical representation of fluid instability, but so far they have been unsuccessful in every case.... Indeed, Orr remarks... 'It would seem improbable that any sharp criterion for stability of fluid motion will ever be arrived at mathematically'.

Orr may have been pessimistic, but the instability of fluid motion continues, to this day, to pose formidable problems. Moreover, the flow between rotating cylinders remains an extraordinarily rich subject for research, and many other aspects of the problem have come to light since Taylor's original study.

It is important to recognize, for example, that with any realistic conditions at the ends of the apparatus the purely rotary flow (9.28) will not be an exact solution of the problem, even when Ω_1 is very small. In the case of stationary rigid boundaries at $z = 0$ and $z = L$, say, eqn (9.28) fails to satisfy the no-slip condition at either end, and even at very small values of Ω_1 the (modified) rotary flow is then accompanied by a weak secondary circulation. In an attempt to minimize such end-effects Taylor used cylinders that were 90 cm long, while in one set of experiments the gap width $d = r_2 - r_1$ was only 0.235 cm. Subsequent experiments with long cylinders have confirmed that as Ω_1 is increased from zero in small steps there is indeed a very rapid development of the vortices as Ω_1 approaches and moves through the critical value corresponding to the instability of the 'infinite-cylinder' flow (9.28). That development is nevertheless an essentially smooth process, and the vortices can be seen spreading from the two ends of the apparatus until they link up to form a continuous chain, as in Fig. 9.8. Furthermore, the end conditions play a crucial part in determining the precise number, and sense of spin, of the vortices that are observed as Ω_1 is gradually increased through this quasi-critical range. More significantly still, it is possible, by the use of more exotic 'switch-on' procedures than a gradual increase in Ω_1, to produce many *different* steady Taylor vortex flows satisfying the same steady boundary conditions, and in such cases the ends of the apparatus again play an essential role in determining what is actually observed, even if the apparatus is very long (see §9.7 and Benjamin and Mullin 1982).

A quite different complication arises from the possibility of time-dependent *wavy vortex* flows (Fig. 9.9). Suppose that Ω_1 is increased in small steps beyond the stage at which the axisymmetric Taylor vortices make their appearance. Then at some critical value of Ω_1 those vortices become unstable to θ-dependent disturbances, and take on the appearance of waves which travel round the apparatus. Non-uniqueness is again in evidence; by different switch-on procedures it is possible to produce several different wavy vortex flows at a single, sufficiently large, value of Ω_1 (Coles 1965).

Extensive reviews of the whole complicated problem have been given by Di Prima and Swinney (1985) and by Stuart (1986). Some excellent photographs of both steady and wavy vortices may be found in Joseph (1976, p. 131), van Dyke (1982, pp. 76–77), Thompson (1982, p. 139), Craik (1985, p. 247), Tritton (1988, p. 259), and, not least, in Taylor's original paper.

9.5. Instability of parallel shear flow

The inviscid theory

Consider the two-dimensional flow of an inviscid fluid between two flat plates $y = -L$ and $y = L$. The basic equations are

$$\frac{\partial u}{\partial t} + u \frac{\partial u}{\partial x} + v \frac{\partial u}{\partial y} = -\frac{1}{\rho} \frac{\partial p}{\partial x},$$

$$\frac{\partial v}{\partial t} + u \frac{\partial v}{\partial x} + v \frac{\partial v}{\partial y} = -\frac{1}{\rho} \frac{\partial p}{\partial y},$$

$$\frac{\partial u}{\partial x} + \frac{\partial v}{\partial y} = 0.$$

The parallel shear flow

$$\boldsymbol{u}_0 = [U(y), 0, 0] \tag{9.46}$$

Instability 321

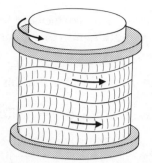

Fig. 9.9. Wavy Taylor vortices.

is an exact solution of these equations for any $U(y)$. The corresponding pressure is constant, p_0.

Let us now consider the linear stability of this flow to two-dimensional disturbances, writing

$$\boldsymbol{u} = [U(y) + u_1, v_1, 0],$$

where u_1 and v_1 are small functions of x, y, and t. Similarly, we write p_1 for the perturbation to the pressure field. Then putting these expressions into the equations and linearizing we have

$$\frac{\partial u_1}{\partial t} + U \frac{\partial u_1}{\partial x} + v_1 U' = -\frac{1}{\rho} \frac{\partial p_1}{\partial x},$$

$$\frac{\partial v_1}{\partial t} + U \frac{\partial v_1}{\partial x} = -\frac{1}{\rho} \frac{\partial p_1}{\partial y},$$

$$\frac{\partial u_1}{\partial x} + \frac{\partial v_1}{\partial y} = 0,$$

where a prime denotes differentiation with respect to y.

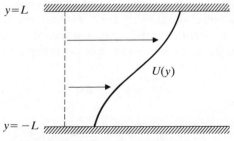

Fig. 9.10. An inviscid shear flow.

322 Instability

The above equations have coefficients which depend on y alone. We therefore explore modes of the form

$$v_1 = \mathcal{R}[\hat{v}(y)e^{i(kx-\omega t)}], \tag{9.47}$$

with similar expressions for u_1 and p_1. In this way we obtain

$$-i(\omega - Uk)\hat{u} + U'\hat{v} = -\frac{1}{\rho}ik\hat{p},$$

$$-i(\omega - Uk)\hat{v} = -\frac{1}{\rho}\hat{p}',$$

$$ik\hat{u} + \hat{v}' = 0.$$

On eliminating $\hat{p}$ and $\hat{u}$ we find

$$\hat{v}'' + \left(\frac{kU''}{\omega - Uk} - k^2\right)\hat{v} = 0, \tag{9.48}$$

subject to

$$\hat{v} = 0 \quad \text{at } y = \pm L, \tag{9.49}$$

as our eigenvalue problem for ω.

At first sight, perhaps, we may proceed no further unless we take special cases, settling on particular velocity profiles $U(y)$ and then solving the eigenvalue problem numerically in each case. We may, however, obtain a sufficient condition for stability by a clever argument due to Lord Rayleigh (1880).

Take eqn (9.48), multiply it by $\bar{v}$, the complex conjugate of $\hat{v}$ and integrate between $-L$ and L to obtain

$$\int_{-L}^{L} \bar{v}\hat{v}'' \, dy + \int_{-L}^{L} \left(\frac{kU''}{\omega - Uk} - k^2\right)|\hat{v}|^2 \, dy = 0. \tag{9.50}$$

The merit of this manoeuvre is that we can obviously say something definite about $|\hat{v}|^2$, even without solving for $\hat{v}$: it is greater than or equal to zero. Now, it is true that the first integral in eqn (9.50) looks troublesome, but on integrating by parts we obtain

$$[\hat{v}'\bar{v}]_{-L}^{L} - \int_{-L}^{L} |\hat{v}'|^2 \, dy + \int_{-L}^{L} \left(\frac{kU''}{\omega - Uk} - k^2\right)|\hat{v}|^2 \, dy = 0.$$

The first term vanishes, because $\hat{v}$ is zero at $y = \pm L$, and so therefore is $\bar{v}$. Now let us write $\omega = \omega_R + i\omega_I$, so that

$$-\int_{-L}^{L} |\hat{v}'|^2 \, dy + \int_{-L}^{L} \left[\frac{(\omega_R - Uk - i\omega_I)kU''}{|\omega - Uk|^2} - k^2 \right] |\hat{v}|^2 \, dy = 0. \quad (9.51)$$

The real and imaginary parts of the left-hand side must individually be zero, and the imaginary part yields

$$\omega_I k \int_{-L}^{L} \frac{U'' |\hat{v}|^2}{|\omega - Uk|^2} \, dy = 0. \quad (9.52)$$

Let us suppose, then, that there is at least one mode which has $\omega_I > 0$, corresponding to exponential growth of the amplitude with time. According to eqn (9.52) this is impossible unless $U''(y)$ changes sign somewhere in the interval, for otherwise the integral cannot vanish. This gives us the following:

Rayleigh's Inflection Point Theorem. A necessary condition for the linear instability of an inviscid shear flow $U(y)$ is that $U''(y)$ should change sign somewhere in the flow.

Note that the presence of an inflection point in the velocity profile is a *necessary* condition for instability to infinitesimal disturbances; there is no claim here that any velocity profile with an inflection point is unstable.

The viscous theory

If the fluid is viscous, the above analysis may easily be modified as far as eqn (9.48) to give

$$i\nu(\hat{\psi}'''' - 2k^2\hat{\psi}'' + k^4\hat{\psi}) + (Uk - \omega)(\hat{\psi}'' - k^2\hat{\psi}) - U''k\hat{\psi} = 0. \quad (9.53)$$

instead. Here the velocity perturbations u_1, v_1 have been written in terms of a perturbation stream function:

$$u_1 = \partial \psi / \partial y, \qquad v_1 = -\partial \psi / \partial x,$$

and

$$\psi = \mathcal{R}[\hat{\psi}(y) e^{i(kx - \omega t)}].$$

The boundary conditions now include no slip; the basic flow

$U(y)$ must satisfy this, and so must the perturbations, so

$$\hat{\psi} = \hat{\psi}' = 0 \quad \text{at } y = \pm L. \tag{9.54}$$

In the case of plane Poiseuille flow, for which

$$U(y) = U_{\max}\left(1 - \frac{y^2}{L^2}\right), \tag{9.55}$$

(see Exercise 2.3), the fourth-order eigenvalue problem consisting of eqns (9.53) and (9.54) leads to a curve of marginal stability as shown in Fig. 9.11(a), so instability occurs for some band of wavenumbers k if

$$R = U_{\max}L/\nu > 5772. \tag{9.56}$$

Now this is interesting, for according to a strictly inviscid theory the velocity profile (9.55) should be stable, as it has no point of inflection. Viscosity therefore plays a dual role: eqn (9.56) shows it to be stabilizing, in the sense that the critical velocity increases with ν; yet if ν were precisely zero there would be no instability at all. Figure 9.11(a) displays the sense in which the inviscid and viscous theories agree, after a fashion, as $R \to \infty$, for the width of the band of unstable wavenumbers k tends to zero in that limit.

For comparison, we show in Fig. 9.11(b) a typical marginal-stability curve, according to the viscous theory, for a velocity

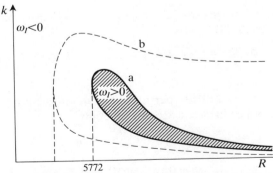

Fig. 9.11. Marginal stability curves for (a) plane Poiseuille flow and (b) a typical velocity profile having an inflection point.

profile which is unstable according to inviscid theory (and which therefore has an inflection point). The curve does not close up in the same way as $R \to \infty$, for viscous effects are no longer crucial to the instability mechanism. The critical Reynolds number is typically a great deal lower than those for profiles with no inflection point.

Experimental results

The criterion (9.56) for plane Poiseuille flow has been confirmed experimentally by Nishioka *et al.* (1975), but to obtain that confirmation they had to take extraordinary pains, keeping the background turbulence below about 0.05 per cent of $U_{\max}$. When $R = 5000$, for example, the flow was indeed stable to sufficiently small disturbances, but they found a definite threshold amplitude of only about 1 per cent of $U_{\max}$ above which disturbances grew.

9.6. A general theorem on the stability of viscous flow

We have been mainly concerned in §§9.2–9.5 with aspects of linear stability theory. We have identified in several systems a critical value R_c of some parameter R, above which infinitesimal disturbances do not remain infinitesimally small as time proceeds. This demonstrates instability when $R > R_c$.

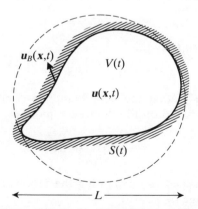

Fig. 9.12. Definition sketch for Serrin's general theorem on the stability of viscous flow. u_B denotes the velocity of the boundary, and may be purely tangential, as in Fig. 9.13.

To pronounce confidently that a system is stable, on the other hand, we need to know the fate of finite-amplitude, as well as infinitesimal disturbances, and we now prove a general theorem due to Serrin (1959):

THEOREM. *Let incompressible viscous fluid occupy a region $V(t)$ which may be enclosed within a sphere of diameter L. Let there be a solution $u(x, t)$ to the Navier–Stokes equations in $V(t)$ satisfying the boundary condition $u = u_B(x, t)$ on $S(t)$, the boundary of $V(t)$. Let u_M be an upper bound to $|u|$ in $V(t)$ for all time t. Let there be another solution $u^*(x, t)$ which satisfies the same boundary condition, but suppose that u and u^* satisfy different initial conditions at $t = 0$. Then the kinetic energy $\mathscr{E}$ of the 'difference flow' $v = u^* - u$ satisfies*

$$\mathscr{E} \leq \mathscr{E}_0 e^{(u_M^2 - 3\pi^2 \nu^2/L^2)t/\nu}, \tag{9.57}$$

where $\mathscr{E}_0$ is its initial value. Thus if

$$R = \frac{u_M L}{\nu} < \pi\sqrt{3} \tag{9.58}$$

then $\mathscr{E} \to 0$ as $t \to \infty$, and the flow u is stable.

To prove the theorem we consider the difference motion

$$v = u^* - u \tag{9.59}$$

which has the property that

$$v = 0 \quad \text{on } S(t), \tag{9.60}$$

because u and u^* satisfy the same boundary conditions. Define a 'kinetic energy' based on the difference motion

$$\mathscr{E} = \tfrac{1}{2}\rho \int_{V(t)} v^2 \, dV. \tag{9.61}$$

The analysis proceeds by exploiting the following expression for the rate of change of $\mathscr{E}$:

$$\frac{1}{\rho}\frac{d\mathscr{E}}{dt} = \int_{V(t)} \left[\frac{\partial v_i}{\partial x_j} v_j u_i - \nu \left(\frac{\partial v_i}{\partial x_j} \right)^2 \right] dV. \tag{9.62}$$

Proof of eqn (9.62)

Both u and u^* are solutions of the Navier–Stokes equations, so

$$\frac{\partial u}{\partial t} + (u \cdot \nabla)u = -\frac{1}{\rho}\nabla p + \nu\nabla^2 u,$$

$$\frac{\partial u^*}{\partial t} + (u^* \cdot \nabla)u^* = -\frac{1}{\rho}\nabla p^* + \nu\nabla^2 u^*.$$

Subtracting, and writing $u^* = u + v$, we have

$$\frac{\partial v}{\partial t} + (v \cdot \nabla)u + (u^* \cdot \nabla)v = -\nabla P + \nu\nabla^2 v,$$

where $P = (p^* - p)/\rho$. In suffix notation this becomes

$$\frac{\partial v_i}{\partial t} + v_j\frac{\partial u_i}{\partial x_j} + u_j^*\frac{\partial v_i}{\partial x_j} = -\frac{\partial P}{\partial x_i} + \nu\frac{\partial^2 v_i}{\partial x_j^2},$$

where the summation convention is understood. Multiplying by v_i gives

$$\frac{\partial}{\partial t}(\tfrac{1}{2}v_i^2) = -v_i v_j\frac{\partial u_i}{\partial x_j} - v_i u_j^*\frac{\partial v_i}{\partial x_j} - v_i\frac{\partial P}{\partial x_i} + \nu v_i\frac{\partial^2 v_i}{\partial x_j^2}.$$

We now try to write as much of this as possible in divergence form:

$$\frac{\partial}{\partial t}(\tfrac{1}{2}v_i^2) = \frac{\partial}{\partial x_j}\left[-v_i v_j u_i - \tfrac{1}{2}v_i^2 u_j^* - v_j P + \nu v_i\frac{\partial v_i}{\partial x_j}\right]$$

$$+ \frac{\partial v_i}{\partial x_j}v_j u_i + v_i\frac{\partial v_j}{\partial x_j}u_i + \tfrac{1}{2}v_i^2\frac{\partial u_j^*}{\partial x_j} + \frac{\partial v_j}{\partial x_j}P - \nu\left(\frac{\partial v_i}{\partial x_j}\right)^2.$$

The middle three of the last five terms vanish, for $\nabla \cdot u = \nabla \cdot u^* = \nabla \cdot v = 0$. On integrating over $V(t)$ and applying the divergence theorem we have

$$\int_{V(t)}\frac{\partial}{\partial t}(\tfrac{1}{2}v^2)\,dV = \int_{S(t)}\left[-v_i v_j u_i - \tfrac{1}{2}v_i^2 u_j^* - v_j P + \nu v_i\frac{\partial v_i}{\partial x_j}\right]n_j\,dS$$

$$+ \int_{V(t)}\left[\frac{\partial v_i}{\partial x_j}v_j u_i - \nu\left(\frac{\partial v_i}{\partial x_j}\right)^2\right]dV.$$

328 Instability

But $v = 0$ on $S(t)$, so the surface integral is zero. Furthermore, Reynolds's transport theorem (6.6a) may be written in the form

$$\frac{\mathrm{d}}{\mathrm{d}t}\int_{V(t)} G\,\mathrm{d}V = \int_{V(t)} \frac{\partial G}{\partial t}\,\mathrm{d}V + \int_{S(t)} G u_n\,\mathrm{d}S,$$

where $G(\mathbf{x}, t)$ is any scalar function and u_n denotes the normal velocity of the points of the boundary $S(t)$. Setting $G = \frac{1}{2}v^2$, and using the fact that $v = 0$ on $S(t)$, we establish eqn (9.62).

Proof of the theorem

Start by observing that for any A_{ij}

$$\left(\frac{\partial v_i}{\partial x_j}\right)^2 + 2A_{ij}\frac{\partial v_i}{\partial x_j} + A_{ij}^2 = \left(\frac{\partial v_i}{\partial x_j} + A_{ij}\right)^2 \geq 0. \quad (9.63)$$

Choosing $A_{ij} = -u_i v_j / \nu$ gives

$$u_i v_j \frac{\partial v_i}{\partial x_j} \leq \frac{1}{2\nu} u_i^2 v_j^2 + \tfrac{1}{2}\nu\left(\frac{\partial v_i}{\partial x_j}\right)^2,$$

and on substituting in eqn (9.62) we obtain

$$\frac{\mathrm{d}\mathscr{E}}{\mathrm{d}t} \leq \frac{\rho}{2\nu}\int_{V(t)}\left[u_i^2 v_j^2 - \nu^2\left(\frac{\partial v_i}{\partial x_j}\right)^2\right]\mathrm{d}V$$

$$\leq \frac{1}{\nu}\left[u_M^2 \mathscr{E} - \frac{\rho}{2}\int_{V(t)} \nu^2\left(\frac{\partial v_i}{\partial x_j}\right)^2 \mathrm{d}V\right]. \quad (9.64)$$

Let us now return to eqn (9.63) and put $A_{ij} = h_j v_i$, where we hope to choose the vector function $\mathbf{h}(\mathbf{x}, t)$ to advantage. Then

$$\left(\frac{\partial v_i}{\partial x_j}\right)^2 \geq -2h_j v_i \frac{\partial v_i}{\partial x_j} - h_j^2 v_i^2$$

$$= -\frac{\partial}{\partial x_j}(h_j v_i^2) - h_j^2 v_i^2 + v_i^2 \frac{\partial h_j}{\partial x_j}.$$

Integrating over $V(t)$, and using the divergence theorem with $v_i = 0$ on the boundary $S(t)$, we obtain

$$\int_{V(t)} \left(\frac{\partial v_i}{\partial x_j}\right)^2 \mathrm{d}V \geq \int_{V(t)} (\nabla \cdot \mathbf{h} - \mathbf{h}^2) v_i^2\,\mathrm{d}V.$$

Suppose, then, that we manage to pick a vector function $\boldsymbol{h}(\boldsymbol{x}, t)$ such that

$$\nabla \cdot \boldsymbol{h} - \boldsymbol{h}^2 \geq C, \qquad \text{say, in } V(t), \qquad (9.65)$$

where C is some positive constant. Then

$$\int_{V(t)} \left(\frac{\partial v_i}{\partial x_j}\right)^2 dV \geq \frac{2C}{\rho} \mathscr{E},$$

and, by eqn (9.64),

$$\frac{d\mathscr{E}}{dt} \leq \frac{1}{\nu}(u_M^2 - C\nu^2)\mathscr{E}. \qquad (9.66)$$

Our final task is to find a function $\boldsymbol{h}$ such that eqn (9.65) holds, for some $C > 0$. It is at this stage in the argument that the linear dimensions of $V(t)$ enter, and we are supposing it is possible to choose an origin such that $V(t)$ always lies inside a sphere, centre the origin, of diameter L (see Fig. 9.12). The simplest kind of $\boldsymbol{h}$ to contemplate is $\boldsymbol{h} = h(r)\boldsymbol{e}_r$, in which case

$$\nabla \cdot \boldsymbol{h} - \boldsymbol{h}^2 = \frac{1}{r^2}\frac{d}{dr}(r^2 h) - h^2.$$

One satisfactory $h(r)$ is $4r/L^2$, for then

$$\nabla \cdot \boldsymbol{h} - \boldsymbol{h}^2 = \frac{4}{L^2}\left(3 - \frac{4r^2}{L^2}\right),$$

so in $r < \frac{1}{2}L$ we have

$$\nabla \cdot \boldsymbol{h} - \boldsymbol{h}^2 > 8/L^2.$$

We may therefore put $C = 8/L^2$ in eqn (9.66). But a better $h(r)$ is

$$h(r) = \frac{\pi}{L}\tan\left(\frac{\pi r}{L}\right),$$

for this is differentiable in $r < \frac{1}{2}L$ and gives

$$\nabla \cdot \boldsymbol{h} - \boldsymbol{h}^2 = \left(\frac{\pi}{L}\right)^2\left(1 + \frac{2}{x}\tan x\right),$$

where $x = \pi r/L$. On minimizing this over x we find that in $r < \tfrac{1}{2}L$

$$\nabla \cdot \boldsymbol{h} - \boldsymbol{h}^2 \geq 3(\pi/L)^2.$$

We may therefore put $C = 3\pi^2/L^2$ in eqn (9.66), which makes for a stronger bound on $d\mathscr{E}/dt$, and from it we deduce immediately that

$$\mathscr{E} \leq \mathscr{E}_0 e^{(u_M^2 - 3\pi^2 \nu^2/L^2)t/\nu}, \qquad (9.67)$$

which proves the theorem.

9.7. Uniqueness and non-uniqueness of steady viscous flow

An immediate corollary of the theorem in §9.6 is the following result, also due to Serrin (1959):

THEOREM. *Let a fixed fluid region V be of such size that it may be enclosed within a sphere of diameter L. Let $\boldsymbol{u}$ and $\boldsymbol{u}^*$ be two steady solutions of the Navier–Stokes equations in V, having the same velocity $\boldsymbol{u}_B(\boldsymbol{x})$ on the boundary of V. Let u_M denote an upper bound to $|\boldsymbol{u}|$ in V. Then if*

$$R = \frac{u_M L}{\nu} < \pi\sqrt{3} \qquad (9.68)$$

the two flows must be identical, i.e. $\boldsymbol{u} = \boldsymbol{u}^*$.

In other words, if we have steady viscous flow in V satisfying eqn (9.68) and the boundary conditions, it is the *only* steady viscous flow in V satisfying those conditions.

The proof is extraordinarily simple, but rests on the unusual step of thinking about the steady flows $\boldsymbol{u}$ and $\boldsymbol{u}^*$ as time proceeds. Precisely because these do not change at all, the kinetic energy $\mathscr{E}$ of the difference motion $\boldsymbol{u}^* - \boldsymbol{u}$ must be constant. But $\mathscr{E}$ must also satisfy eqn (9.57). If eqn (9.68) is satisfied, the only way both these constraints can be satisfied as $t \to \infty$ is by $\mathscr{E}$ being zero, which implies $\boldsymbol{u} = \boldsymbol{u}^*$.

An example of non-uniqueness of steady flow

Let us consider again the Taylor experiment of §9.4 in which viscous fluid occupies the gap $r_1 < r < r_2$ between two cylinders.

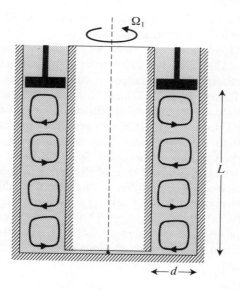

Fig. 9.13. A Taylor-vortex apparatus with variable aspect ratio $\Gamma = L/d$.

Suppose that the inner cylinder rotates with angular velocity Ω_1 but that the outer cylinder is fixed. Let the two plane ends of the apparatus, $z = 0$ and $z = L$, also be fixed, but suppose that the top end $z = L$ is adjustable (Fig. 9.13) so that the length L of the apparatus may be varied. We may characterize the system by three dimensionless parameters, namely the radius ratio r_1/r_2, a Reynolds number

$$R = \Omega_1 r_1 d/\nu, \qquad (9.69)$$

and an aspect ratio

$$\Gamma = L/d, \qquad (9.70)$$

where $d = r_2 - r_1$ is the gap width.

Now, for the particular values

$$r_1/r_2 = 0.6, \quad \Gamma = 12.61, \quad R = 359, \qquad (9.71)$$

Benjamin and Mullin (1982) demonstrated experimentally no fewer than 20 different stable steady flows in this apparatus, and they inferred on theoretical grounds the existence of a further 19 steady flows which were unstable and consequently not observed. All the 20 flows observed were of an axisymmetric cellular nature, as in Fig. 9.8, but they were distinguished by having different numbers of cells and/or a different sense of rotation within each individual cell. Which flow was observed depended on how the steady boundary conditions (9.71) were achieved from an initial state of rest. If R was gradually increased in small steps from 0 to 359, the same flow consisting of 12 cells was always observed. Some of the other flows could be produced by sudden starts of the rotation rate of the inner cylinder, once the various transients had died down. Others were produced in a still more devious manner, by first setting Γ at a different value from that in eqn (9.71), then increasing R to 359, and then changing Γ in small steps to its final value of 12.61. Benjamin and Mullin provide excellent photographs of these flows.

Hysteresis

With all these different flows around, a point of major interest is, of course, how one flow evolves into another as the parameters of the problem are changed. We illustrate this with reference to some earlier experiments by Benjamin (1978) on very short cylinders, with $\Gamma = L/d$ at most 4 or 5. The transition between two-cell and four-cell modes may be indicated schematically on a diagram of the kind in Fig. 9.14, which catastrophe theory has now made so familiar (see, e.g., Thompson 1982). The fold in the surface implies a multiplicity of solutions in certain parts of the R–Γ plane, the character of these solutions being broadly as described on the sketch. The middle sheet of the fold corresponds to an unstable solution, which is consequently not observed.

For a good example of how the observed steady solution may depend on the starting-up process, suppose that Γ is 3.8, i.e. greater than the value corresponding to the point B in Fig. 9.14, and suppose that the inner cylinder is initially at rest. As R is gradually increased from zero to 100, say, the state of the system progresses smoothly along the upper sheet in Fig. 9.14, and ends up at the top right-hand corner as a clear four-cell mode (Fig. 9.13). We may, however, first trick the system

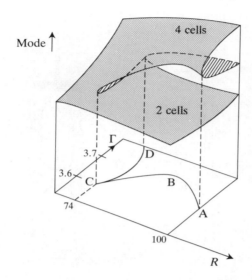

Fig. 9.14. State diagram of the two-cell/four-cell transition in the experiments of Benjamin (1978), for which $r_1/r_2 = 0.615$.

into producing a two-cell mode by starting with Γ below the value corresponding to C on the diagram, then increasing R from 0 to 100, then increasing Γ to a value of 3.8. The last step will simply have the effect of stretching the two-cell mode in such a way that we end up on the lower solution surface in Fig. 9.14, as in Fig. 9.15.

Now, if we reverse that sequence of boundary conditions the sequence of steady flows also reverses. More generally, we expect *hysteresis*. By way of example, fix Γ somewhere between the values corresponding to the points C and B in Fig. 9.14. As R is increased from zero the flow at first shows traits of both a two-cell and a four-cell structure. It develops continuously until the curve CB is reached, at which stage we drop over the edge in Fig. 9.14, so to speak, and there is an abrupt transition to a clear two-cell structure. If R is then reduced, the new two-cell form changes continuously until the curve CD is reached, at which stage the system jumps back to the state it originally had at that particular value of R.

334 *Instability*

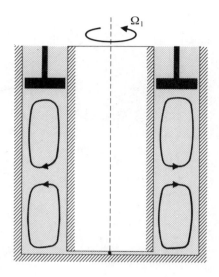

Fig. 9.15. Another Taylor-vortex flow satisfying the same boundary conditions as that in Fig. 9.13.

9.8. Instability, chaos, and turbulence

We begin our short treatment of this enormous topic with the opening remarks from Lorenz's highly influential paper 'Deterministic non-periodic flow' (1963):

Certain hydrodynamical systems exhibit steady-state flow patterns, while others oscillate in a regular periodic fashion. Still others vary in an irregular, seemingly haphazard manner, and, even when observed for long periods of time, do not appear to repeat their previous history.

Lorenz had in mind, in particular, some experiments on thermally driven motions in a rotating annulus. Here the inner and outer cylinders rotate with the same angular velocity Ω, and the fluid would rotate as a solid body were it not for the fact that the outer cylinder is heated and the inner cylinder is cooled. This, then, is the atmosphere stripped to its bare essentials, namely a basic rotation and some differential heating. At sufficiently small values of Ω a weak differential rotation is observed, as in Fig. 9.16(*a*). As Ω is increased in small steps past

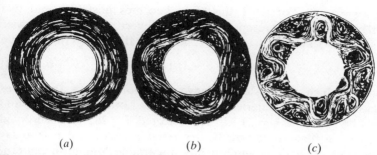

(a)　　　　　　　(b)　　　　　　　(c)

Fig. 9.16. Three types of flow in a differentially heated rotating annulus, viewed from the rotating frame (after Hide 1977).

a critical value (which depends on the temperature difference between the cylinders), this flow succumbs to *baroclinic instability*, which takes the form of amplifying, non-axisymmetric waves (see Hide 1977). As Ω is increased further the amplitude of these waves increases, and a distinctive meandering jet structure emerges (Fig. 9.16(b)), reminiscent of the jet stream in the atmosphere. The amplitude, shape, or wavenumber of this jet may be steady or may vary in a periodic manner. But at higher values of Ω still these variations become irregular, and the waves show complicated aperiodic fluctuations (Fig. 9.16(c)). It was this type of behaviour that interested Lorenz.

The analysis in his 1963 paper was, however, for a thermally convecting system of the kind in §9.3. By means of some drastic approximations he obtained three ordinary differential equations:

$$\dot{x} = \sigma(y - x)$$
$$\dot{y} = rx - y - xz, \quad (9.72)$$
$$\dot{z} = -bz + xy.$$

Here $x(t)$ is proportional to the intensity of the convective motion, while $y(t)$ and $z(t)$ represent certain broad features of the temperature field in the fluid. The parameter r denotes the ratio of the actual Rayleigh number to its critical value for the onset of convection, the parameter b acts as a measure of the horizontal extent of the convection cells (and is not really externally controllable, of course), while σ denotes the ratio ν/κ (see §9.3). For $r < 1$ the only steady state is that of no motion,

336 Instability

$x = y = z = 0$. For $r > 1$ this state becomes unstable, and two others appear, representing steady convection rolls (clockwise and anticlockwise). If $\sigma > b + 1$, however, there is a critical value

$$r_c = \sigma(\sigma + b + 3)/(\sigma - b - 1) \qquad (9.73)$$

above which these steady convective motions are themselves unstable. In his numerical computations Lorenz took $b = \frac{8}{3}$ and $\sigma = 10$, so that $r_c = 24.74$. He selected $r = 28$, and observed behaviour that would now be described as *chaotic*, that is to say irregular oscillations without any discernible long-term pattern. Moreover, two very slightly different sets of initial conditions would lead, eventually, to completely different behaviour. Lorenz saw this to be a general feature of chaotic, or non-periodic dynamics, and realized the implications only too well:

> When our results concerning the instability of non-periodic flow are applied to the atmosphere, which is ostensibly non-periodic, they indicate that prediction of the sufficiently distant future is impossible by any method, unless the present conditions are known exactly. In view of the inevitable inaccuracy and incompleteness of weather observations, precise very-long-range forecasting would seem to be non-existent.

Many other systems of evolution equations possess chaotic solutions. Perhaps the simplest is the non-linear difference equation

$$x_{n+1} = \lambda x_n(1 - x_n), \qquad (9.74)$$

where λ is a constant. This serves as a simple model for biological populations (May 1976). We shall restrict attention to initial values x_0 which lie in the interval $[0, 1]$; it follows that x_n will also lie in that interval if $0 < \lambda < 4$.

For $0 < \lambda < 1$ the solution x_n tends to the steady solution $x = 0$ as $n \to \infty$. For $1 < \lambda < 3$, x_n tends to the steady solution $x = 1 - 1/\lambda$ as $n \to \infty$. If $\lambda > 3$, both these steady solutions are unstable (Exercise 9.6). For $3 < \lambda < 3.449$ there is an oscillatory solution with period 2, i.e. such that $x_{n+2} = x_n$ (Fig. 9.17(*b*)). When λ exceeds 3.449 this oscillatory solution itself becomes unstable, but a period 4 solution then appears, which is stable (Fig. 9.17(*c*)). This *period doubling* continues indefinitely as λ approaches the value $\lambda_\infty = 3.570$, the gap between successive

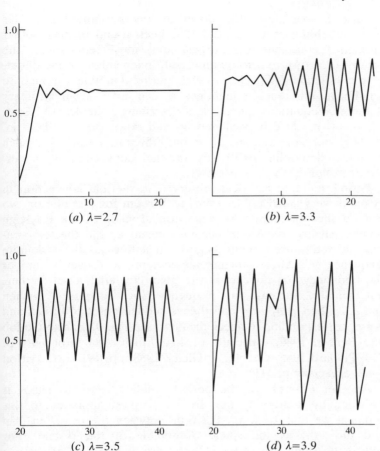

Fig. 9.17. Solutions of the non-linear difference equation $x_{n+1} = \lambda x_n(1-x_n)$, with $x_0 = 0.1$. In case (d) the starting condition $x_0 = 0.100000001$ leads to a completely different result for $n \geq 30$.

period-doublings diminishing rapidly according to the law

$$\frac{(\lambda_m - \lambda_{m-1})}{(\lambda_{m+1} - \lambda_m)} \to 4.669 \quad \text{as } m \to \infty. \tag{9.75}$$

(Feigenbaum 1980). For $\lambda > \lambda_\infty$ chaotic solutions are possible (Fig. 9.17(d)); for a typical starting value x_0 the subsequent behaviour does not settle down to any periodic form. This period-doubling route to chaos has been found in many other difference equations $x_{n+1} = f(x_n)$ and, most remarkably, the result (9.75) turns out to be universal, independent of the detail of the function $f(x)$. Nor is such period-doubling confined to non-linear difference equations; it can arise from ordinary differential equations, such as that describing a simple pendulum the pivot of which is oscillated up and down (Moon 1987, pp 79–80), and from systems of partial differential equations, such as those describing oscillatory thermal convection in a salt stratified fluid (Moore et al. 1983).

Period-doubling has been observed in thermal convection in boxes of very small aspect ratio, with room for just one or two cells in the horizontal. At some critical value of the Rayleigh number steady convection becomes unstable, and the temperature at some fixed point begins oscillating at some definite frequency ω_1. After a further increase of the Rayleigh number the frequency $\frac{1}{2}\omega_1$ appears in the spectrum, after a yet further but smaller, increase in frequency $\frac{1}{4}\omega_1$ appears..., then $\frac{1}{8}\omega_1$..., then $\frac{1}{16}\omega_1$..., and then a sudden onset of broadband noise, corresponding to aperiodic flow (see Fig. 9.18; also Gollub and Benson 1980, especially p. 464; Miles 1984, especially p 210; Pippard 1985, Chapter 4; Gleick 1988, pp. 191–211; Tritton 1988, especially p. 411).

Another example of the period-doubling route to chaos is provided by a dripping tap. In the simplest approach to this problem we may treat x_n, the time between the $(n-1)$th drip and the nth, as the single observable of the system. For sufficiently small flow rates, Q, the dripping is regular, with a single period. As Q is increased the dripping sequence takes to repeating itself after two drips, then at higher Q still after four and so on, with chaos eventually setting in at some definite value of Q (see Moon 1987, pp. 116–117; Gleick 1988, pp. 262–267 Tritton 1988, p. 409).

But successive period-doubling is not the only route to chaos. Consider again Taylor vortex flow between two cylinders, and suppose Ω_1 to be sufficiently large that the system is in the 'wavy vortex' regime of Fig. 9.9. At higher values of Ω_1 still this wavy vortex flow becomes unstable, and in the experiments of Fenstermacher et al. (1979) a second frequency appeared in the

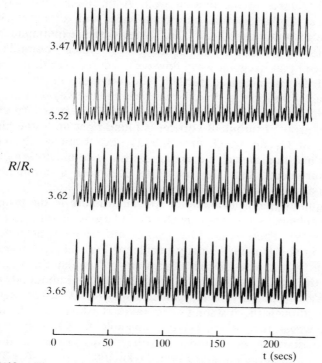

Fig. 9.18. The thermal convection experiment of Libchaber *et al.* (1982): direct time recordings of temperature for various stages of the period doubling cascade.

spectrum, incommensurate with the first. In that particular experiment this happened when Ω_1 was about 10 times the value at which Taylor vortices appeared. When Ω_1 was increased to about 12 times that value, there was a sudden appearance of broadband noise. This route to chaos is not a period-doubling one, then, but appears instead to be in keeping with one proposed by Ruelle and Takens (see Ruelle 1980; Lanford 1985), which again emerged from studies of finite-dimensional systems of ODEs such as eqn (9.72).

So far we have been solely concerned with the question of how irregular fluctuations in time may appear in a fluid flow. Yet in looking to understand how *turbulence* arises we seek to explain spatial features of the flow as well. A turbulent flow may, for instance, display disorder on certain scales and yet a remarkable degree of order on others. The Taylor vortex experiment again provides an excellent, but not necessarily typical, example. We have seen that irregular wavy flow occurs at some critical value of Ω_1. If Ω_1 is increased further, the waviness of the vortices eventually disappears, but the vortices themselves do not, even though each one is in an increasingly turbulent state. An evenly spaced array of turbulent vortices is found right up to the highest values of Ω_1 for which experiments have been conducted. A major question in such circumstances is not so much 'Why is the flow turbulent?' but 'How on earth does such a turbulent flow retain a large-scale spatially periodic structure?'.

A quite different example of spatial structure in the transition to turbulence is provided by the boundary layer on a flat plate (Fig. 9.19). There is no adverse pressure gradient, no inflection point in the laminar velocity profile of Fig. 8.8, yet instability occurs by the viscous mechanism in §9.5 when the boundary layer thickness δ grows to the point that $U\delta/\nu$ is about 500 or so, which corresponds to a Reynolds number Ux/ν in the region of 10^5. Although the instability first takes the form of 2-D waves, these waves themselves become unstable to 3-D disturbances, and a startling development further downstream is the appearance of turbulent *spots* (van Dyke 1982, pp. 62–65; Tritton

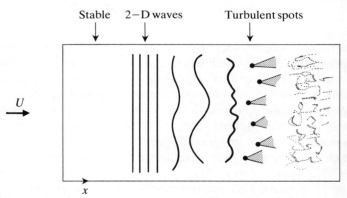

Fig. 9.19. Transition to turbulence in the boundary layer on a flat plate (the plate is in the plane of the paper).

1988, p. 282). These eventually coalesce to form a fully developed turbulent boundary layer.

Fully developed turbulence has a rich spatial structure of its own, and is characterized not only by rapid, irregular velocity fluctuations at any particular point in space, but by motions on many different length scales at once. It is in general a fully three-dimensional affair, with energy being transferred by non-linear processes from large-scale motions to smaller-scale eddies. These processes include the tortuous stretching and twisting of vortex lines, as described by Helmholtz's theorems (§5.3); viscosity is typically important only for the smallest-scale eddies, where it dissipates the energy that has been passed down from larger scales.

Until comparatively recently the notion of two-dimensional turbulence was largely dismissed as a theoretical abstraction, but experiments on rapidly rotating fluids (cf. §8.5), on electrically conducting fluids subject to strong magnetic fields, and on thin liquid films have renewed interest in the subject (see Couder and Basdevant (1986) for references). The most distinctive feature of two-dimensional turbulence is the way in which energy can be transferred from small scales to large scales. Random small-scale forcing can then lead to the emergence of comparatively large-scale flow features, by a process akin to the vortex merging described towards the end of §5.8.

9.9. Instability at very low Reynolds number

The powerful theorem of §9.6 guarantees stability of flow, at sufficiently low Reynolds number, when u is prescribed at some known, but possibly varying boundary. It does not extend to flows in which free boundaries are involved, and there are several known instabilities at low Reynolds number involving free boundaries. In these closing pages we present just two examples.

Viscous fingering in a Hele-Shaw cell

Take two sheets of transparent plastic, and drill a hole in one of them to accommodate the nozzle of a small syringe. Put a generous blob of golden syrup on the other one, and press one

342 Instability

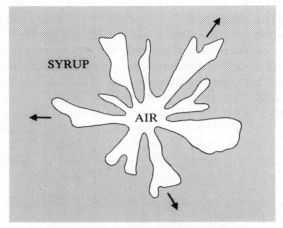

Fig. 9.20. The Saffman–Taylor instability.

sheet down on top of the other, using coins to keep the sheets about 2 mm apart, so forming a rudimentary Hele-Shaw cell (see §7.7). Now inject air by pressing down on the syringe. In principle one might expect the air to displace the golden syrup in a symmetrical manner, the interface between the two fluids remaining circular as its radius grows with time. Such a flow is, however, unstable, and the air/syrup interface develops ripples which grow rapidly into large-amplitude fingers (Fig. 9.20).

This kind of behaviour is liable to happen whenever a more viscous fluid is displaced by a less viscous one, and it is known as the Saffman–Taylor instability. Homsy (1987) gives a good review with photographs, and further excellent photographs (some of them showing fractal fluid behaviour) may be found in Walker (1987) and Chen (1989).

The buckling of viscous jets

Take a pot of golden syrup, spoon out a generous helping, and let it drain slowly back into the pot. If the height H of the falling jet is less than some critical value H_c the jet will be stable and will remain more or less symmetric about a vertical axis, as in Fig. 9.21(a). If $H > H_c$, on the other hand, the jet will be unstable and will buckle, coiling up at the base as in Fig. 9.21(b). Some

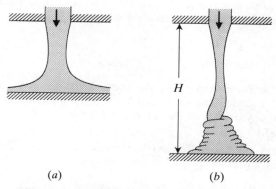

Fig. 9.21. The buckling of a viscous jet.

excellent photographs of this may be found in Cruickshank (1988).

This instability is particularly strange, when viewed against the background of the rest of this chapter, in that it occurs—other things being equal—only if the Reynolds number is *less* than some critical value.

Exercises

9.1. Consider the rotary flow $U_\theta(r)\mathbf{e}_\theta$ of an inviscid fluid. Examine its stability to small-amplitude axisymmetric disturbances by writing

$$\mathbf{u} = [u_r^*, U_\theta(r) + u_\theta^*, u_z^*], \qquad p = p_0(r) + p^*,$$

in Euler's equations, where u_r^*, u_θ^*, and u_z^* are functions of r, z, and t. Linearize the resulting equations, and examine modes of the form

$$u_r^* = \hat{u}_r(r) e^{inz+st},$$

with similar expressions for the other perturbation variables, the real part being understood. Show by elimination that

$$(r\hat{u}_r')' - \left[\frac{n^2}{s^2 r^3}(r^2 U_\theta^2)' + n^2 + \frac{1}{r^2}\right] r\hat{u}_r = 0.$$

If the fluid is contained between two circular cylinders, $r = r_1$ and $r = r_2$, then

$$\hat{u}_r = 0 \qquad \text{at } r = r_1, r_2,$$

344 *Instability*

and we have an eigenvalue problem for s. By a method similar to that used on eqn (9.48), show that the flow is

stable if $(U_\theta^2 r^2)' > 0$ in $r_1 \leq r \leq r_2$,

unstable if $(U_\theta^2 r^2)' < 0$ in $r_1 \leq r \leq r_2$.

[It is in fact possible to invoke the Sturm–Liouville theory of ordinary differential equations to show that the flow is unstable if $(U_\theta^2 r^2)'$ is negative in any portion of the interval $r_1 \leq r \leq r_2$ (see Drazin and Reid 1981, p. 78).]

9.2. It may be shown that small-amplitude 2-D disturbances to the shear flow $U(y)$ of an inviscid stratified fluid are governed by

$$\hat{v}'' + \frac{\rho_0'}{\rho_0}\hat{v}' + \left[\frac{N^2}{(c-U)^2} + \frac{U''}{c-U} + \frac{\rho_0'}{\rho_0}\frac{U'}{(c-U)} - k^2\right]\hat{v} = 0,$$

where $c = \omega/k$, and N denotes the buoyancy frequency, defined by

$$N^2 = -\frac{g}{\rho_0}\frac{d\rho_0}{dy}.$$

Verify that this equation reduces to eqn (3.84) in the case $U = 0$ and to eqn (9.48) in the case of constant density. Verify too that if the density $\rho_0(y)$ varies with height much more slowly than either $U(y)$ or $\hat{v}(y)$ then the equation reduces to the Taylor–Goldstein equation

$$\hat{v}'' + \left[\frac{N^2}{(c-U)^2} + \frac{U''}{c-U} - k^2\right]\hat{v} = 0.$$

Let the shear flow take place between two plane rigid boundaries, so that

$$\hat{v} = 0 \quad \text{at } y = \pm L.$$

Make the change of variable

$$\hat{v} = (U-c)^n q,$$

where n is a parameter at our disposal, rewrite the Taylor–Goldstein equation as an equation for q, and apply the method of §9.5 to show that

$$\int_{-L}^{L} (U-c)^{2n}[|q'|^2 + k^2|q|^2]\, dy$$

$$= \int_{-L}^{L} [\{N^2 + n(n-1)U'^2\}(U-c)^{2n-2} + (n-1)U''(U-c)^{2n-1}]|q|^2\, dy$$

Write $c = c_R + ic_I$, and by choosing n suitably show that c_I must be zero, so that the flow is stable, if

$$\frac{N^2}{(dU/dy)^2} > \tfrac{1}{4} \quad \text{in } -L \leq y \leq L.$$

By making a suitably different choice of n, show that if, on the other hand, the flow is unstable, then the wave speed c_R of any amplifying mode must lie between the least and greatest values of $U(y)$ in the interval $-L \leq y \leq L$.

9.3. *Salt fingering.* Suppose that a layer of hot, salty water lies on top of a layer of cold, fresh water. Suppose too that the effect of the difference in temperature outweighs that of the difference in salinity, so that the density of the upper fluid is less than that of the lower fluid. Even though the system is bottom-heavy it may be unstable, and tall, thin convection cells known as 'salt fingers' may develop at the interface (see Turner 1973, pp. 251–259; Tritton 1988, pp. 378–385).

Try to explain the instability, by considering the fate of a small fluid parcel which is displaced from the lower layer into the upper layer, allowing for the fact that heat and salt diffuse in water at very different rates.

9.4. In §9.1 the Reynolds experiment illustrated how a system which is stable to infinitesimal disturbances may yet be unstable to disturbances of finite amplitude. In §9.7 we used the Taylor vortex flow between two cylinders to illustrate non-uniqueness and hysteresis.

Some feeling for how all these ideas are related may be obtained from a very simple set of experiments with a length of net-curtain wire. Hold it vertically with, say, a pair of pliers, so that a length L of the wire extends vertically upward from the support. Observe that:

(i) If L is increased in small steps, and disturbances are kept to a minimum, the wire is stable in its vertical position until L reaches some value L_c, when it suddenly flops down into a new steady state.

(ii) If L is then decreased in small steps, the wire does not immediately revert to the vertical position; the new state changes continuously until L becomes less than some value $L_E < L_c$, at which stage the wire suddenly springs back to the vertical position.

(iii) For values of L such that $L_E < L < L_c$ the vertical position is stable to disturbances of small magnitude, but not to sufficiently large disturbances.

[The wire does not typically confine its movement to one vertical plane, of course, and the experiment may be conducted in a more controlled manner by passing both ends of the wire through holes in board, so that the stability of a vertical arch is investigated (see Joseph (1985) for a fuller discussion).]

9.5. *The Landau equation.* Suppose that a fluid system becomes unstable according to linear theory as some parameter R is increased beyond a critical value R_c. Then, in certain special circumstances, provided $R - R_c$ is small, the evolution of the disturbance amplitude $|A|$ is governed by

$$d|A|^2/dt = \beta(R - R_c)|A|^2 - l|A|^4.$$

Here β is a positive constant, such that $\frac{1}{2}\beta(R - R_c)$ represents the growth/decay rate according to linear theory, and l is a constant which may be positive or negative, depending on whether finite-amplitude effects are stabilizing or destabilizing.

Solve this equation for $|A|^2$, given that $|A| = A_0$ at $t = 0$. If $l > 0$ show that:

(i) $|A| \to 0$ as $t \to \infty$ if $R < R_c$;

(ii) $|A| \to [\beta(R - R_c)/l]^{\frac{1}{2}}$ as $t \to \infty$ if $R > R_c$.

Show also that if $l < 0$ and $R < R_c$ then:

(i) $|A| \to 0$ as $t \to \infty$ if $A_0 < [\beta(R_c - R)/|l|]^{\frac{1}{2}}$;

(ii) $|A|$ becomes infinite in a finite time if

$$A_0 > [\beta(R_c - R)/|l|]^{\frac{1}{2}};$$

although the last result really implies only that $|A|$ will grow until the approximations leading to the Landau equation break down.

What sign do we expect l to take in the case of (a) thermal instability (§9.3), (b) the instability of a viscous shear flow (§9.5), and (c) the instability of Exercise 9.4?

[The Landau equation is the simplest evolution equation that arises in weakly non-linear stability theory (Drazin and Reid 1981, pp. 370–423), although the explicit calculation of l in any particular case may, even then, be a complicated matter. The 'special circumstances' for the equation's validity include (i) a certain symmetry to the system (otherwise cubic terms in A appear) and (ii) $A = 0$ constituting, for a time, *a* solution of the problem. In the case of thermal convection, for example, the first condition is broken if variations of viscosity with temperature are taken into account (see, e.g., Busse 1985), while the second is broken by the presence of side walls, for the state of no motion is typically not then a solution of the problem (see, e.g., Hall and Walton 1977).]

9.6. *Period-doubling.* Show that the difference equation

$$x_{n+1} = \lambda x_n(1 - x_n), \qquad \lambda > 0$$

has two steady, or constant, solutions, $x = 0$ and $x = 1 - \lambda^{-1}$. Show that if $\lambda < 1$ the first of these is stable, so that if x_1 is small then $x_n \to 0$ as $n \to \infty$. By writing $x_n = 1 - \lambda^{-1} + \varepsilon_n$ and assuming ε_n to be small, show likewise that the second steady solution is stable if $1 < \lambda < 3$ but unstable if $\lambda > 3$.

Show that, provided that $\lambda > 3$, there is a period 2 solution in which x_n alternates between the two values given by

$$2x = 1 + \frac{1}{\lambda} \pm \left[\left(1 - \frac{1}{\lambda}\right)^2 - \frac{4}{\lambda^2}\right]^{\frac{1}{2}}.$$

Suppose that there is a period 2 solution X_n of the difference equation

$$x_{n+1} = f(x_n),$$

so that $X_{n+2} = X_n$. Show that such a solution is unstable if

$$|f'(X_n)f'(X_{n+1})| > 1.$$

Hence deduce that the period 2 solution to $x_{n+1} = \lambda x_n(1 - x_n)$ loses its stability when

$$\lambda > 1 + \sqrt{6} \doteq 3.449.$$

Appendix

A.1. Vector identities

$$(a \wedge b) \wedge c = (a \cdot c)b - (b \cdot c)a, \tag{A.1}$$

$$\nabla \wedge \nabla \phi = 0, \qquad \nabla \cdot (\nabla \wedge F) = 0, \tag{A.2, A.3}$$

$$\nabla \cdot (\phi F) = \phi \nabla \cdot F + F \cdot \nabla \phi, \tag{A.4}$$

$$\nabla \wedge (\phi F) = \phi \nabla \wedge F + (\nabla \phi) \wedge F, \tag{A.5}$$

$$\nabla \wedge (F \wedge G) = (G \cdot \nabla)F - (F \cdot \nabla)G + F(\nabla \cdot G) - G(\nabla \cdot F), \tag{A.6}$$

$$\nabla \cdot (F \wedge G) = G \cdot (\nabla \wedge F) - F \cdot (\nabla \wedge G), \tag{A.7}$$

$$\nabla (F \cdot G) = F \wedge (\nabla \wedge G) + G \wedge (\nabla \wedge F) + (F \cdot \nabla)G + (G \cdot \nabla)F, \tag{A.8}$$

$$(F \cdot \nabla)F = (\nabla \wedge F) \wedge F + \nabla(\tfrac{1}{2}F^2), \tag{A.9}$$

$$\nabla^2 F = \nabla(\nabla \cdot F) - \nabla \wedge (\nabla \wedge F). \tag{A.10}$$

A.2. Two properties of the gradient operator ∇

Let $\phi(x)$ be some scalar function of x, and let $d\phi/ds$ be its rate of change, with distance s, in the direction of some unit vector t. Then

$$d\phi/ds = t \cdot \nabla \phi. \tag{A.11}$$

For this very reason, the line integral of $\nabla \phi$ along some curve C is equal to the difference in ϕ between the two end-points of the curve:

$$\int_C \nabla \phi \cdot dx = [\phi]_C. \tag{A.12}$$

A.3. The divergence theorem

Let the region V be bounded by a simple closed surface S with unit outward normal $\mathbf{n}$. Then

$$\int_S \mathbf{F} \cdot \mathbf{n} \, \mathrm{d}S = \int_V \nabla \cdot \mathbf{F} \, \mathrm{d}V. \tag{A.13}$$

In suffix notation, and using the summation convention, this takes the form

$$\int_S F_j n_j \, \mathrm{d}S = \int_V \frac{\partial F_j}{\partial x_j} \, \mathrm{d}V.$$

There are many identities which may be derived from the divergence theorem. The identity

$$\int_S \phi \mathbf{n} \, \mathrm{d}S = \int_V \nabla \phi \, \mathrm{d}V \tag{A.14}$$

is particularly valuable, and may be written

$$\int_S \phi n_j \, \mathrm{d}S = \int_V \frac{\partial \phi}{\partial x_j} \, \mathrm{d}V. \tag{A.15}$$

The following are immediate consequences:

$$\int_S F_i n_j \, \mathrm{d}S = \int_V \frac{\partial F_i}{\partial x_j} \, \mathrm{d}V, \quad \int_S T_{ij} n_j \, \mathrm{d}S = \int_V \frac{\partial T_{ij}}{\partial x_j} \, \mathrm{d}V,$$
$$\tag{A.16, A.17}$$

$$\int_S u_i v_j n_j \, \mathrm{d}S = \int_V \frac{\partial}{\partial x_j} (u_i v_j) \, \mathrm{d}V. \tag{A.18}$$

Other identities derivable from the divergence theorem include:

$$\int_S \mathbf{F} \wedge \mathbf{n} \, \mathrm{d}S = -\int_V \nabla \wedge \mathbf{F} \, \mathrm{d}V, \quad \int_S \mathbf{n} \cdot \nabla \phi \, \mathrm{d}S = \int_V \nabla^2 \phi \, \mathrm{d}V,$$
$$\tag{A.19, A.20}$$

$$\int_S \phi \frac{\partial \psi}{\partial n} \, \mathrm{d}S = \int_V (\phi \nabla^2 \psi + \nabla \phi \cdot \nabla \psi) \, \mathrm{d}V, \tag{A.21}$$

$$\int_S \left(\phi \frac{\partial \psi}{\partial n} - \psi \frac{\partial \phi}{\partial n} \right) \mathrm{d}S = \int_V (\phi \nabla^2 \psi - \psi \nabla^2 \phi) \, \mathrm{d}V. \tag{A.22}$$

A.4. Stokes's theorem

Let C be a simple closed curve spanned by a surface S with unit normal $\mathbf{n}$. Then

$$\int_C \mathbf{F} \cdot d\mathbf{x} = \int_S (\nabla \wedge \mathbf{F}) \cdot \mathbf{n} \, dS, \tag{A.23}$$

where the line integral is taken in an appropriate sense, according to that of $\mathbf{n}$ (see Fig. A.1).

Green's theorem in the plane may be viewed as a special case of Stokes's therorem, with $\mathbf{F} = [u(x, y), v(x, y), 0]$. If C is a simple closed curve in the x–y plane, and S denotes the region enclosed by C, then

$$\int_C u \, dx + v \, dy = \int_S \left(\frac{\partial v}{\partial x} - \frac{\partial u}{\partial y}\right) dx \, dy. \tag{A.24}$$

A useful identity derivable from Stokes's theorem is

$$\int_C \phi \, d\mathbf{x} = -\int_S (\nabla \phi) \wedge \mathbf{n} \, dS. \tag{A.25}$$

A.5. Orthogonal curvilinear coordinates

Let u, v, and w denote a set of orthogonal curvilinear coordinates, and let $\mathbf{e}_u$, $\mathbf{e}_v$ and $\mathbf{e}_w$ denote unit vectors parallel to the coordinate lines and in the directions of increase of u, v, and w respectively. Then

$$\mathbf{e}_u = \mathbf{e}_v \wedge \mathbf{e}_w, \qquad \text{etc.,}$$

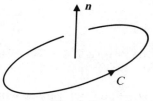

Fig. A.1.

and
$$\delta \boldsymbol{x} = h_1\, \delta u \boldsymbol{e}_u + h_2\, \delta v \boldsymbol{e}_v + h_3\, \delta w \boldsymbol{e}_w,$$
where
$$h_1 = |\partial \boldsymbol{x}/\partial u|, \quad \text{etc.}$$
Furthermore,
$$\nabla \phi = \frac{1}{h_1} \frac{\partial \phi}{\partial u} \boldsymbol{e}_u + \frac{1}{h_2} \frac{\partial \phi}{\partial v} \boldsymbol{e}_v + \frac{1}{h_3} \frac{\partial \phi}{\partial w} \boldsymbol{e}_w, \tag{A.26}$$

$$\nabla \cdot \boldsymbol{F} = \frac{1}{h_1 h_2 h_3} \left[\frac{\partial}{\partial u}(h_2 h_3 F_u) + \frac{\partial}{\partial v}(h_3 h_1 F_v) + \frac{\partial}{\partial w}(h_1 h_2 F_w) \right], \tag{A.27}$$

$$\nabla \wedge \boldsymbol{F} = \frac{1}{h_1 h_2 h_3} \begin{vmatrix} h_1 \boldsymbol{e}_u & h_2 \boldsymbol{e}_v & h_3 \boldsymbol{e}_w \\ \dfrac{\partial}{\partial u} & \dfrac{\partial}{\partial v} & \dfrac{\partial}{\partial w} \\ h_1 F_u & h_2 F_v & h_3 F_w \end{vmatrix}. \tag{A.28}$$

For *cylindrical polar coordinates* (Fig. A.2)
$$u = r, \quad v = \theta, \quad w = z,$$
$$h_1 = 1, \quad h_2 = r, \quad h_3 = 1.$$

For *spherical polar coordinates* (Fig. A.3)
$$u = r, \quad v = \theta, \quad w = \phi,$$
$$h_1 = 1, \quad h_2 = r, \quad h_3 = r \sin \theta.$$

A.6. Cylindrical polar coordinates

Cylindrical polar coordinates (r, θ, z) are such that
$$x_1 = r \cos \theta, \quad x_2 = r \sin \theta, \quad x_3 = z,$$
as in Fig. A.2. Clearly,
$$\delta \boldsymbol{x} = \delta r \boldsymbol{e}_r + r\, \delta \theta \boldsymbol{e}_\theta + \delta z \boldsymbol{e}_z$$
and
$$\boldsymbol{e}_r = \cos \theta\, \boldsymbol{e}_1 + \sin \theta\, \boldsymbol{e}_2, \qquad \boldsymbol{e}_\theta = -\sin \theta\, \boldsymbol{e}_1 + \cos \theta\, \boldsymbol{e}_2, \qquad \boldsymbol{e}_z = \boldsymbol{e}_3.$$
The unit vectors do not change with r or z, but
$$\frac{\partial \boldsymbol{e}_r}{\partial \theta} = \boldsymbol{e}_\theta, \qquad \frac{\partial \boldsymbol{e}_\theta}{\partial \theta} = -\boldsymbol{e}_r, \qquad \frac{\partial \boldsymbol{e}_z}{\partial \theta} = 0. \tag{A.29}$$

Appendix

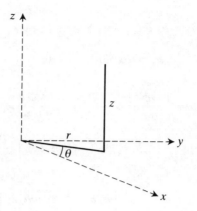

Fig. A.2 Cylindrical polar coordinates.

Also,
$$\nabla \phi = \frac{\partial \phi}{\partial r} \boldsymbol{e}_r + \frac{1}{r} \frac{\partial \phi}{\partial \theta} \boldsymbol{e}_\theta + \frac{\partial \phi}{\partial z} \boldsymbol{e}_z, \tag{A.30}$$

$$\nabla \cdot \boldsymbol{F} = \frac{1}{r} \frac{\partial}{\partial r} (rF_r) + \frac{1}{r} \frac{\partial F_\theta}{\partial \theta} + \frac{\partial F_z}{\partial z}, \tag{A.31}$$

$$\nabla \wedge \boldsymbol{F} = \frac{1}{r} \begin{vmatrix} \boldsymbol{e}_r & r\boldsymbol{e}_\theta & \boldsymbol{e}_z \\ \frac{\partial}{\partial r} & \frac{\partial}{\partial \theta} & \frac{\partial}{\partial z} \\ F_r & rF_\theta & F_z \end{vmatrix}, \tag{A.32}$$

$$\nabla^2 = \frac{1}{r} \frac{\partial}{\partial r} \left(r \frac{\partial}{\partial r} \right) + \frac{1}{r^2} \frac{\partial^2}{\partial \theta^2} + \frac{\partial^2}{\partial z^2}, \tag{A.33}$$

$$\boldsymbol{u} \cdot \nabla = u_r \frac{\partial}{\partial r} + \frac{u_\theta}{r} \frac{\partial}{\partial \theta} + u_z \frac{\partial}{\partial z}. \tag{A.34}$$

Appendix

The Navier–Stokes equations in cylindrical polar coordinates are:

$$\frac{\partial u_r}{\partial t} + (\boldsymbol{u}\cdot\nabla)u_r - \frac{u_\theta^2}{r} = -\frac{1}{\rho}\frac{\partial p}{\partial r} + \nu\left(\nabla^2 u_r - \frac{u_r}{r^2} - \frac{2}{r^2}\frac{\partial u_\theta}{\partial \theta}\right),$$

$$\frac{\partial u_\theta}{\partial t} + (\boldsymbol{u}\cdot\nabla)u_\theta + \frac{u_r u_\theta}{r} = -\frac{1}{\rho r}\frac{\partial p}{\partial \theta} + \nu\left(\nabla^2 u_\theta + \frac{2}{r^2}\frac{\partial u_r}{\partial \theta} - \frac{u_\theta}{r^2}\right),$$

(A.35)

$$\frac{\partial u_z}{\partial t} + (\boldsymbol{u}\cdot\nabla)u_z = -\frac{1}{\rho}\frac{\partial p}{\partial z} + \nu\,\nabla^2 u_z,$$

$$\frac{1}{r}\frac{\partial}{\partial r}(ru_r) + \frac{1}{r}\frac{\partial u_\theta}{\partial \theta} + \frac{\partial u_z}{\partial z} = 0.$$

The components of the rate-of-strain tensor are given by:

$$e_{rr} = \frac{\partial u_r}{\partial r}, \qquad e_{\theta\theta} = \frac{1}{r}\frac{\partial u_\theta}{\partial \theta} + \frac{u_r}{r}, \qquad e_{zz} = \frac{\partial u_z}{\partial z},$$

$$2e_{\theta z} = \frac{1}{r}\frac{\partial u_z}{\partial \theta} + \frac{\partial u_\theta}{\partial z}, \qquad 2e_{zr} = \frac{\partial u_r}{\partial z} + \frac{\partial u_z}{\partial r}, \qquad (A.36)$$

$$2e_{r\theta} = r\frac{\partial}{\partial r}\left(\frac{u_\theta}{r}\right) + \frac{1}{r}\frac{\partial u_r}{\partial \theta}.$$

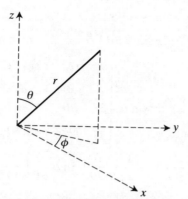

Fig. A.3. Spherical polar coordinates.

A.7. Spherical polar coordinates

Spherical polar coordinates (r, θ, ϕ) are such that

$$x_1 = r \sin \theta \cos \phi, \qquad x_2 = r \sin \theta \sin \phi, \qquad x_3 = r \cos \theta,$$

as in Fig. A.3. Clearly,

$$\delta \boldsymbol{x} = \delta r \boldsymbol{e}_r + r \delta \theta \boldsymbol{e}_\theta + r \sin \theta \, \delta \phi \boldsymbol{e}_\phi$$

and

$$\boldsymbol{e}_r = \sin \theta \cos \phi \, \boldsymbol{e}_1 + \sin \theta \sin \phi \, \boldsymbol{e}_2 + \cos \theta \, \boldsymbol{e}_3,$$

$$\boldsymbol{e}_\theta = \cos \theta \cos \phi \, \boldsymbol{e}_1 + \cos \theta \sin \phi \, \boldsymbol{e}_2 - \sin \theta \, \boldsymbol{e}_3,$$

$$\boldsymbol{e}_\phi = -\sin \phi \, \boldsymbol{e}_1 + \cos \phi \, \boldsymbol{e}_2.$$

The unit vectors do not change with r, but

$$\partial \boldsymbol{e}_r / \partial \theta = \boldsymbol{e}_\theta, \qquad \partial \boldsymbol{e}_\theta / \partial \theta = -\boldsymbol{e}_r, \qquad \partial \boldsymbol{e}_\phi / \partial \theta = 0,$$

$$\partial \boldsymbol{e}_r / \partial \phi = \sin \theta \, \boldsymbol{e}_\phi, \qquad \partial \boldsymbol{e}_\theta / \partial \phi = \cos \theta \, \boldsymbol{e}_\phi, \qquad \text{(A.37)}$$

$$\partial \boldsymbol{e}_\phi / \partial \phi = -\sin \theta \, \boldsymbol{e}_r - \cos \theta \, \boldsymbol{e}_\theta.$$

Also,

$$\nabla \Phi = \frac{\partial \Phi}{\partial r} \boldsymbol{e}_r + \frac{1}{r} \frac{\partial \Phi}{\partial \theta} \boldsymbol{e}_\theta + \frac{1}{r \sin \theta} \frac{\partial \Phi}{\partial \phi} \boldsymbol{e}_\phi, \qquad \text{(A.38)}$$

$$\nabla \cdot \boldsymbol{F} = \frac{1}{r^2} \frac{\partial}{\partial r}(r^2 F_r) + \frac{1}{r \sin \theta} \frac{\partial}{\partial \theta}(F_\theta \sin \theta) + \frac{1}{r \sin \theta} \frac{\partial F_\phi}{\partial \phi}, \qquad \text{(A.39)}$$

$$\nabla \wedge \boldsymbol{F} = \frac{1}{r^2 \sin \theta} \begin{vmatrix} \boldsymbol{e}_r & r\boldsymbol{e}_\theta & r \sin \theta \boldsymbol{e}_\phi \\ \dfrac{\partial}{\partial r} & \dfrac{\partial}{\partial \theta} & \dfrac{\partial}{\partial \phi} \\ F_r & rF_\theta & r \sin \theta F_\phi \end{vmatrix}, \qquad \text{(A40)}$$

$$\nabla^2 = \frac{1}{r^2} \frac{\partial}{\partial r}\left(r^2 \frac{\partial}{\partial r}\right) + \frac{1}{r^2 \sin \theta} \frac{\partial}{\partial \theta}\left(\sin \theta \frac{\partial}{\partial \theta}\right) + \frac{1}{r^2 \sin^2 \theta} \frac{\partial^2}{\partial \phi^2},$$
(A.41)

$$\boldsymbol{u} \cdot \nabla = u_r \frac{\partial}{\partial r} + \frac{u_\theta}{r} \frac{\partial}{\partial \theta} + \frac{u_\phi}{r \sin \theta} \frac{\partial}{\partial \phi}. \qquad \text{(A.42)}$$

Appendix 355

The Navier–Stokes equations in spherical polar coordinates are:

$$\frac{\partial u_r}{\partial t} + (\boldsymbol{u} \cdot \nabla)u_r - \frac{u_\theta^2}{r} - \frac{u_\phi^2}{r}$$

$$= -\frac{1}{\rho}\frac{\partial p}{\partial r} + \nu\left[\nabla^2 u_r - \frac{2u_r}{r^2} - \frac{2}{r^2 \sin\theta}\frac{\partial}{\partial \theta}(u_\theta \sin\theta) - \frac{2}{r^2 \sin\theta}\frac{\partial u_\phi}{\partial \phi}\right],$$

$$\frac{\partial u_\theta}{\partial t} + (\boldsymbol{u} \cdot \nabla)u_\theta + \frac{u_r u_\theta}{r} - \frac{u_\phi^2 \cot\theta}{r}$$

$$= -\frac{1}{\rho r}\frac{\partial p}{\partial \theta} + \nu\left[\nabla^2 u_\theta + \frac{2}{r^2}\frac{\partial u_r}{\partial \theta} - \frac{u_\theta}{r^2 \sin^2\theta} - \frac{2\cos\theta}{r^2 \sin^2\theta}\frac{\partial u_\phi}{\partial \phi}\right],$$

$$\frac{\partial u_\phi}{\partial t} + (\boldsymbol{u} \cdot \nabla)u_\phi + \frac{u_\phi u_r}{r} + \frac{u_\theta u_\phi \cot\theta}{r}$$

$$= -\frac{1}{\rho r \sin\theta}\frac{\partial p}{\partial \phi} + \nu\left[\nabla^2 u_\phi + \frac{2}{r^2 \sin\theta}\frac{\partial u_r}{\partial \phi} + \frac{2\cos\theta}{r^2 \sin^2\theta}\frac{\partial u_\theta}{\partial \phi} - \frac{u_\phi}{r^2 \sin^2\theta}\right],$$

$$\frac{1}{r^2}\frac{\partial}{\partial r}(r^2 u_r) + \frac{1}{r \sin\theta}\frac{\partial}{\partial \theta}(u_\theta \sin\theta) + \frac{1}{r \sin\theta}\frac{\partial u_\phi}{\partial \phi} = 0. \quad \text{(A.43)}$$

The components of the rate-of-strain tensor are given by:

$$e_{rr} = \frac{\partial u_r}{\partial r}, \qquad e_{\theta\theta} = \frac{1}{r}\frac{\partial u_\theta}{\partial \theta} + \frac{u_r}{r},$$

$$e_{\phi\phi} = \frac{1}{r \sin\theta}\frac{\partial u_\phi}{\partial \phi} + \frac{u_r}{r} + \frac{u_\theta \cot\theta}{r},$$

$$2e_{\theta\phi} = \frac{\sin\theta}{r}\frac{\partial}{\partial \theta}\left(\frac{u_\phi}{\sin\theta}\right) + \frac{1}{r \sin\theta}\frac{\partial u_\theta}{\partial \phi}, \quad \text{(A.44)}$$

$$2e_{\phi r} = \frac{1}{r \sin\theta}\frac{\partial u_r}{\partial \phi} + r\frac{\partial}{\partial r}\left(\frac{u_\phi}{r}\right),$$

$$2e_{r\theta} = r\frac{\partial}{\partial r}\left(\frac{u_\theta}{r}\right) + \frac{1}{r}\frac{\partial u_r}{\partial \theta}.$$

Hints and answers for exercises

Chapter 1

1.1 The rate of flow of mass out of S is $\int_S \rho \boldsymbol{u} \cdot \boldsymbol{n}\, dS$, and this must be equal to $-\int_V (\partial \rho/\partial t)\, dV$, the rate at which mass is decreasing in the region enclosed by S.

Use (A.4).

$D\rho/Dt = 0$ does *not* mean that ρ is a constant; it means that ρ is conserved by each individual fluid element, and this makes sense, as each element conserves both its mass and (if $\nabla \cdot \boldsymbol{u} = 0$) its volume.

1.2. The flow is not irrotational, as $\nabla \wedge \boldsymbol{u} = (0, 0, 2\Omega)$, so the theorem following eqn (1.17) does not apply. The flow *is* steady, so the Bernoulli streamline theorem applies, but there is then no telling how p might vary from one streamline to another.

Free surface: $z = (\Omega^2/2g)(x^2 + y^2) + \text{constant}$.

1.3. The preceding exercise implies

$$p/\rho = \tfrac{1}{2}\Omega^2 r^2 - gz + c_1 \qquad \text{for } r < a.$$

For $r > a$ the flow is irrotational, so the Bernoulli theorem following eqn (1.17) gives

$$\frac{p}{\rho} = -\tfrac{1}{2}\frac{\Omega^2 a^4}{r^2} - gz + c_2 \qquad \text{for } r > a.$$

Continuity of p at $r = a$ implies that $c_2 - c_1 = \Omega^2 a^2$ etc.

1.4. Take the Euler equation in the form (1.14), multiply by ρ, take the dot product of both sides with $\boldsymbol{u}$, and then use eqn (A.4) to obtain

$$\frac{\partial}{\partial t}(\tfrac{1}{2}\rho \boldsymbol{u}^2) = -\nabla \cdot [(p' + \tfrac{1}{2}\rho \boldsymbol{u}^2)\boldsymbol{u}].$$

Then integrate both sides over V and use the divergence theorem.

1.5. Much as in §1.5, but use eqn (A.5) in dealing with

$$\nabla \wedge \left(\frac{1}{\rho}\nabla p\right).$$

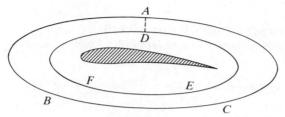

Fig. H.1.

Use conservation of mass to replace $\nabla \cdot \boldsymbol{u}$, when it appears, by

$$-\frac{1}{\rho}\frac{D\rho}{Dt}.$$

If $p = p(\rho)$, then $\nabla p = p'(\rho)\nabla\rho$, so $\nabla p \wedge \nabla\rho = 0$.

1.6. Consider the circulation round the closed circuit ABCADEFDA in Fig. H.1, which does not enclose the wing, and thereby show that

$$\int_{\text{ABCA}} \boldsymbol{u} \cdot d\boldsymbol{x} = \int_{\text{DFED}} \boldsymbol{u} \cdot d\boldsymbol{x}.$$

1.7. See Fig. 4.1 for the streamlines.

No, because $Dc/Dt = 0$.

The results are true in general because holding $\boldsymbol{X}$ (as opposed to $\boldsymbol{x}$) constant corresponds to restricting attention to a particular fluid element.

$c(X, Y, t) = \beta X^2 Y$, and this gives a slightly different perspective on why it is that c does not change with time for a particular fluid element.

1.8 The streamlines are

$$y = \frac{kt}{u_0}x + \text{constant}, \qquad z = \text{constant},$$

obtained by integrating eqn (1.5) at fixed time t.

A particle path is obtained by integrating

$$(\partial x/\partial t)_X = u_0, \qquad (\partial y/\partial t)_X = kt, \qquad (\partial z/\partial t)_X = 0$$

(see Exercise 1.7), so

$$x = u_0 t + F_1(\boldsymbol{X}), \qquad y = \tfrac{1}{2}kt^2 + F_2(\boldsymbol{X}), \qquad z = F_3(\boldsymbol{X}).$$

The arbitrary functions of X, Y, and Z are determined by the condition that $\boldsymbol{x} = \boldsymbol{X}$ when $t = 0$, so

$$x = u_0 t + X, \qquad y = \tfrac{1}{2}kt^2 + Y, \qquad z = Z.$$

Eliminating t gives the particle paths.

Chapter 2

2.1. (i) 4×10^7; (ii) 10^3; (iii) 0.003; (iv) 10^{-3}.
The boundary layer thickness is of order 1 mm.

2.2.
$$(\mathbf{u}' \cdot \nabla')\mathbf{u}' = -\nabla'p' + \frac{1}{R}\nabla'^2\mathbf{u}', \qquad \nabla' \cdot \mathbf{u}' = 0,$$

subject to

$$\mathbf{u}' = 0 \quad \text{on } x'^2 + y'^2 = 1; \qquad \mathbf{u}' \to (1, 0, 0) \quad \text{as } x'^2 + y'^2 \to \infty.$$

Solving this will give $\mathbf{u}'$ as some specific function (or functions—the solution may not be unique) of $\mathbf{x}'$ and R. Thus, in particular, the *direction* of $\mathbf{u}'$ depends, at a given $\mathbf{x}' = \mathbf{x}/a$, not on ν, a, or U individually but only on $R = Ua/\nu$.

(The same argument may evidently be used for flow past a body of any shape.)

2.3. (i) Seek a solution to the Navier–Stokes equations of the form $\mathbf{u} = [u(y), 0, 0]$, taking care that: (1) $\nabla \cdot \mathbf{u} = 0$; (2) all three components of the momentum equation (2.3) are satisfied; and (3) $u(y)$ satisfies the no-slip condition on $y = \pm h$.

(ii) Likewise, assuming $\mathbf{u} = [0, 0, u_z(r)]$, and using eqn (2.22). The condition that u_z is finite at $r = 0$ is needed.

2.4. Extend the single-layer analysis in the text to show that $\partial p/\partial x = 0$ in both layers. The interface conditions are

$$u_1 = u_2, \qquad \mu_1 \frac{du_1}{dy} = \mu_2 \frac{du_2}{dy} \quad \text{at } y = h_1,$$

because the tangential stress exerted on the lower layer by the upper layer is $\mu_2 \, du_2/dy$, that exerted on the upper layer by the lower layer is $-\mu_1 \, du_1/dy$, and the two must be equal and opposite.

The upper layer is not accelerating, and there is no tangential stress on it from above, so the tangential stress on it from below must exactly cancel the net gravitational force on it in the x-direction, which is proportional to its mass. The (equal and opposite) tangential stress on the lower layer thus depends on h_2, but not on ν_2.

2.5. The boundary conditions are homogeneous, but the equation is not, so write $u = u_0(y) + u_1(y, t)$, where

$$u_0(y) = (P/2\mu)(h^2 - y^2)$$

is the steady solution satisfying the boundary conditions (but not the initial conditions), as in Exercise 2.3. Then solve the resulting problem for $u_1(y, t)$ as in the text.

$$u = \frac{P}{2\mu}\left[h^2 - y^2 - h^2 \sum_{N=0}^{\infty} \frac{4(-1)^N}{(N+\frac{1}{2})^3 \pi^3} e^{-(N+\frac{1}{2})^2 \pi^2 vt/h^2} \cos(N+\frac{1}{2})\frac{\pi y}{h}\right].$$

2.6. When $v_0 h/v \ll 1$, $e^{-v_0 y/v} \doteq 1 - v_0 y/v$ etc., so

$$u \doteq U\left(1 - \frac{y}{h}\right).$$

When $v_0 h/v \gg 1$, $e^{-v_0 h/v}$ is extremely small, and so too is $e^{-v_0 y/v}$ throughout most of the range $0 < y < h$, though *not* within a distance of order v/v_0 of the lower boundary $y = 0$. In this boundary layer

$$u \doteq U e^{-v_0 y/v}.$$

2.8. There will at first be a thin boundary layer of negative vorticity on $r = a$, and this will gradually diffuse outwards until, as $t \to \infty$, the vorticity tends to zero at any finite r, however large.

On the outer cylinder $r = b$ there will at first be a boundary layer with positive vorticity. This vorticity will diffuse inwards, gradually cancelling the outward-diffusing negative vorticity, so that $\omega \to 0$ as $t \to \infty$.

2.9. The result for u_r comes directly from $\nabla \cdot \boldsymbol{u} = 0$. The general solution of the equation is

$$u_\theta = \frac{A}{r} + \frac{B}{r^{R-1}},$$

provided that $R \neq 2$. When $R > 2$ a free parameter is left in the solution, so there are in fact infinitely many flows satisfying the conditions.

2.10. We have

$$f'' + \left(\tfrac{1}{2}\eta - \frac{1}{\eta}\right)f' = 0,$$

which is a first-order equation for f'.

The vorticity is concentrated in the (expanding) region $r < O(vt)^{\frac{1}{2}}$, and it decreases with time (as t^{-1}) at $r = 0$.

2.11. If $\boldsymbol{u} = u_\theta(r, z)\boldsymbol{e}_\theta$, the r- and z-components of the Navier–Stokes equations (2.22) together imply that $u_\theta(r, z)$ is independent of z, but this is incompatible with the no-slip boundary conditions on $z = 0$ and $z = h$.

2.12 Multiply both sides by ru_θ, integrate both sides between $r = 0$ and

$r = a$, and then use integration by parts to obtain

$$\frac{d}{dt}\int_0^a \tfrac{1}{2}u_\theta^2 r\, dr = -\nu \int_0^a r\left(\frac{\partial u_\theta}{\partial r}\right)^2 dr - \nu \int_0^a \frac{u_\theta^2}{r}\, dr.$$

The second term is less than or equal to zero; the third term needs to be compared with

$$-\nu \int_0^a \frac{r u_\theta^2}{a^2}\, dr.$$

$E \to 0$ as $t \to \infty$ because $Ee^{2\nu t/a^2}$ is a decreasing function of t.

2.14. Substitute into the Navier–Stokes equations, integrate the y-component with respect to y, and deduce that $\partial p/\partial x$ is a function of x only. Turn to the x-component, deduce that $x^{-1}\,\partial p/\partial x$ is a constant, because the rest of the equation is a function of y alone, etc.

$$p = -\tfrac{1}{2}\rho\alpha^2 x^2 - \rho\nu\alpha(f' + \tfrac{1}{2}f^2) + \text{constant}.$$

2.15. The supposition that the main, inviscid flow is not much disturbed requires the existence of a thin boundary layer on the plate in order to satisfy the no-slip condition. But the mainstream flow speed αy at the edge of such a boundary layer would decrease rapidly with distance along the plate from the leading edge. By Bernoulli's theorem there would therefore be a substantial *increase* in pressure p along the plate in the flow direction (as is evident from Fig. 2.13), and, as explained in §2.1, this is exactly the circumstance in which boundary-layer separation occurs.

Chapter 3

3.1. The new boundary condition is $\partial\phi/\partial y = 0$ on $y = -h$. The analysis is valid only if $\eta \ll \lambda$ and $\eta \ll h$. The particle paths are ellipses that become flatter with depth.

3.2. The condition that $p_1 = p_2$ at the interface gives, on using eqn (3.19) in each layer and linearizing:

$$\rho_1 \frac{\partial \phi_1}{\partial t} + \rho_1 g \eta = \rho_2 \frac{\partial \phi_2}{\partial t} + \rho_2 g \eta \quad \text{on } y = 0.$$

Seek suitable solutions of $\partial^2\phi/\partial x^2 + \partial^2\phi/\partial y^2 = 0$ in each layer, ensuring that $\phi_1 \to 0$ as $y \to -\infty$ and $\phi_2 \to 0$ as $y \to \infty$ (at which stage $|k|$ enters the analysis).

3.3. By the argument of §3.4, $p_2 - p_1 = T\,\partial^2\eta/\partial x^2$ at $y = \eta(x, t)$. In each layer, seek solutions of Laplace's equation of the form $\phi = f(x)g(y)\sin(\omega t + \varepsilon)$; the conditions $\partial\phi/\partial x = 0$ at $x = 0$ and at $x = a$ help to determine $f(x)$.

Hints and answers for exercises 361

3.4. If $\rho_2 > \rho_1$ then *some* of the normal 'frequencies' ω_N may be pure imaginary, so that $\omega_N = \pm i p_N$, say, in which case

$$\cos \omega_N t = \tfrac{1}{2}(e^{p_N t} + e^{-p_N t}).$$

As a result, $\eta(x, t)$ does not remain small as time proceeds. The only way this may be avoided is if

$$(\rho_1 - \rho_2)g + T\frac{N^2\pi^2}{a^2} > 0,$$

for *all* values of $N = 1, 2, \ldots$.

3.5. Using eqn (3.19), the pressure condition at the free surface may be written

$$\tfrac{1}{2}\left[\left(U + \frac{\partial\phi}{\partial x}\right)^2 + \left(\frac{\partial\phi}{\partial y}\right)^2\right] + g\eta = \text{constant} \qquad \text{on } y = \eta(x),$$

and on linearization this becomes

$$U\frac{\partial\phi}{\partial x} + g\eta = 0 \qquad \text{on } y = 0.$$

In similar fashion, the kinematic condition (3.18) takes the form

$$U\frac{d\eta}{dx} = \frac{\partial\phi}{\partial y} \qquad \text{on } y = 0,$$

and this same expression may be used to derive the lower boundary condition, by replacing η by $\varepsilon \cos kx$.

Seek a solution to Laplace's equation of the form $\phi = f(y)\sin kx$.

The free surface displacement is

$$\eta(x) = \varepsilon \cos kx \left/ \left[\cosh(kh) - \frac{g}{U^2 k}\sinh(kh)\right]\right..$$

3.6. The linearized interface conditions are

$$\frac{\partial\phi_1}{\partial y} = \frac{\partial\eta}{\partial t}, \qquad \frac{\partial\phi_2}{\partial y} = \frac{\partial\eta}{\partial t} + U\frac{\partial\eta}{\partial x},$$

$$\rho_1\left(\frac{\partial\phi_1}{\partial t} + g\eta\right) - \rho_2\left(\frac{\partial\phi_2}{\partial t} + U\frac{\partial\phi_2}{\partial x} + g\eta\right) = T\frac{\partial^2\eta}{\partial x^2} \qquad \text{on } y = 0.$$

The dispersion relation has one root with a positive imaginary part to ω unless

$$\rho_1\rho_2 U^2 < (\rho_1 + \rho_2)\left[|k|\,T + (\rho_1 - \rho_2)\frac{g}{|k|}\right].$$

362 Hints and answers for exercises

Thus *some* Fourier components of a general disturbance (3.30) will grow in amplitude with time, corresponding to instability, unless the above inequality is satisfied *for all* k.

3.7. According to property (ii) following eqn (3.29), waves of wavelength $\lambda = 2\pi/k$ are eventually to be found at a distance $c_g t$ from the source, but eqn (3.47) has a minimum with respect to k when

$$k = \left(\frac{2}{\sqrt{3}} - 1\right)^{\frac{1}{2}} \left(\frac{\rho g}{T}\right)^{\frac{1}{2}},$$

so no waves at all are to be found in $x < c_g^{\min} t$.

3.8. A group of waves of wavelength λ travels at velocity c_g (but consists of different crests as time proceeds). Using $\omega^2 = gk = 2\pi g/\lambda$ we may think instead of waves of a certain *period* $T = 2\pi/\omega$ moving with the group velocity, so that waves of period T arrive at the coast from distance d in time $4\pi d/gT$.

d is roughly 5000 km.

3.9. Obtain the exact dispersion relation for plane waves with finite depth and surface tension, and expand the expression for $c = \omega/k$ in powers of kh (assumed small).

3.10. Modify the analysis leading to eqn (3.42), using eqn (3.49) in place of eqn (3.39). The simplest way to follow a particular crest is to hold the phase function θ constant; the wavelength then changes as $t^{\frac{2}{3}}$. The slowly varying wavetrain assumption is only valid if $Tt^2 \ll \rho x^3$.

3.11. Putting $k_1 = k - k_0$ gives

$$\eta(x, 0) = a_0 e^{ik_0 x} e^{-x^2/4\sigma} \int_{-\infty}^{\infty} e^{-\sigma(k_1 - ix/2\sigma)^2} \, dk_1.$$

The integral is the same as

$$\int_{-\infty}^{\infty} e^{-\sigma s^2} \, ds = (\pi/\sigma)^{\frac{1}{2}},$$

but the 'substitution' $s = k_1 - (ix/2\sigma)$ is not good enough to demonstrate this, and a more careful argument is needed, which involves the application of Cauchy's theorem to

$$\int e^{-\sigma z^2} \, dz$$

round a rectangular contour in the complex z-plane (see, e.g., Priestley 1985, pp. 128 and 148).

3.12. Make use of eqns (3.23), (3.25), and (3.26), and the expressions for u and v derived from eqn (3.25). The expression for the perturbation

Hints and answers for exercises 363

pressure comes from eqn (3.19), on recognizing that χ is constant at a fixed point in space and that u^2 is quadratically small and therefore negligible compared with $\partial \phi / \partial t$.

3.13. Seek oscillatory solutions to eqn (3.65) by putting

$$h(r, t) = f(r)\cos \omega t,$$

and note that (i) p_1 must be finite at $r = 0$ and (ii) u_r must be zero for all t on $r = L$, which implies, via eqn (3.61), a further boundary condition on p_1.

3.14. The normal force on a length δs of the upper surface is $p \, \delta s$, and if γ is the local angle of slope of the surface the x-component of this force is $p \, \delta s \sin \gamma = p \, \delta y$. The net drag on the upper surface is therefore

$$\int_0^L (p_0 + p_1)_{y=f(x)} f'(x) \, \mathrm{d}x.$$

The p_0 term gives no contribution, and as $f(x)$ is small the drag is essentially

$$\int_0^L p_1(x, 0) f'(x) \, \mathrm{d}x.$$

Then use eqns (3.70) and (3.77).

3.15.
$$u_1 = -\frac{Al}{k} \cos \xi, \qquad p_1 = -\frac{\rho_0 A \omega l}{k^2} \cos \xi,$$

where $\xi = kx + ly - \omega t$.

3.16. Internal gravity waves are remarkably different to sound waves in that, given N, the frequency ω fixes the ratio l^2/k^2, via eqn (3.89). This in turn fixes the direction of the group velocity (3.91), in which direction all the energy of the disturbance will be found.

$$\cos \alpha = \omega / N.$$

If $\omega > N$ no waves exist to carry energy to great distances, and the disturbance produced by the cylinder will be a localized one.

3.17. In eliminating u_1, v_1, and w_1 it can be helpful to first show that

$$2\Omega \left(\frac{\partial u_1}{\partial y} - \frac{\partial v_1}{\partial x} \right) = -\frac{1}{\rho} \nabla^2 p_1,$$

and then show that

$$\frac{\partial}{\partial t} \left(\frac{\partial u_1}{\partial y} - \frac{\partial v_1}{\partial x} \right) = 2\Omega \left(\frac{\partial u_1}{\partial x} + \frac{\partial v_1}{\partial y} \right).$$

To show that c_g is perpendicular to $\mathbf{k}$,

$$2\omega \frac{\partial \omega}{\partial k} = \frac{-4\Omega^2 m^2 2k}{(k^2 + l^2 + m^2)^2} \quad \text{etc.}$$

[For a remarkable photograph of *axisymmetric* waves produced by an oscillating disc in a rotating tank see Greenspan 1968, p. 3 or Tritton 1988, p. 232.]

3.18. If we assume that $u + 2c = 2c_0$, then on the plate curve $x = Vt$ we have $u = V$ and $c = c_0 - \frac{1}{2}V$. Thus $u + 2c = 2c_0$ along the characteristic $dx/dt = u + c$ coming off the plate curve, and $u - 2c = 2V - 2c_0$ along characteristics $dx/dt = u - c$ coming off that curve, so wherever these characteristics intersect $u = V$ and $c = c_0 - \frac{1}{2}V$. Throughout the region in which these characteristics intersect, then, the characteristics are straight lines

$$dx/dt = V \pm (c_0 - \tfrac{1}{2}V).$$

The first set has $dx/dt > V$ and the second has $dx/dt < V$, so the intersection region is bounded by the member of the second set which comes from the origin, i.e. $x = (\tfrac{3}{2}V - c_0)t$.

3.19. The characteristic curve is the curve $x = x(s)$, $t = t(s)$, going through $(x_0, 0)$ such that

$$dt/ds = 1, \qquad dx/ds = z.$$

Along it, $dz/ds = 0$, so $z = g(x_0)$ along it; therefore it takes the form $x = g(x_0)t + \text{constant}$, etc.

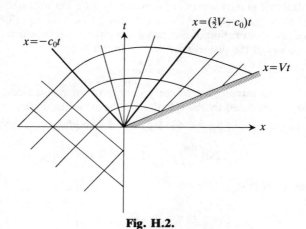

Fig. H.2.

Hints and answers for exercises 365

At the time t at which two characteristics from x_0 and $x_0 + \delta x_0$ cross, the value of x is the same, so the relevant time t is that at which $(\partial x/\partial x_0)_t = 0$.

3.20 The first term represents the rate of working by pressure forces on the cross-section; the second and third represent the rates at which kinetic and potential energy are being swept through the cross-section. To obtain the first result use the fact that the pressure is hydrostatic, i.e. $p = -\rho g(y - h_1)$, because the flow is uniform. To obtain the second result, consider the difference in energy fluxes and use eqns (3.124) and (3.125).

3.21. Write down the Euler momentum equation (1.12) with $g = 0$, use eqn (A.9) to rewrite the $(u \cdot \nabla)u$ term, and then write $p = c\rho^\gamma$, where c is a constant, etc.

3.22. Write $p = c\rho^\gamma$, where c is a constant, and eliminate p. It can then be helpful to establish the relationship

$$\frac{2}{a}\frac{\partial a}{\partial x} = \frac{\gamma - 1}{\rho}\frac{\partial \rho}{\partial x},$$

which also holds if $\partial/\partial x$ is replaced by $\partial/\partial t$.

The final part involves adding and subtracting the equations for u and a.

3.23. Use eqn (3.136), writing $u = F(\xi)$, where $\xi = x - [\frac{1}{2}(\gamma + 1)u + a_0]t$. Therefore

$$\frac{\partial u}{\partial x} = \frac{F'(\xi)}{1 + \frac{1}{2}(\gamma + 1)tF'(\xi)} \quad (\text{cf. eqn (3.115)}),$$

and, of course, $F(\xi) = \frac{1}{2}U[1 - \tanh(\xi/L)]$, by virtue of the initial conditions.

3.24. After one integration

$$(a_0 - V)f + \frac{1}{4}(\gamma + 1)f^2 = \frac{2}{3}vf' + c,$$

where c is a constant. Now $f \to 0$ as $\xi \to \infty$, so $\frac{2}{3}vf' \to -c$ as $\xi \to \infty$, and the only way this is compatible with $f \to 0$ as $\xi \to \infty$ is by c being zero. Similar considerations as $\xi \to -\infty$ give the shock speed V.

Chapter 4

4.1. (i) Use Stokes's theorem (A.23).
(ii) Use Green's theorem in the plane (A.24), with u in place of v and $-v$ in place of u.

366 *Hints and answers for exercises*

4.2. Treat similarly to the line vortex flow (4.18).

No contradiction: does $\partial u/\partial x + \partial v/\partial y = 0$—on which Exercise 4.1(ii) depends—hold *everywhere* for a line source flow?

$$w = \frac{Q}{2\pi}\log(z-d) + \frac{Q}{2\pi}\log(z+d).$$

On wall $x=0$, $v = Qy/\pi(y^2+d^2)$ and $p + \tfrac{1}{2}\rho v^2 = $ constant.

4.3. Use eqn (4.11) to obtain ϕ, and then eqn (4.12) to obtain the complex potential of the flow when the cylinder is absent, $w = \tfrac{1}{2}A(z-c)^2$. Then use the circle theorem (4.29).

$$\psi = Ay\left[x\left\{1+\frac{a^2}{x^2+y^2}\right\} - c\right]\left(1-\frac{a^2}{x^2+y^2}\right)$$

(so $\psi = 0$ on $x^2+y^2 = a^2$, as desired).

$$F_x = -2\pi\rho A^2 c a^2, \qquad F_y = 0.$$

4.5. $\delta\mathcal{N} = x\delta F_y - y\delta F_x = \mathcal{R}[(x+iy)(\delta F_y + i\delta F_x)]$.

4.7. Note that $Z = X + iY = 2a\cos\theta$ on the plate.

$\int_{-1}^{1}[(1-s)/(1+s)]^{\tfrac{1}{2}}\,ds$ can be evaluated by putting $s = \cos 2\gamma$.

For the torque, re-work the calculation for the ellipse in §4.10, but with

$$w = U\left(ze^{-i\alpha} + \frac{a^2}{z}e^{i\alpha}\right) - \frac{i\Gamma}{2\pi}\log z.$$

The terms involving Γ do give a contribution to the integral, but this disappears when the real part is taken.

4.8. Note as a check, or as part of the argument in obtaining Fig. 4.16, that $dZ/dz = 0$ (but $d^2Z/dz^2 \neq 0$) at $z = a$, so that angles between two line elements through $z = a$ are doubled by the mapping, as in Fig. H.3 (see §4.6).

$$W(Z) = U\left[z - ia\cot\beta + \frac{a^2\cosec^2\beta}{z-ia\cot\beta}\right] - \frac{i\Gamma}{2\pi}\log(z-ia\cot\beta),$$

where

$$z = \tfrac{1}{2}Z + (\tfrac{1}{4}Z^2 - a^2)^{\tfrac{1}{2}}.$$

Fig. H.3.

Hints and answers for exercises   367

4.9. (i) The image sources are at $z = \pm 2nib$, $n = 1, 2, 3 \ldots$. To sum, use the argument leading to eqn (5.29), and the identity quoted there.

$$w = \frac{Q}{2\pi} \log(e^{\pi z/2b} - e^{-\pi z/2b}).$$

(ii) Write $Z = Re^{i\Theta} = e^{\alpha(x+iy)}$ to find the corresponding fluid region in the Z-plane. The choice $\alpha = \pi/b$ opens that region up so that there is just a barrier along the negative real Z-axis, which does not affect the flow caused by the line source at $Z = 1$. But there is a subtlety. In the z-plane there is a volume flux $Q/2$ in both positive and negative x-directions. As $x \to \pm\infty$, then, the flow becomes uniform with velocity $(\pm Q/4b, 0)$. What happens, under the given mapping, to the uniform flow at large negative x?

4.10. Apply the x-component of eqn (4.70) to the region ABCDA in Fig. 4.10, showing that the right-hand side is zero, while the left-hand side is $(p_1 - p_2)d - F_x$, where p_1 is the pressure far upstream and p_2 the pressure far downstream. Then apply Bernoulli's streamline theorem and use eqn (4.73).

No: the Kutta–Joukowski theorem is for a single aerofoil, and in any case $F_x \to 0$ as $d \to \infty$ (and $v_2 \to 0$) for fixed Γ.

Chapter 5

5.1. $\boldsymbol{u} = (\partial \boldsymbol{x}/\partial t)_s = (a\alpha \sin s, 0, 0)$. (Note that holding s constant is exactly like holding X constant in Exercise 1.7.)

The integrand $\boldsymbol{u} \cdot \partial \boldsymbol{x}/\partial s$ is time-dependent, as expected, but $\Gamma = -\pi a^2 \alpha$.

5.2.

$$\frac{\partial}{\partial t}\left(\boldsymbol{u} \cdot \frac{\partial \boldsymbol{x}}{\partial s}\right) = \left(\frac{\partial \boldsymbol{u}}{\partial t}\right)_s \cdot \left(\frac{\partial \boldsymbol{x}}{\partial s}\right)_t + \boldsymbol{u} \cdot \left(\frac{\partial}{\partial t}\right)_s \left(\frac{\partial \boldsymbol{x}}{\partial s}\right)_t.$$

Now, $(\partial \boldsymbol{u}/\partial t)_s$ is the acceleration of a fluid element, otherwise written $D\boldsymbol{u}/Dt$ (cf. Exercise 1.7). Also,

$$\left(\frac{\partial}{\partial t}\right)_s\left(\frac{\partial \boldsymbol{x}}{\partial s}\right)_t = \left(\frac{\partial}{\partial s}\right)_t\left(\frac{\partial \boldsymbol{x}}{\partial t}\right)_s = \left(\frac{\partial \boldsymbol{u}}{\partial s}\right)_t,$$

and

$$\boldsymbol{u} \cdot \frac{\partial \boldsymbol{u}}{\partial s} = \frac{\partial}{\partial s}(\tfrac{1}{2}\boldsymbol{u}^2),$$

which integrates to zero, as $\boldsymbol{u}$ is a single-valued function of position.

[Note that the partial derivatives commute only because $\boldsymbol{x}$ is viewed consistently as a function of s and t. Suppose the original $\partial/\partial t$ had been the 'normal' one in this text, namely $\partial/\partial t$ *holding x constant*. Then typically,

$$\left(\frac{\partial}{\partial t}\right)_x \left(\frac{\partial x}{\partial s}\right)_t \neq \left(\frac{\partial}{\partial s}\right)_t \left(\frac{\partial x}{\partial t}\right)_x,$$

because the right-hand side is zero, because $(\partial x/\partial t)_x$ is trivially zero, but the left-hand side is typically not zero. The reason that these differential operators do not commute is that in each case the dependent variable is being viewed as a function of x and t during one differentiation but as a function of s and t during the other.]

5.3. See note (c) following the proof of the theorem in §5.1.

5.4. After Stokes's theorem use (A.5), and note that if $p = f(\rho)$ then $\nabla p = f'(\rho)\,\nabla\rho$. The unnecessary assumption is the same as that in Exercise 5.3. Alternative:

$$-\frac{1}{\rho}f'(\rho)\,\nabla\rho = \nabla h,$$

where

$$h(\rho) = -\int_0^\rho \frac{1}{\alpha} f'(\alpha)\,\mathrm{d}\alpha.$$

Then $\mathrm{d}\Gamma/\mathrm{d}t = [h]_C = 0$, as ρ is a single-valued function of position.

In the thin vortex tube argument replace conservation of volume by (the more generally valid) conservation of mass.

5.5. Apply the divergence theorem to $\boldsymbol{\omega} = \nabla \wedge \boldsymbol{u}$, the region V being a portion of the vortex tube of finite length.

5.6. Proceed as in Exercise 5.2 until

$$\frac{\mathrm{d}\mathscr{C}}{\mathrm{d}t} = \int_0^1 \left[\left(\frac{\partial \boldsymbol{a}}{\partial t}\right)_s \cdot \frac{\partial \boldsymbol{x}}{\partial s} + \boldsymbol{a} \cdot \frac{\partial \boldsymbol{u}}{\partial s}\right] \mathrm{d}s.$$

Then take the right-hand side of the desired result, recognizing that $\partial \boldsymbol{a}/\partial t$ there means $(\partial \boldsymbol{a}/\partial t)_x$, and write it using the suffix notation and the summation convention as

$$\int_{C(t)} \left[\left(\frac{\partial \boldsymbol{a}}{\partial t}\right)_x + \left(\boldsymbol{e}_i \wedge \frac{\partial \boldsymbol{a}}{\partial x_i}\right) \wedge \boldsymbol{u}\right] \cdot \mathrm{d}\boldsymbol{x}.$$

Expand the triple vector product, and note that

$$(\partial \boldsymbol{a}/\partial t)_s = (\partial \boldsymbol{a}/\partial t)_x + (\boldsymbol{u} \cdot \nabla)\boldsymbol{a},$$

because both sides of this expression denote the rate of change of $\boldsymbol{a}$ following a particular fluid particle (cf. Exercise 5.2).

Hints and answers for exercises   369

5.7. Note that $\boldsymbol{\omega} = \omega(R, z, t)\boldsymbol{e}_\phi$, so that

$$(\boldsymbol{\omega} \cdot \nabla)\boldsymbol{u} = \frac{\omega}{R}\frac{\partial}{\partial \phi}[u_R(R, z, t)\boldsymbol{e}_R + u_z(R, z, t)\boldsymbol{e}_z] = \frac{\omega u_R}{R}\boldsymbol{e}_\phi,$$

by virtue of eqn (A.29), allowing for the difference in notation.

5.8. Let the vortex be at $z_1 = x_1 + iy_1$. The image system consists of *three* vortices at $\bar{z}_1$, $-\bar{z}_1$, and $-z_1$. Proceed as with eqn (5.27) to obtain

$$\frac{dx_1}{dt} - i\frac{dy_1}{dt} = -\frac{i\Gamma}{2\pi}\left[-\frac{1}{z_1 - \bar{z}_1} - \frac{1}{z_1 + \bar{z}_1} + \frac{1}{2z_1}\right];$$

hence
$$dy_1/dx_1 = -(y_1/x_1)^3 \qquad \text{etc.}$$

To understand the behaviour of the trailing vortices, view the whole of $y \geq 0$ in the above problem as occupied by fluid, with a single boundary at $y = 0$.

5.9. See eqn (3.19).

5.10. The net force on the wall is zero.

If the vortex were somehow fixed at $(d, 0)$, the $\partial \phi/\partial t$ term would be absent and there would be a net force $\rho\Gamma^2/4\pi d$ on the wall, directed towards the vortex.

5.12.
$$w(z) = -\frac{i\Gamma}{2\pi}[\log(z - z_2) - \log(z - \bar{z}_2) - \log(z - z_1) + \log(z - \bar{z}_1)] - \tfrac{1}{2}\alpha z^2.$$

5.13. Let $P = \bar{\varepsilon}_2 + \varepsilon_1$ and $Q = \bar{\varepsilon}_2 - \varepsilon_1$, and deduce that

$$\dot{P} = -\tfrac{1}{4}\bar{P}, \qquad \dot{Q} = -\tfrac{1}{2}iQ.$$

Then write $P = P_R + iP_I$ and solve for P_R and P_I.

5.14. Let the n vortices be at

$$z_m = ae^{i2\pi m/n}, \qquad m = 0, 1, \ldots, n-1.$$

The complex potential due to all these vortices is

$$w = -\frac{i\Gamma}{2\pi}\log(z^n - a^n),$$

and that due to all except the one at $z = a$ can be written

$$w = -\frac{i\Gamma}{2\pi}\log\left[1 + \frac{z}{a} + \left(\frac{z}{a}\right)^2 + \ldots + \left(\frac{z}{a}\right)^{n-1}\right].$$

5.15. For last part, multiply the equation by Γ_s and sum from 1 to n.

5.16.
$$\frac{\mathrm{D}}{\mathrm{D}t}(\boldsymbol{u}\cdot\boldsymbol{\omega}) = \frac{\mathrm{D}\boldsymbol{u}}{\mathrm{D}t}\cdot\boldsymbol{\omega} + \boldsymbol{u}\cdot\frac{\mathrm{D}\boldsymbol{\omega}}{\mathrm{D}t}.$$

Then use Euler's equation (1.12), the vorticity equation (5.7), and the fact that $\nabla\cdot(\nabla\wedge\boldsymbol{u})=0$ to show that

$$\frac{\mathrm{D}}{\mathrm{D}t}(\boldsymbol{u}\cdot\boldsymbol{\omega}) = \nabla\cdot\left[\left(-\frac{p}{\rho}-\chi+\tfrac{1}{2}u^2\right)\boldsymbol{\omega}\right].$$

Then use the divergence theorem.

5.17. Bring $\nabla\cdot[(\boldsymbol{\omega}\wedge\boldsymbol{u})\wedge\nabla\lambda]$ into play using eqn (A.7), expand the triple vector product using eqn (A.1), then use eqn (A.4).

Having said this, the problem lends itself to a much more straightforward approach using suffix notation and the summation convention:

$$\frac{\mathrm{D}}{\mathrm{D}t}\left(\omega_i\frac{\partial\lambda}{\partial x_i}\right) = \frac{\mathrm{D}\omega_i}{\mathrm{D}t}\frac{\partial\lambda}{\partial x_i} + \omega_i\left(\frac{\partial}{\partial t} + u_k\frac{\partial}{\partial x_k}\right)\frac{\partial\lambda}{\partial x_i}$$

$$= \omega_j\frac{\partial u_i}{\partial x_j}\frac{\partial\lambda}{\partial x_i} + \omega_i\frac{\partial}{\partial x_i}\left(\frac{\partial\lambda}{\partial t}\right) + \omega_i u_k\frac{\partial^2\lambda}{\partial x_k\,\partial x_i},$$

where we have used the vorticity equation in its form (5.7). By reversing the order of partial differentiation in the final term, and then changing the dummy suffices, it may be written $\omega_j u_i\,\partial^2\lambda/\partial x_j\,\partial x_i$, with summation understood, of course, over $i=1,2,3$ and $j=1,2,3$. Thus

$$\frac{\mathrm{D}}{\mathrm{D}t}\left(\omega_i\frac{\partial\lambda}{\partial x_i}\right) = \omega_i\frac{\partial}{\partial x_i}\left(\frac{\partial\lambda}{\partial t}\right) + \omega_j\frac{\partial}{\partial x_j}\left(u_i\frac{\partial\lambda}{\partial x_i}\right),$$

which is the result.

5.18. The flow has two elements: a *uniform rotation*, angular velocity $\Omega e^{\alpha t}$, which steadily increases with time as a result of a *secondary flow* of the kind in Fig. 5.17 which keeps stretching the vortex lines (if $\alpha>0$).

5.19. The condition that u_θ be finite at $r=0$ is needed.

5.20. The suitable vector identities are eqns (A.9) and (A.10).

5.21. X_1, X_2, and X_3 are three scalar quantities that are (rather trivially) conserved by an individual fluid element, and $X_i=x_i$ at $t=0$, so Ertel's theorem gives

$$\omega_j\frac{\partial X_i}{\partial x_j} = \omega_{0i}.$$

These are three linear algebraic equations for ω_1, ω_2, and ω_3. They can be solved by writing

$$\omega_{0i}\frac{\partial x_k}{\partial X_i} = \omega_j \frac{\partial X_i}{\partial x_j}\frac{\partial x_k}{\partial X_i} = \omega_j \frac{\partial x_k}{\partial x_j} = \omega_j \delta_{jk} = \omega_k,$$

where we have used the chain rule in the second step (note that summation over $i = 1, 2, 3$ is understood).

5.22. Let $x_i = x_i\{X(s), t\}$ denote the current position of the particle that was, at $t = 0$, a distance s along the initial curve. Then

$$\left(\frac{\partial x_i}{\partial s}\right)_t = \left(\frac{\partial x_i}{\partial X_j}\right)_t \frac{dX_j}{ds} = \frac{\partial x_i}{\partial X_j}\frac{\omega_{0j}}{|\boldsymbol{\omega}_0|} = \frac{\omega_i}{|\boldsymbol{\omega}_0|}.$$

5.23.

$$T = \tfrac{1}{2}\rho \int (\nabla\phi) \cdot (\nabla\phi)\, dV = \tfrac{1}{2}\rho \int [\nabla \cdot (\phi\,\nabla\phi) - \phi\nabla \cdot (\nabla\phi)]\, dV,$$

by eqn (A.4), and then use the divergence theorem.

5.24. Suppose there are two different irrotational flows, i.e. $\boldsymbol{u}_1 = \nabla\phi_1$ and $\boldsymbol{u}_2 = \nabla\phi_2$, where

$$\nabla^2\phi_1 = 0 \quad \text{in } V, \qquad \partial\phi_1/\partial n = f \quad \text{on } S;$$
$$\nabla^2\phi_2 = 0 \quad \text{in } V, \qquad \partial\phi_2/\partial n = f \quad \text{on } S.$$

Consider the problem for $\phi' = \phi_2 - \phi_1$, and use Exercise 5.23 to show that $\nabla\phi'$ must be zero, whereupon $\boldsymbol{u}_1 = \boldsymbol{u}_2$, contradicting the original supposition.

5.25. Let $\nabla\phi$ be the unique irrotational flow satisfying the boundary condition, and write any other incompressible flow doing so as $\boldsymbol{u} = \nabla\phi + \boldsymbol{u}'$, where, consequently,

$$\nabla \cdot \boldsymbol{u}' = 0 \quad \text{in } V, \qquad \text{and} \qquad \boldsymbol{u}' \cdot \boldsymbol{n} = 0 \quad \text{on } S.$$

Then expand $T = \tfrac{1}{2}\rho \int (\nabla\phi + \boldsymbol{u}')^2\, dV$, and use eqn (A.4) and the divergence theorem.

5.26. If z denotes distance downstream from the centre of the sphere, then $\phi \sim Uz$ as $r \to \infty$, i.e. $\phi \sim Ur\cos\theta$ as $r \to \infty$. The other boundary condition is $\partial\phi/\partial r = 0$ on $r = a$. By trying $\phi = f(r)\cos\theta$—or by a more formal application of the method of separation of variables—obtain

$$\phi = U\left(r + \frac{a^3}{2r^2}\right)\cos\theta.$$

To find the pressure distribution on the sphere, use Bernoulli's theorem.

372 Hints and answers for exercises

5.27. Use Exercise 5.23 to find the kinetic energy, $T = \frac{1}{4}MU^2$, where $M = \frac{4}{3}\pi a^3 \rho$ is the mass of fluid displaced by the sphere. Then use the argument leading to eqn (4.77).

5.28. The boundary conditions are

$$\frac{1}{r}\frac{\partial \phi}{\partial \theta} = \pm \Omega r \quad \text{at } \theta = \pm \Omega t,$$

and trying an appropriate separable solution gives

$$\phi = -(\Omega r^2 \cos 2\theta)/(2 \sin 2\Omega t).$$

The streamlines are $xy = \text{constant}$. At $\theta = \pm \Omega t$ use Exercise 5.9 to obtain

$$\frac{p}{\rho} = \text{constant} - \tfrac{1}{2}\Omega^2 r^2 \left[\frac{3}{\sin^2 2\Omega t} - 2 \right].$$

5.29. The last result in Exercise 5.23 assumes that ϕ is a single-valued function of position, which is guaranteed only in a simply connected region. Here, $\phi = \Gamma \theta / 2\pi$. Remedy: make the region simply connected by a *cut* (Fig. H.4), and apply the last result of Exercise 5.23 to the surface ABCDEFA, which encloses a simply connected region V. Check your answer by using the second result in Exercise 5.23, which is a more straightforward method.

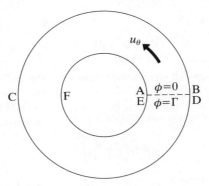

Fig. H.4.

Chapter 6

6.1.

$$n \wedge (\nabla \wedge u) = n \wedge \left(e_i \wedge \frac{\partial u}{\partial x_i}\right)$$

$$= \left(n \cdot \frac{\partial u}{\partial x_i}\right) e_i - (n \cdot e_i) \frac{\partial u}{\partial x_i}$$

$$= n_j \frac{\partial u_j}{\partial x_i} e_i - n_i \frac{\partial}{\partial x_i} (u_j e_j)$$

$$= n_j \frac{\partial u_j}{\partial x_i} e_i - n_j \frac{\partial u_i}{\partial x_j} e_i \quad \text{etc.},$$

where in the last term we have switched the dummy suffices i and j, summation being understood over both $i = 1, 2, 3$ and $j = 1, 2, 3$.

6.2. The vector identities needed are eqns (A.14), (A.19), (A.20), and (A.10).

6.4. The torque is

$$\int_0^{2\pi} r_1 t_\theta r_1 \, d\theta = \frac{4\pi\mu(\Omega_2 - \Omega_1) r_1^2 r_2^2}{r_2^2 - r_1^2}$$

per unit length in the z-direction. It is positive if $\Omega_2 > \Omega_1$, as we would expect.

6.5. Put $G = \rho$ in eqn (6.6a).

6.7.

$$\frac{\partial}{\partial s_k} (e_{ij} s_i s_j) = e_{ij} (\delta_{ik} s_j + s_i \delta_{jk})$$

$$= e_{kj} s_j + e_{ik} s_i.$$

But $e_{ik} = e_{ki}$, so this is equal to $2 e_{kj} s_j$ etc.

6.8.

$$u_Q = u_P + \tfrac{1}{2}(\beta s_2, -\beta s_1, 0) + \tfrac{1}{2}(\beta s_2, \beta s_1, 0).$$

The principal axes of e_{ij} are (everywhere) at an angle of $\pi/4$ to the coordinate axes, which is why T'_{12} turned out to be zero in the analysis leading to eqn (6.15). (More usually, the principal axes will vary with the position of the point P.)

6.10.

$$2 e_{\theta\phi} = [(e_\theta \cdot \nabla) u] \cdot e_\phi + [(e_\phi \cdot \nabla) u] \cdot e_\theta \quad \text{etc.},$$

using eqns (A.37) and (A.38).

6.11. Direct substitution gives $T^D_{ij} = A\delta_{ij}e_{kk} + Be_{ij} + Ce_{ji}$, and $e_{ij} = e_{ji}$ by eqn (6.16). In the second part, note that $e_{ii} = \nabla \cdot \boldsymbol{u}$ and $\delta_{ii} = 3$.

6.12. In the momentum equation, the term $\mu \nabla^2 \boldsymbol{u}$ or $-\mu \nabla \wedge (\nabla \wedge \boldsymbol{u})$ represents the *net* viscous force on a small fluid element (cf. eqn (2.11)). While this is zero for an irrotational flow, it does not follow at all that 'the viscous forces are zero'. There will typically be viscous stresses all around a fluid element, even if the *resultant* force is zero; these stresses will be zero all around the element only if it is not being *deformed*, and that is certainly not the case here (see Fig. 1.5).

6.13. $DJ/Dt = (\partial J/\partial t)_X$, and differentiating the determinant gives the sum of three determinants, in each of which only one row is differentiated. The top row of the first of these is

$$\left(\frac{\partial}{\partial t}\right)_X \frac{\partial x_1}{\partial X_1} \quad \left(\frac{\partial}{\partial t}\right)_X \frac{\partial x_1}{\partial X_2} \quad \left(\frac{\partial}{\partial t}\right)_X \frac{\partial x_1}{\partial X_3},$$

and on changing the order of partial differentiation this is

$$\frac{\partial u_1}{\partial X_1} \quad \frac{\partial u_1}{\partial X_2} \quad \frac{\partial u_1}{\partial X_3}.$$

On using the chain rule this can be written

$$\frac{\partial u_1}{\partial x_i}\frac{\partial x_i}{\partial X_1} \quad \frac{\partial u_1}{\partial x_i}\frac{\partial x_i}{\partial X_2} \quad \frac{\partial u_1}{\partial x_i}\frac{\partial x_i}{\partial X_3},$$

where summation over $i = 1, 2, 3$ is understood in each case. The $i = 1$ terms give a contribution $J \partial u_1/\partial x_1$ to this first determinant. The terms $i = 2, 3$ give no contribution to it, because in each case two rows of the resulting determinant are multiples of one another.

To prove Reynolds's transport theorem,

$$\frac{d}{dt}\int_{V(t)} G\, dV = \frac{d}{dt}\int_{V(t)} G\, dx_1\, dx_2\, dx_3$$

$$= \frac{d}{dt}\int_{V(0)} GJ\, dX_1\, dX_2\, dX_3$$

$$= \int_{V(0)} \left[\left(\frac{\partial G}{\partial t}\right)_X J + G\left(\frac{\partial J}{\partial t}\right)_X\right] dX_1\, dX_2\, dX_3 \quad \text{etc.,}$$

the Jacobian determinant J entering when we make a change of variables, $V(0)$ denoting the region *initially* occupied by the 'dyed' blob

Hints and answers for exercises 375

Chapter 7

7.1. $u_\theta = \Omega r z/h$. $t_\theta = -T_{\theta z}$, as $\boldsymbol{n} = (0, 0, -1)$ on the upper boundary. The torque exerted by the fluid on the upper disc is
$$\int_0^{2\pi}\int_0^a r t_\theta r \, dr \, d\theta.$$

7.2. $u_\phi = \Omega r \sin\theta$ on $r = a$. Try $u_\phi = f(r)\sin\theta$. The torque exerted by the fluid on the sphere is
$$\int_0^{2\pi}\int_0^\pi a \sin\theta \, t_\phi a^2 \sin\theta \, d\theta \, d\phi.$$

7.3. The boundary conditions are
$$u_r = 0, \quad t_\theta = 0 \quad \text{on } r = a,$$
so
$$\partial\Psi/\partial\theta = 0, \quad T_{\theta r} = 0 \quad \text{on } r = a,$$
as $\boldsymbol{n} = \boldsymbol{e}_r$. The last of these implies $e_{\theta r} = 0$ on $r = a$, and using eqn (A.44) this means
$$r\frac{\partial}{\partial r}\left(\frac{u_\theta}{r}\right) = r\frac{\partial}{\partial r}\left\{-\frac{1}{r^2 \sin\theta}\frac{\partial\Psi}{\partial r}\right\} = 0 \quad \text{on } r = a$$
etc.
$$t_r = T_{rr} = -p + 2\mu e_{rr} = -p + 2\mu\frac{\partial u_r}{\partial r},$$
where we have again used eqn (A.44). To find p, proceed as in §7.2:
$$E^2\Psi = \frac{Ua}{r}\sin^2\theta, \quad p = p_\infty - \frac{\mu Ua}{r^2}\cos\theta.$$

7.5. Boundary conditions:
$$\frac{1}{r}\frac{\partial\psi}{\partial\theta} = 0, \quad -\frac{\partial\psi}{\partial r} = \pm\Omega r \quad \text{at } \theta = \pm\Omega t.$$

7.6. $\psi = f(\theta)$; the equation for $f(\theta)$ is
$$f'''' + 4f'' = 0.$$

7.7. The boundary conditions imply
$$(\lambda - 2)\tan\lambda\alpha = \lambda \tan(\lambda - 2)\alpha;$$
then put $p = \lambda - 1$ etc.
In Fig. 7.5, draw a horizontal line from the local maximum that appears on the right of the figure.

7.8. One boundary condition is that $y=0$ is a streamline, so we may as well put $\psi = 0$ at $y = 0$. In the non-dimensionalized problem, with

$$x' = kx, \qquad x'_0 = kx_0 \qquad \text{etc.},$$

the other boundary condition is, on dropping primes,

$$\partial\psi/\partial y = -\cos x_0 \qquad \text{when } x = x_0 + \varepsilon \sin x_0 \text{ and } y = 0.$$

This may be expanded to

$$\left.\frac{\partial\psi}{\partial y}\right|_{x_0} + \varepsilon \sin x_0 \left.\frac{\partial^2\psi}{\partial x\, \partial y}\right|_{x_0} + \ldots = -\cos x_0 \qquad \text{on } y = 0$$

etc.

7.9. The thin-film approximation here implies

$$\frac{1}{r}\frac{\partial u_r}{\partial \theta} \gg \frac{\partial u_r}{\partial r}, \qquad \text{etc.},$$

and from $\nabla \cdot \mathbf{u} = 0$ we infer $u_\theta/u_r = O(\alpha) \ll 1$. Equations (A.35) reduce on neglecting the terms $(\mathbf{u}\cdot\nabla)\mathbf{u}$ and $\partial\mathbf{u}/\partial t$, to

$$0 = -\frac{\partial p}{\partial r} + \frac{\mu}{r^2}\frac{\partial^2 u_r}{\partial\theta^2},$$

$$0 = -\frac{1}{r}\frac{\partial p}{\partial\theta} + \frac{\mu}{r^2}\left(\frac{\partial^2 u_\theta}{\partial\theta^2} + 2\frac{\partial u_r}{\partial\theta}\right);$$

hence

$$\frac{1}{r}\frac{\partial p}{\partial\theta} \ll \frac{\partial p}{\partial r},$$

so $p = p(r)$ to a first approximation etc.

The pressure is

$$p = \frac{6\mu}{\alpha^3}\frac{d\alpha}{dt}\log\left(\frac{r}{a}\right) + p_0,$$

where p_0 is atmospheric pressure. The logarithmic singularity in p at $r = 0$ is inevitable in this simple model of a 'peeling' process; the torque (per unit length in the z-direction) is nonetheless finite.

7.10.

$$\frac{2g\sin\alpha}{\nu}h(x,t) = -\frac{1}{\beta t} + \left[\frac{1}{\beta^2 t^2} + \frac{4g\sin\alpha}{\nu}\left(\frac{x}{t}\right)\right]^{\frac{1}{2}}.$$

7.11.

$$\frac{\partial h}{\partial t} + \frac{gh^2}{\nu}\frac{\partial h}{\partial x} = 0,$$

with solution
$$h = f\left(x - \frac{gh^2}{\nu}t\right),$$
so h is constant along curves
$$x - \frac{gh^2}{\nu}t = \text{constant};$$
therefore all the characteristics are straight lines. On this basis we may construct Fig. H.5 (see §3.9), assuming $h = 0$ at $x = 0$ for $t > 0$.

7.12. Use conical coordinates (x, z, ϕ), where ϕ denotes the angle which the x–z plane makes with some fixed plane through the symmetry axis of the cone. These orthogonal curvilinear coordinates have $h_1 = 1$, $h_2 = 1$, and h_3 equal to the distance of a general point P from the symmetry axis, which is $x \cos \alpha + z \sin \alpha$. According to eqn (A.27), $\nabla \cdot \boldsymbol{u} = 0$ implies
$$\frac{\partial}{\partial x}(h_3 u) + \frac{\partial}{\partial z}(h_3 w) + \frac{\partial}{\partial \phi}(0) = 0,$$
as there is no velocity component in the ϕ-direction. But as $z \ll x$ we have $h_3 \doteqdot x \cos \alpha$, and the result then follows.

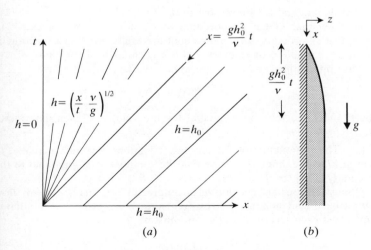

Fig. H.5.

378 Hints and answers for exercises

The structure of the evolution equation for $h(x, t)$ allows a similarity solution of the same form as before:

$$h = \left(\frac{3\nu}{5g \sin \alpha}\right)^{\frac{1}{2}}\left(\frac{x}{t}\right)^{\frac{1}{2}}.$$

The volume V of the blob as a whole is

$$\int_0^{x_N(t)} 2\pi x \cos \alpha \, h(x, t) \, dx = V.$$

If V is constant this leads to $x_N \propto t^{\frac{1}{5}}$.

7.13.

$$\nu \frac{\partial^2 u}{\partial z^2} = g \frac{\partial h}{\partial x}; \quad u = \frac{g}{\nu}\frac{\partial h}{\partial x}\left(\frac{z^2}{2} - hz\right) \qquad \text{etc.}$$

Axisymmetric case: exactly the same until the incompressibility condition

$$\frac{\partial w}{\partial z} = -\frac{1}{r}\frac{\partial}{\partial r}(ru_r) \qquad \text{etc.}$$

7.14. Rewrite the conservation of volume

$$2\pi \int_0^{r_N(t)} rh(r, t) \, dr = V,$$

so that the integral is with respect to η.

To show that $k(t)$ is proportional to $t^{\frac{1}{8}}$, substitute as suggested; in order that an ordinary differential equation results, for F as a function of the single variable η, it emerges that $k^7 k'(t)$ must be a constant.

After integration, and two applications of $F(\eta_N) = 0$, we have

$$F^3 = \tfrac{3}{16}(\eta_N^2 - \eta^2);$$

substitution into the above integral for the volume V then determines η_N.

7.15. The formulation of the thin-film equations is identical, really, to that in §7.9, the difference being that the component of $\mathbf{g}$ parallel to the boundary is $-g \cos \theta$, and thus varies with θ.

Q must be constant because of incompressibility: this is a steady flow with a fixed (but unknown) stress-free surface $z = h(\theta)$, so the volume flux across each section $\theta = $ constant must be the same.

$$\bar{h} = \frac{1}{2\pi}\int_0^{2\pi} h(\theta) \, d\theta, \qquad \text{etc.}$$

Hints and answers for exercises 379

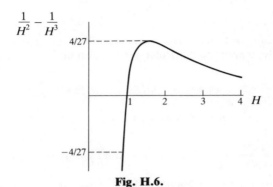

Fig. H.6.

7.16. The integrals
$$I_n = \int_0^{2\pi} \frac{d\theta}{(1 - \lambda \cos \theta)^n}, \qquad n = 2, 3$$

are best done by contour integration round the unit circle in the z-plane, so that $\cos \theta = \frac{1}{2}(z + z^{-1})$. There is just one pole, of order n, inside the circle at
$$z_0 = \frac{1}{\lambda} - \left(\frac{1}{\lambda^2} - 1\right)^{\frac{1}{2}}.$$

Straightforward, but slightly messy, applications of the residue calculus give
$$I_2 = 2\pi/(1 - \lambda^2)^{\frac{3}{2}}, \qquad I_3 = \pi(2 + \lambda^2)/(1 - \lambda^2)^{\frac{5}{2}}.$$

7.17. The reason the no-slip condition on the obstacle cannot be satisfied is that the highest derivatives with respect to x and y have been neglected in the thin-film approximation
$$\nu\left(\frac{\partial^2 u}{\partial x^2} + \frac{\partial^2 u}{\partial y^2} + \frac{\partial^2 u}{\partial z^2}\right) \doteqdot \nu \frac{\partial^2 u}{\partial z^2}.$$

The thickness δ of the boundary layer on the obstacle must be such as to restore the importance of those terms, so
$$U/\delta^2 = 0(U/h^2),$$

i.e. $\delta = O(h)$. Notably, the boundary layer thickness here has nothing to do with the viscosity ν, let alone the Reynolds number.

380 *Hints and answers for exercises*

Chapter 8

8.2. To show that $g(x)$ is a constant, follow the routine in §8.4. Subsequently, there is no particular need to introduce a stream function; thus
$$\partial v/\partial y = -\partial u/\partial x = -\alpha f'(\eta),$$
so
$$v = -(v\alpha)^{\frac{1}{2}}f(\eta) + k(x), \quad \text{etc.}$$

8.3.
$$u = \frac{\partial \psi}{\partial y} = \frac{F(x)}{g(x)} f'(\eta), \qquad u \to U(x) \text{ as } \eta \to \infty, \quad \text{etc.}$$

On substituting into the boundary layer equations, take out a factor UU', and two terms proportional to $f'f''$ cancel.

To obtain an ordinary differential equation for $f(\eta)$ we must have

$$g^2 U' = \text{constant} \quad \text{and} \quad \frac{U}{U'}\frac{g'}{g} = \text{constant},$$

for these will otherwise be functions of x alone. Combining these leads to the restricted possibilities for $U(x)$ that are given.

8.4. In dealing with $\int_{-\infty}^{\infty} v(\partial u/\partial y)\,\mathrm{d}y$, integrate by parts and use the incompressibility condition.

Substitution into the jet equation (8.64) gives

$$-\tfrac{1}{2} g^{\frac{1}{2}} g'(f'^2 + ff'') = v(2\rho/3M)^{\frac{1}{2}} f''',$$

whence $g^{\frac{1}{2}} g'$ must be a constant.

The equation for f can be written

$$f''' + (ff')' = 0.$$

Inspection of the final result for u shows that, at given x, u changes substantially when η changes by an $O(1)$ amount; the thickness of the jet, δ, is thus obtained by putting $\eta = O(1)$, so $\delta = O[g(x)]$; i.e.

$$\delta \sim (\rho v^2/M)^{\frac{1}{3}} x^{\frac{2}{3}}.$$

The boundary-layer-type treatment rests on the assumption that the jet is thin, i.e. $\delta \ll x$. It is valid, then, if

$$(Mx/\rho v^2)^{\frac{1}{3}} \gg 1.$$

The analysis inevitably breaks down very close to the slit.

8.5. The usual boundary-layer-type argument leads to

$$(U + u_1)\frac{\partial u_1}{\partial x} + v_1 \frac{\partial u_1}{\partial y} = v \frac{\partial^2 u_1}{\partial y^2},$$

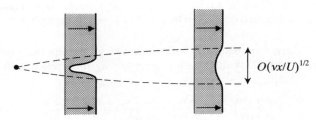

Fig. H.7.

$\mathrm{d}p/\mathrm{d}x$ being zero because the flow outside the (thin) wake is uniform, speed U. Then neglect quadratically small terms. The rest of the analysis is similar to, but easier than, that in Exercise 8.4; the symmetry condition $\partial u_1/\partial y = 0$ at $y = 0$, i.e. $f'(0) = 0$, is needed.

8.6. Multiply by F', integrate once, and apply $F(0) = 0$ to obtain

$$\tfrac{1}{2}F'^2 + F - \tfrac{1}{3}F^3 = \tfrac{1}{2}[F'(0)]^2.$$

Now, $\lim_{\eta \to \infty} F(\eta) = -1$. The above equation then implies that $\lim_{\eta \to \infty} F'(\eta) = c$; yet that constant c must be zero, for otherwise $F(\eta)$ itself would not tend to a finite limit as $\eta \to \infty$. A contradiction then follows: in the limit $\eta \to \infty$ the left-hand side of the above equation is negative, but the right-hand side is positive.

8.7. By substituting $u_r = F(\theta)/r$ into eqn (A.35) and eliminating the pressure by cross-differentiating the r and θ components,

$$\nu\left(\frac{\mathrm{d}^3 F}{\mathrm{d}\theta^3} + 4\frac{\mathrm{d}F}{\mathrm{d}\theta}\right) + 2F\frac{\mathrm{d}F}{\mathrm{d}\theta} = 0.$$

Integrate eqn (8.68) once, multiply by f' and integrate again, then apply conditions $f'(0) = 0$ and $f(0) = 1$.

Results (8.71) and (8.72) are obtained by observing that as α, R, f, and c are all positive

$$\tfrac{2}{3}\alpha R(f^2 + f) + 4\alpha^2 f + c$$

is greater than or equal to both $\tfrac{2}{3}\alpha R(f^2 + f)$ and $4\alpha^2 f$ throughout the interval $0 \leq \eta \leq 1$.

8.8 The r component of eqn (A.35) gives

$$f^2 + hf' - g^2 = -\frac{1}{\rho\Omega^2}\frac{1}{r}\frac{\partial p}{\partial r} + f'',$$

whence $r^{-1} \partial p/\partial r$ must be a function of z only. But integration of the

382   Hints and answers for exercises

z-component shows it to be a function of r only, so it must be a constant, c. The conditions $f \to 0$, $g \to 0$ as $\xi \to \infty$ then imply that c must be zero.

8.10. Differentiate eqn (8.1) with respect to y, use eqn (8.2) to cancel two terms, and differentiate with respect to y again. Using eqn (8.2) this gives

$$\tfrac{1}{2}\frac{\partial}{\partial x}\left(\frac{\partial u}{\partial y}\right)^2 + u\frac{\partial^3 u}{\partial x\,\partial y^2} - \frac{\partial u}{\partial x}\frac{\partial^2 u}{\partial y^2} + v\frac{\partial^3 u}{\partial y^3} = v\frac{\partial^4 u}{\partial y^4}.$$

But $u = v = 0$ on $y = 0$, so $\partial u/\partial x$ is also zero on $y = 0$, and therefore

$$\tfrac{1}{2}\frac{d}{dx}\left[\left(\frac{\partial u}{\partial y}\right)^2\bigg|_{y=0}\right] = v\frac{\partial^4 u}{\partial y^4}\bigg|_{y=0} \quad \text{etc.}$$

Chapter 9

9.1. The linearized equations are

$$\frac{\partial u_r^*}{\partial t} - \frac{2U_\theta u_\theta^*}{r} = -\frac{1}{\rho}\frac{\partial p^*}{\partial r},$$

$$\frac{\partial u_\theta^*}{\partial t} + \left(U_\theta' + \frac{U_\theta}{r}\right)u_r^* = 0,$$

$$\frac{\partial u_z^*}{\partial t} = -\frac{1}{\rho}\frac{\partial p^*}{\partial z}, \qquad \frac{1}{r}\frac{\partial}{\partial r}(ru_r^*) + \frac{\partial u_z^*}{\partial z} = 0.$$

In the last part, s is a constant, so

$$s^2 \int_{-L}^{L} \left\{r\,|\hat{u}_r'|^2 + r\left(n^2 + \frac{1}{r^2}\right)|\hat{u}_r|^2\right\}dr = -\int_{-L}^{L} \frac{n^2}{r^2}(r^2 U_\theta^2)'\,|\hat{u}_r|^2\,dr.$$

9.2. The reduction to the Taylor–Goldstein equation follows on comparing (i) the middle two terms in square brackets and (ii) the first two terms in the equation.

Cast the Taylor–Goldstein equation into the form

$$[(U-c)^{2n}q']' + [\ldots]q = 0$$

before multiplying by the c.c. of q and integrating.

Choose $n = \tfrac{1}{2}$ in the first case and $n = 1$ in the second, focusing on the imaginary part of the resulting equation in each case.

9.3. Consider first the case in which there is no diffusion of heat or salt. When the parcel is displaced upwards it retains its initial density, find itself more dense than its surroundings, and is forced back towards its equilibrium position. But consider now the effects of diffusion, recognizing that *heat diffuses much faster than salt*. If the parcel moves sufficiently slowly it can continually adjust its temperature to (almost) that of its surroundings. It then finds itself with essentially the same temperature as its surroundings, but with essentially its original salt content. In this manner it will be *less* dense than its new surroundings, and buoyancy forces will keep it moving upward, further still from its equilibrium position.

9.5. The Landau equation may be rewritten as a linear equation in the variable $|A|^{-2}$, so

$$\frac{1}{|A|^2} = \frac{l}{\beta(R - R_c)} + \left[\frac{1}{A_0^2} - \frac{l}{\beta(R - R_c)}\right] e^{-\beta(R - R_c)t}.$$

(a) $l > 0$, because a state of steady convection is reached at slightly supercritical values of the Rayleigh number, and there is also stability to disturbances of arbitrary magnitude at subcritical values of the Rayleigh number.

(b) $l < 0$, because of the experiments of Nishioka *et al.* (1975).

(c) $l < 0$.

9.6. The disturbances ε_n to the $1 - \lambda^{-1}$ steady solution satisfy $\varepsilon_{n+1} = (2 - \lambda)\varepsilon_n$ if quadratically small terms are ignored, so $|\varepsilon_n|$ does not remain small if $|2 - \lambda| > 1$.

To examine the period 2 solution note that

$$x_{n+2} = \lambda^2 x_n(1 - x_n)[1 - \lambda x_n(1 - x_n)].$$

Seek a constant solution X to this, and factorize the resulting cubic, knowing that $X - 1 + \lambda^{-1}$ must be a factor.

To examine the stability of a period 2 solution to $x_{n+1} = f(x_n)$ write $x_n = X_n + \varepsilon_n$ and obtain, for small ε_n,

$$\varepsilon_{n+1} = f'(X_n)\varepsilon_n.$$

Thus $\varepsilon_{n+2} = f'(X_{n+1})\varepsilon_{n+1}$ etc.

Bibliography

Aref, H. (1983). *Ann. Rev. Fluid Mech.* **15,** 345–89.
Aref, H. (1986). *J. Fluid Mech.* **173,** 15–41.
Atiyah, M. F. (1988). *Q. J. R. Astron. Soc.* **29,** 287–99.
Batchelor, G. K. (1956). *J. Fluid Mech.* **1,** 177–90.
Batchelor, G. K. (1967). *An introduction to fluid dynamics.* Cambridge University Press.
Benjamin, T. B. (1978). *Proc. R. Soc. Lond.* A **359,** 27–43.
Benjamin, T. B. and Mullin, T. (1982). *J. Fluid Mech.* **121,** 219–30.
Bouard, R. and Coutanceau, M. (1980). *J. Fluid Mech.* **101,** 583–607.
Bourne, D. E. and Kendall, P. C. (1977). *Vector analysis and Cartesian tensors.* Nelson, London.
Busse, F. H. (1985). In Swinney and Gollub (1985), pp. 97–137.
Campbell, L. J. and Ziff, R. M. (1979). *Phys. Rev.* B **20,** 1886–902.
Carrier, G. F., Krook, M., and Pearson, C. E. (1966). *Functions of a complex variable.* McGraw-Hill, New York.
Cattaneo, F. and Hughes, D. W. (1988). *J. Fluid Mech.* **196,** 323–44.
Chaiken, J., Chevray, R., Tabor, M., and Tan, Q. M. (1986). *Proc. R. Soc. Lond.* A **408,** 165–74.
Chen, J.-D. (1989). *J. Fluid Mech.* **201,** 223–42.
Childress, S. (1981). *Mechanics of swimming and flying.* Cambridge University Press.
Coles, D. (1965). *J. Fluid Mech.* **21,** 385–425.
Couder, Y. and Basdevant, C. (1986). *J. Fluid Mech.* **173,** 225–51.
Coutanceau, M. and Bouard, R. (1977). *J. Fluid Mech.* **79,** 231–56.
Craik, A. D. D. (1985). *Wave interactions and fluid flows.* Cambridge University Press.
Crapper, G. D. (1984). *Introduction to water waves.* Ellis Horwood, Chichester.
Cruickshank, J. O. (1988). *J. Fluid Mech.* **193,** 111–27.
Cvitanović, P. (1986). *Universality in chaos.* Adam Hilger, Bristol.
Dalton, S. (1977). *The miracle of flight.* McGraw-Hill, New York.
Di Prima, R. C. and Swinney, H. L. (1985). In Swinney and Gollub (1985), pp. 139–80.
Donnelly, R. J. (1988). *Scient. Am.* **259**(5), 66–74.
Drazin, P. G. and Johnson, R. S. (1989). *Solitons: an introduction.* Cambridge University Press.
Drazin, P. G. and Reid, W. H. (1981). *Hydrodynamic stability.* Cambridge University Press.

Dritschel, D. G. (1985). *J. Fluid Mech.* **157,** 95–134.
Dritschel, D. G. (1986), *J. Fluid Mech.* **172,** 157–82.
Euler, L. (1755). See the translation and commentary in Truesdell (1954), especially pp. LXXXIV–XCI.
Feigenbaum, M. J. (1980). *Los Alamos Science* **1,** 4–27. (Also in Cvitanović (1986), pp. 49–84.)
Fenstermacher, P. R., Swinney, H. L., and Gollub, J. P. (1979), *J. Fluid Mech.* **94,** 103–29.
Fohl, T. and Turner, J. S. (1975). *Phys. Fluids* **18,** 433–6.
Fornberg, B. (1985). *J. Comput. Phys.* **61,** 297–320.
Gleick, J. (1988). *Chaos.* Heinemann, London.
Goldstein, S. (1938). *Modern developments in fluid dynamics.* Clarendon Press, Oxford. (Paperback edn: Dover, New York, 1965.)
Goldstein, S. (1969). *Ann. Rev. Fluid Mech.* **1,** 1–28.
Gollub, J. P. and Benson, S. V. (1980). *J. Fluid Mech.* **100,** 449–70.
Greenspan, H. P. (1968). *The theory of rotating fluids.* Cambridge University Press.
Greenspan, H. P. and Howard, L. N. (1963). *J. Fluid Mech.* **17,** 385–404.
Hall, P. and Walton, I. C. (1977). *Proc. R. Soc. Lond.* A **358,** 199–221.
Hasimoto, H. and Sano, O. (1980). *Ann. Rev. Fluid Mech.* **12,** 335–63.
Hele-Shaw, H. J. S. (1898). *Nature* **58,** 34–6.
Helmholtz, H. von (1858). See translation in *Phil. Mag.* (4) **33,** 485–512 (1867).
Hide, R. (1977), *Q. J. R. Met. Soc.* **103,** 1–28.
Higdon, J. J. L. (1985). *J. Fluid Mech.* **159,** 195–226.
Homsy, G. M. (1987). *Ann. Rev. Fluid Mech.* **19,** 271–311.
Huppert, H. E. (1982). *J. Fluid Mech.* **121,** 43–58.
Huppert, H. E. (1986). *J. Fluid Mech.* **173,** 557–94.
Jimenez, J. (1987). *J. Fluid Mech.* **178,** 177–94.
Joseph, D. D. (1976). *Stability of fluid motions.* Springer-Verlag, New York.
Joseph, D. D. (1985). In Swinney and Gollub (1985), pp. 27–76.
Kelvin, Lord (1869). *Trans. R. Soc. Edinb.* **25,** 217–260. (*Math. Phys. Papers* **4,** 13–66.)
Korteweg, D. J. and de Vries, G. (1895). *Phil. Mag.* **39,** 422–43.
Kutta, W. M. (1902). *Illust. Aeronaut. Mitt.* **6,** 133–5.
Lamb, H. (1932). *Hydrodynamics* (6th edn). Cambridge University Press.
Lanchester, F. W. (1907). *Aerodynamics.* Constable, London.
Landau, L. D. and Lifshitz, E. M. (1987). *Fluid mechanics* (2nd edn). Pergamon, Oxford.
Lanford, O. E. (1985). In Swinney and Gollub (1985), pp. 7–26.

Libchaber, A., Laroche, C., and Fauve, S. (1982). *J. Phys. Lett.* **43**, L211–16. (Also in Cvitanović (1986), pp. 137–42.)
Lighthill, J. (1973). *J. Fluid Mech.* **60**, 1–17.
Lighthill, J. (1978). *Waves in fluids*. Cambridge University Press.
Lighthill, J. (1986). *An informal introduction to theoretical fluid mechanics*. Clarendon Press, Oxford.
Loc, T. P. and Bouard, R. (1985). *J. Fluid Mech.* **160**, 93–117.
Lorenz, E. N. (1963). *J. Atmos. Sci.* **20**, 130–41.
Maxworthy, T. (1972). *J. Fluid Mech.* **51**, 15–32.
May, R. M. (1976). *Nature* **261**, 459–67.
Meiron, D. I., Saffman, P. G., and Schatzman, J. C. (1984). *J. Fluid Mech.* **147**, 187–212.
Mellor, G. L., Chapple, P. J., and Stokes, V. K. (1968). *J. Fluid Mech.* **31**, 95–112.
Miles, J. (1984). *Adv. Appl. Mech.* **24**, 189–214.
Moffatt, H. K. (1964). *J. Fluid Mech.* **18**, 1–18.
Moffatt, H. K. (1969). *J. Fluid Mech.* **35**, 117–29.
Moffatt, H. K. (1977a). *J. Mécanique* **16**, 651–73.
Moffatt, H. K. (1977b). Six lectures on fluid dynamics. In *Fluid dynamics*, Proc. Summer School Theor. Phys., Les Houches, France, July 1973. pp. 149–234. Gordon & Breach, London.
Moffatt, H. K. and Duffy, B. R. (1980). *J. Fluid Mech.* **96**, 299–313.
Moon, F. C. (1987). *Chaotic vibrations*. Wiley, Chichester.
Moore, D. R., Toomre, J., Knobloch, E., and Weiss, N. O. (1983) *Nature* **303**, 348–52.
Nakayama, Y. (ed.) (1988). *Visualized flow*. Pergamon, Oxford.
Newton, I. (1687). *Principia*. Translated into English by Andrew Motte in 1729. The translation revised, and supplied with an historical and explanatory appendix, by Florian Cajori. University of California Press, Berkeley (1934).
Nishioka, M., Iida, S. and Ichikawa, Y. (1975). *J. Fluid Mech.* **72** 731–51.
Ockendon, H. and Tayler, A. B. (1983). *Inviscid fluid flows* Springer-Verlag, New York.
Oshima, Y. (1978). *J. Phys. Soc. Japan* **45**, 660–64.
Oshima, Y. and Asaka, S. (1975). *Nat. Sci. Rep. Ochanomizu Univ.* **26** 31–7.
Ottino, J. M. (1989a). *Scient. Am.* **260**(1), 40–9.
Ottino, J. M. (1989b). *The kinematics of mixing: stretching, chaos and transport*. Cambridge University Press.
Perry, A. E., Chong, M. S., and Lim, T. T. (1982). *J. Fluid Mech.* **116** 77–90.
Pippard, A. B. (1985). *Response and stability*. Cambridge University Press.

Prandtl, L. (1905). *Verhandlungen des III Internationalen Mathematiker-Kongresses* (Heidelberg 1904), Leipzig, pp. 484-91. (Also in Prandtl, L. (1961). *Gesammelte Abhandlungen* **2**, 575-84. Springer-Verlag, Berlin.)

Prandtl, L. (1952). *The essentials of fluid dynamics*. Blackie, London.

Prandtl, L. and Tietjens, O. G. (1934). *Applied hydro- and aeromechanics*. McGraw-Hill, New York. (Paperback edn: Dover, New York, 1957.)

Priestley, H. A. (1985). *Introduction to complex analysis*. Clarendon Press, Oxford.

Proudman, I. and Pearson, J. R. A. (1957). *J. Fluid Mech.* **2**, 237-62.

Rayleigh, Lord (1880). *Proc. Lond. Math. Soc.* **11**, 57-70. (Also *Scientific Papers* **1**, 474-87 (1899). Cambridge University Press.)

Rayleigh, Lord (1910). *Proc. R. Soc. Lond.* A **84**, 247-84. (Also *Scientific Papers* **5**, 573-610 (1912). Cambridge University Press.)

Rayleigh, Lord (1916a). *Phil. Mag.* **32**, 529-46. (Also *Scientific Papers* **6**, 432-46 (1920). Cambridge University Press.)

Rayleigh, Lord (1916b). *Proc. R. Soc. Lond.* A **93**, 148-54. (Also *Scientific Papers* **6**, 447-53 (1920). Cambridge University Press.)

Reed, H. L. (1987). *Phys. Fluids* **30**, 2597-606.

Reichenbach, H. (1983). *Ann. Rev. Fluid Mech.* **15**, 1-28.

Reynolds, O. (1883). *Phil. Trans. R. Soc. Lond.* **174**, 935-82. (Also *Scientific Papers* **2**, 51-105 (1901). Cambridge University Press.)

Roberts, P. H. and Donnelly, R. J. (1974). *Ann. Rev. Fluid Mech.* **6**, 179-225.

Rosenhead, L. (ed.) (1963). *Laminar boundary layers*. Clarendon Press, Oxford. (Paperback edn: Dover, New York, 1988.)

Rosensweig, R. E. (1985). *Ferrohydrodynamics*. Cambridge University Press.

Rouse, H. (1946). *Elementary mechanics of fluids*. Wiley, New York. (Paperback edn: Dover, New York, 1978.)

Ruelle, D. (1980). *Math. Intellig.* **2**, 126-37. (Also in Cvitanović (1986), pp. 37-48.)

Russell, J. S. (1845). *Rep. 14th Meet. Br. Assoc. Adv. Sci., York*, pp. 311-90. John Murray, London.

Schlichting, H. (1979). *Boundary-layer theory* (7th edn). McGraw-Hill, New York.

Schneider, W. (1981). *J. Fluid Mech.* **108**, 55-65.

Schneider, W. (1985). *J. Fluid Mech.* **154**, 91-110.

Scorer, R. S. (1972). *Clouds of the world: a colour encyclopaedia*. Lothian, David & Charles, Melbourne.

Serrin, J. (1959). Mathematical principles of classical fluid mechanics. In *Handbuch der Physik* VIII/1, pp. 125-263. Springer-Verlag, Berlin.

Smith, F. T. (1977). *Proc. R. Soc. Lond.* A **356**, 443-63.

Smith, F. T. (1979). *J. Fluid Mech.* **92,** 171–205.
Smith, R. C. and Smith, P. (1968). *Mechanics.* Wiley, Chichester.
Sobey, I. J. and Drazin, P. G. (1986). *J. Fluid Mech.* **171,** 263–87.
Spedding, G. R. and Maxworthy, T. (1986). *J. Fluid Mech.* **165,** 247–72.
Stewartson, K. (1981). *SIAM Rev.* **23,** 308–43.
Stokes, G. G. (1845). *Trans. Camb. Phil. Soc.* **8,** 287–305. (Also *Math. Phys. Papers* **1,** 75–129. Cambridge University Press.)
Stokes, G. G. (1848). *Phil. Mag.* **33,** 349–56. (Also *Math. Phys. Papers* **2,** 51–8. Cambridge University Press.)
Stokes, G. G. (1851). *Trans. Camb. Phil. Soc.* **9,** 8–106. (Also *Math. Phys. Papers* **3,** 1–141. Cambridge University Press.)
Stuart, J. T. (1986). *SIAM Rev.* **28,** 315–42.
Swinney, H. L. and Gollub, J. P. (eds) (1985). *Hydrodynamic instabilities and the transition to turbulence.* Springer-Verlag, New York.
Taneda, S. (1965). *J. Phys. Soc. Japan* **20,** 1714–21.
Taneda, S. (1979). *J. Phys. Soc. Japan* **46,** 1935–42.
Tanner, R. I. (1988). *Engineering rheology.* Clarendon Press, Oxford.
Tayler, A. B. (1986). *Mathematical models in applied mechanics.* Clarendon Press, Oxford.
Taylor, G. I. (1910). *Proc. R. Soc. Lond.* A **84,** 371–7.
Taylor, G. I. (1923). *Phil. Trans. R. Soc. Lond.* A **223,** 289–343.
Thompson, J. M. T. (1982). *Instabilities and catastrophes in science and engineering.* Wiley, Chichester.
Thompson, S. P. (1910). *The life of William Thomson, Baron Kelvin of Largs.* Macmillan, London.
Thomson, J. J. (1883). *A treatise on the motion of vortex rings.* Macmillan, London.
Thorpe, S. A. (1969). *J. Fluid Mech.* **39,** 25–48.
Thorpe, S. A. (1971). *J. Fluid Mech.* **46,** 299–319.
Tritton, D. J. (1988). *Physical fluid dynamics* (2nd edn). Clarendon Press, Oxford.
Truesdell, C. A. (1954). Rational fluid mechanics (1687–1765). Editor's introduction to *Euleri Opera Omnia*, Series II, Vol. 12. Orell Füssli, Zurich.
Truesdell, C. A. (1968). *Essays in the history of mechanics.* Springer-Verlag, New York.
Turner, J. S. (1973). *Buoyancy effects in fluids.* Cambridge University Press.
van Dyke, M. (1982). *An album of fluid motion.* Parabolic Press, Stanford, California.
von Kármán, T. (1954). *Aerodynamics: selected topics in the light of their historical development.* Cornell University Press, Ithaca, New York.
Walker, J. (1987). *Scient. Am.* **257**(5), 114–7.

Weis-Fogh, T. (1973). *J. Exp. Biol.* **59,** 169–230.
Weis-Fogh, T. (1975). *Scient. Am.* **233**(5), 80–7.
Widnall, S. E. and Tsai, C.-Y. (1977). *Phil. Trans. R. Soc. Lond.* A **287,** 273–305.
Yamada, H. and Matsui, T. (1978). *Phys. Fluids* **21,** 292–4.
Yarmchuk, E. J., Gordon, M. J. V. and Packard, R. E. (1979). *Phys. Rev. Lett.* **43,** 214–17.
Zandbergen, P. J. and Dijkstra, D. (1987). *Ann. Rev. Fluid Mech.* **19,** 465–91.
Zauner, E. (1985). *J. Fluid Mech.* **154,** 111–19.

Index

acceleration of fluid particle 4
aerofoil
  drag 59, 143, 151
  flow round 20, 60, 139
  generation of circulation round 23, 158
  lift 21, 59, 120, 143
  stall 30
  attached vortices 180, 194
axisymmetric flow
  irrotational 174, 199
  stream function 173
  viscous 223, 253

baroclinic instability 335
barotropic flow 192
Bernoulli equation
  for compressible flow 118
  for steady irrotational flow 10
  for unsteady irrotational flow 66, 193
Bernoulli streamline theorem 9
biharmonic equation 230, 254
Blasius's theorem 140
bluff-body flows 28, 150, 180, 262, 264, 290
body force 8
bore 97, 100
boundary conditions
  for inviscid fluid 199
  for viscous fluid 26, 30, 265
  at free surface 65, 67
boundary layer 26
  adverse pressure gradient 29, 261, 263
  approximation 260, 266
  converging channel 275
  equations 260, 266, 268
  flat plate 271
  instability 275, 290, 340

Prandtl's paper 260
reversed flow 179, 262, 287, 293
rotating fluid 278
separation 28, 122, 150, 160, 169, 180, 261, 287
similarity solutions 271, 275, 292
thickness 32, 50, 268
triple-deck 288
bubble, slow flow past 253
buoyancy (or Brunt–Väisälä)
  frequency 87, 115, 344
buoyancy force
  internal gravity waves 86
  thermal instability 305
  vorticity generation 86, 305
Burgers vortex 187
Burgers' equation 107

capillary waves 76
catastrophe theory 332
Cauchy–Lagrange theorem 161
Cauchy–Riemann equations 124
Cauchy's vorticity formula 198
cavitation 265
centrifugal instability 313
channel flow 51, 324
chaos 334
characteristics, method of 91
circle theorem 129
circular cylinder
  flow due to rotation of 53
  flow past
    development from rest 178, 262
    irrotational 28, 130
    low Reynolds number 226, 253
    high Reynolds number 28, 150, 178, 190, 289
  spin-down within 45, 165
  vortex pair behind 180, 194
  vortex street behind 180, 184, 194

Index

circular flow 12, 43
circulation
  definition 19
  generated by vortex shedding 158
  Kelvin's theorem 157
  Kutta–Joukowski condition 20, 140
    related to lift 21, 120, 143, 147
    related to velocity potential 122
    related to vorticity 19
  round a line vortex 126
  round an aerofoil 19, 121, 139
clap-and-fling lift mechanism 159
coefficient of viscosity 26
complex potential for 2D flow
  defined 125
  examples 125
  flow past aerofoil 139
  flow past circular cylinder 141
  line vortices 126, 178, 183, 193
  relation to flow speed 125
compressible flow
  Bernoulli equation for 118
  equations 79
  past thin aerofoil 59
  shock waves 62, 103
  sound waves 58, 79
  unsteady 1D 102
  viscous 107
conformal mapping 134
conservation of mass 23
conservative force 9
constitutive equations 202, 207
continuity equation
  see under conservation of mass
continuum hypothesis, breakdown of 63
convective derivative
  see under D/Dt
converging channel, flow in
  at low Reynolds number 255
  at high Reynolds number 275
Coriolis force 279
corner eddies 229
Couette flow
  in channel 52
  between rotating cylinders 44, 313
creeping flow
  see under slow flow

D/Dt 4
d'Alembert's paradox 147
dam break problem 92

deformation of fluid element 13, 212
density 6
density variations
  conservation of mass 23, 79
  effect of gravity on 86, 111, 115, 306
differentiation 'following the fluid', *see under* D/Dt
diffusion
  of vorticity 33, 37, 46, 48, 179, 187
  of heat 36, 307, 345
  of salt 345
diffusivity, thermal 307
dimensionless parameters 31, 51, 59, 101, 305, 311, 317, 331
dispersion 56, 64, 69, 108
dissipation of energy due to viscosity 54, 216, 341
divergence theorem 349
diverging channel, flow in 278, 296
double diffusive convection 345
doubly-connected region 19, 122
drag
  coefficient 150
  crisis 290
  at high Reynolds number 150, 261, 274
  at low Reynolds number 226, 253
  in ideal flow 59, 149
  in supersonic flow 61
  on streamlined bodies 151, 274
  due to waves 61
draining plate 256
'dyed' fluid 6

Ekman layer 280
elliptic cylinder, flow round 136, 142
energy
  cascade 341
  dissipation 54, 216, 341
  equation 24, 306
  and group velocity 70, 74, 114
  Kelvin's theorem on minimum 199
  loss in hydraulic jump 100
entropy
  defined 79
  change across a shock 104
equation of state 307
equations of motion
  Cauchy 205
  Euler (inviscid) 8
  Navier–Stokes (viscous) 30, 207
    in cylindrical polar coordinates 42, 353

in spherical polar coordinates 355
relative to a rotating frame 279
Ertel's theorem 196
Euler's equations 8
Euler's principle of linear
   momentum 202
Euler's principle of moment of
   momentum 202

Falkner–Skan equation 292
Feigenbaum number 337
fish, mechanical 235
flat plate
   boundary layer 49, 261, 271, 340
   drag 274
   irrotational flow round 137
force
   on an accelerating body 149, 200
   buoyancy 86, 115, 305
   calculated by Blasius's theorem 140
   centrifugal 164, 318
   Coriolis 279
   pressure 6, 208, 219
   viscous 26, 35, 209, 219
   see also under drag, lift
free streamline theory 289
free surface, conditions at 39, 65, 67, 245
Froude number 101

gas, perfect 79
gravity waves, see under water waves
group velocity 56, 69

Hagen–Poiseuille flow, see under
   Poiseuille flow
heat conduction 36, 307
Hele–Shaw cell 241
helicity 196
helium, superfluid properties of 185
Helmholtz's vortex theorems 162
hexagonal convection cells 312
Hill's spherical vortex 175
homentropic flow 102, 118
hydraulic jump 63, 100
hydrostatic pressure distribution 9
hysteresis 332, 345

ideal fluid 6
images, method of 128, 171

incompressible fluid
   equation 7
   conditions for behaviour as 7, 58
induced drag on a lifting body 23
inertia term 31
inertial waves (in rotating fluid) 116
inner and outer solutions 270
   see also under matched asymptotic
      expansions
insect flight, clap-and-fling
   mechanism 159
instability
   baroclinic 335
   Bénard 313
   boundary layer 290, 340
   centrifugal 313
   and chaos 334
   double diffusive 345
   jet 295
   Kelvin–Helmholtz 113, 303
   line vortex arrays 184
   low Reynolds number 341
   pipe flow 300
   Rayleigh's criterion for circular
      flow 318
   Rayleigh's inflection point
      theorem 323
   Rayleigh–Taylor 112
   Saffman–Taylor 342
   shear flow 320
   stratified shear flow 344
   subcritical 301, 325, 345
   due to surface tension variations 313
   thermal 305
   thermohaline 345
   and turbulence 334
   vortex arrays 184
   vortex rings 172
interface waves 111
internal gravity waves 86
irrotational flow
   axisymmetric 174, 199
   expression for pressure in 10, 66, 193
   kinetic energy 199
   minimum energy of 199
   past a
      aerofoil 138
      circular cylinder 130
      elliptical cylinder 136
      flat plate 137
      sphere 174
   persistence of 161

Index

irrational flow (cont.)
  produced impulsively 179, 199
  uniqueness of 199
  unsteady 66, 149, 193
  velocity potential of 122
isentropic flow 79
isotropic medium 209

Jeffrey–Hamel flow 297
jet 293
Joukowski
  condition at trailing edge 20, 140
  theorem 143
  transformation 136
journal bearing 249

Kármán vortex street 180
Kelvin
  circulation theorem 157
  letter to Helmholtz 168
  minimum energy theorem 199
kinematic condition at free surface 65
kinematic viscosity 26, 28
kinetic energy of irrotational flow 199
Korteweg-de-Vries equation 108
Kutta–Joukowski hypothesis 20, 140
Kutta–Joukowski lift theorem 21, 143

Lagrangian description of flow 25, 191, 197, 198
Lanchester, F. W. 22, 120, 265
Landau equation 346
Laplace's equation 125, 162
leading edge suction 153
length scale, characteristic 31
lift
  on aerofoil 21, 120, 145, 153
  defined 20
  on a cylinder with circulation 133
linear stability theory 303
line source 151
line vortex, *see under* vortex
local motion analysed 13, 209, 212
Lorenz equations 335
lubrication theory 248

Mach lines 59, 85
Mach number 59
  and Froude number 101

marginal stability 304, 316, 324
mass-conservation equation 23, 79
matched asymptotic expansions 227
  see also under inner and outer solutions
material derivative, *see under* D/Dt
mean free path 63
micro-organisms, swimming 33, 235
Milne–Thomson circle theorem 129
minimum energy theorem 199
moment of forces
  on an aerofoil 141, 154
  and moment of momentum 202
momentum equation
  in integral form 145
  inviscid 8
  viscous 30, 208

Navier–Stokes equations 30, 208
  in cylindrical polar coordinates 42, 353
  derivation 34, 207
  simple solutions of 33
  in spherical polar coordinates 355
Newtonian viscous fluid 26, 207
non-Newtonian fluids 26
non-uniqueness
  of irrotational flow in multiply-connected regions 19, 130
  of steady viscous flow 278, 297, 330
normal stresses 208
no-slip condition 30, 265

Orr–Sommerfeld equation 323
oscillating plate 52

particle paths
  and streamlines 4, 25
  in water waves 69
pendulum, chaotic motion of 338
perfect gas 79
period doubling 336
phase function 72
pipe flow 51, 300
piston problem 102
Poiseuille flow 51, 300
polystyrene beads 3
potential flow, *see under* irrotational flow
Prandtl–Batchelor theorem 189

Prandtl number 313
Prandtl's paper 260
predictability 336
pressure 6, 208
pressure gradient
  adverse 29, 287
  hydrostatic 9
  principal axes 214

quantized vortices 185
quarter-chord point 155

radiation condition 82
Rankine vortex 15, 24
Rankine–Hugoniot relations 104
rate of change following the fluid 4
rate of strain tensor 212
  components in curvilinear
    coordinates 214, 353, 355
Rayleigh criterion (circular flow) 318
Rayleigh's inflection point
  theorem 323
Rayleigh number 311
Rayleigh problem 35
Rayleigh–Taylor instability 112
resistance, *see under* drag
reversed flow near solid boundary 29, 179, 251, 287
reversibility 33, 234
Reynolds number
  definition 31
  and dynamic similarity 51
  flow at low 32, 221
  flow at high 31, 49, 150, 190, 260, 300
  and instability 300
  physical significance 31
  typical values 50
Reynolds's transport theorem 206
Richardson number 305
ripple tank 113
rolls, convection 312
rotating cylinders, flow between 44, 313, 330
rotating fluid
  between discs 54, 251, 278, 298
  at low Reynolds number 32, 234, 249, 252
  slow relative motion in 279
  spin-down of 164, 284
  waves in 116

Index 395

Russell, J. S. 63

salt fingers 345
secondary flow 54, 165, 252, 281, 285, 298
separation of boundary layer, *see under* boundary layer
shallow water, waves on 79, 89, 108, 119
shear stress 26, 210
ship waves 57
shock wave 62
  caused by a piston 103
  conditions across 104
  oblique 104
  thickness of 63, 107
similarity solution 36, 247, 258, 261, 271, 275, 292
simple shearing motion 27, 34
sink, *see under* source
slider bearing 248
slope, flow down 38, 245
slow flow equations 221, 233
slowly-varying waves 72
smoke ring 168
solid boundary, conditions at
  for inviscid fluid 199
  for viscous fluid 26, 30, 330
solitary wave 64, 108
soliton 110
sound barrier 61
sound, speed of 58, 81
sound waves
  of infinitesimal amplitude 79
  of finite amplitude 61, 102, 107
source, line 151
source, point 152
specific heat 79, 307
sphere, flow due to a moving
  irrotational 199
  at high Reynolds number 290
  at low Reynolds number 223
spherical vortex, Hill's 175
spin-down 45, 164, 284
spreading drop 257
stability of viscous flow 326
  see also under instability
stagnation point, flow near 48, 55, 126, 291
stall 30
steady flow, definition of 2
steepening
  of sound waves 62, 103

steepening (*cont.*)
  of water waves 63, 98
Stokes flow, *see under* slow flow
  equations
Stokes's law for drag on a moving
  sphere 226
Stokes's stream function 173, 223
Stokes's theorem 350
Stokes waves, *see under* water waves
stratified fluid
  interfacial waves 58, 111
  internal gravity waves 86
  shear flow instability 344
stream function
  in 2D flow 123
  in axisymmetric flow 173
streamlined body 151
streamlines
  definition 3
  and particle paths 4, 25
  and the stream function 124
strength of vortex tube 163
stress 26, 203
stress tensor 203
  for Newtonian viscous fluid 207, 209
  symmetry of 207, 220
stress vector 203
  for Newtonian viscous fluid 209
subcritical and supercritical flow 101
subsonic and supersonic flow 59, 105
suction, delaying separation by 263
suction, flow along channel with 52
suffix notation and summation
  convention 204, 233
superfluid dynamics 185
surface tension 57, 74, 112, 305, 313
surface waves, *see under* water waves
swimming, at low Reynolds
  number 234

tap, dripping 338
Taylor, G. I. 234, 342
Taylor–Goldstein equation 344
Taylor–Proudman theorem 280
Taylor shock 63
Taylor vortices 314, 319, 333, 338
teacup, spin-down in 45, 164, 284
tensor
  isotropic 209, 219
  stress 203
terminal velocity 226
thermal conduction 36, 79, 307

thermal convection 305
thin film flow 222, 238
tornado 164
torque 141, 143, 202, 218, 252
trailing edge
  Kutta–Joukowski condition 20, 140
  separation at 1, 158, 288
transition to turbulence
  boundary layer 290, 340
  chaos 339
  jet 295
  pipe flow 300
  thermal convection 312
  wake 150, 180
transonic flow 105
triple-deck 288
turbulence
  in hydraulic jump 100
  nature of 341
  see also under transition to
    turbulence
turbulent spot 340
two-dimensional flow, definition 2

uniqueness
  of irrotational flow 199
  of steady viscous flow 330
  see also under non-uniqueness

velocity potential 122
viscosity
  coefficient of 26, 207
  kinematic 26
  measured values 28
  and Reynolds number 31
viscous dissipation of energy 54, 216
volume flux 40
vortex
  atoms 169
  Burgers 187
  elliptical 185
  Hill's spherical 175
  line vortex 12, 125
    near corner 193
    near wall 193
  viscous diffusion of 46
  merging 186
  pair 177
  Rankine 15
  rings 168
    collision 171
    instability 172

shedding 1, 150, 159, 181, 288
sheet 38, 290
starting 1, 159, 288
street 180, 194
stretching 164, 187
surface 163
Taylor problem 314, 319, 333, 338
theorems 162
trailing 22
vortex line
 definition 162
 moves with an inviscid fluid 162
vortex tube 162
vorticity
 convection and diffusion 48, 187
 definition 10
 equation
  in axisymmetric flow 167
  in general flow 17
  in 2D flow 17
  viscous 48, 187
 generated at a solid boundary 37, 46, 179, 261
 generation by buoyancy forces 86, 305
 intensification by stretching of vortex lines 164, 166, 187, 285

meter 14
physical interpretation 11, 212
shed into wake 28, 150, 181, 261, 295
theorems for an inviscid fluid 162
in turbulent flow 341
viscous diffusion of 33, 37, 48, 187

wake
 circular cylinder 28, 150, 181, 262, 289
 streamlined body 29, 151, 295
water waves
 dispersion 56, 69
 energy 114
 finite amplitude, in shallow water 89
 finite depth 78
 group velocity 56, 73
 at interface between two fluids 111
 particle paths 69
 surface tension effects on 74
wave drag 61
wave packet 57, 69
Whitehead's paradox 226